Lecture Notes in Economics and Mathematical Systems

600

T0182004

Mark Hickman · Pitu Mirchandani
Stefan Voß

(Editors)

Computer-aided Systems in Public Transport

 Springer

Professor Mark Hickman
Department of Civil Engineering
and Engineering Mechanics
University of Arizona
1209 E. Second Street
Tucson, AZ 85721-0072
USA
mhickman@engr.arizona.edu

Professor Pitu Mirchandani
Department of Systems
and Industrial Engineering
University of Arizona
1127 E. James E. Rogers Way
Tucson, AZ 85721-0020
USA
pitu@sie.arizona.edu

Professor Dr. Stefan Voß
Institute of Information Systems
Department of Business and Economics
University of Hamburg
Von-Melle-Park 5
20146 Hamburg
Germany
stefan.voss@uni-hamburg.de

ISBN 978-3-540-73311-9   e-ISBN 978-3-540-73312-6

DOI 10.1007/978-3-540-73312-6

Lecture Notes in Economics and Mathematical Systems ISSN 0075-8442

Library of Congress Control Number: 2007939763

Production: LE-TEX Jelonek, Schmidt & Vöckler GbR, Leipzig
Cover design: WMX Design GmbH, Heidelberg

Printed on acid-free paper

9 8 7 6 5 4 3 2 1

springer.com

Preface

This proceedings volume consists of selected papers presented at the Ninth International Conference on *Computer-Aided Scheduling of Public Transport* (CASPT 2004), which was held at the Hilton San Diego Resort and Conference Center in San Diego, California, USA, from August 9-11, 2004. The CASPT 2004 conference is the continuation of a series of international workshops and conferences presenting recent research and progress in computer-aided scheduling in public transport. Previous workshops and conferences were held in:

- Chicago (1975)
- Leeds (1980)
- Montreal (1983 and 1990)
- Hamburg (1987)
- Lisbon (1993)
- Cambridge, Mass. (1997)
- Berlin (2000)[1]

---

[1] While there were no formal proceedings for the first workshop (only pre-prints were distributed to participants), the subsequent workshops and conferences were well documented:

Wren, A. (ed.) (1981). *Computer Scheduling of Public Transport*, North-Holland, Amsterdam.

Rousseau, J.-M. (ed.) (1985). *Computer Scheduling of Public Transport 2*, North-Holland, Amsterdam.

Daduna, J.R. and A. Wren (eds.) (1988). *Computer-Aided Transit Scheduling*, Lecture Notes in Economics and Mathematical Systems 308, Springer, Berlin.

Desrochers, M. and J.-M. Rousseau (eds.) (1992). *Computer-Aided Transit Scheduling*, Lecture Notes in Economics and Mathematical Systems 386, Springer, Berlin.

Daduna, J.R., I. Branco, and J.M.P. Paixão (eds.) (1995). *Computer-Aided Transit Scheduling*, Lectures Notes in Economics and Mathematical Systems 430, Springer, Berlin.

Wilson, N.H.M. (ed.) (1999). *Computer-Aided Transit Scheduling*, Lecture Notes in Economics and Mathematical Systems 471, Springer, Berlin.

Voß, S. and J.R. Daduna (eds.) (2001). *Computer-Aided Scheduling of Public Transport*, Lecture Notes in Economics and Mathematical Systems 505, Springer, Berlin.

The scope and purpose of the conference has broadened significantly since 1975, although it retains as its core the primary mission of advancing the state of the art and the state of the practice in computer-aided systems in public transport (which also let us choose the title of this book). Yet, this volume illustrates a greater breadth of subjects in this area. The common theme of these conferences remains on the use of computer-aided methods and operations research techniques to improve:

- Information management
- Network and route planning
- Vehicle and crew scheduling and rostering
- Vehicle monitoring and management
- Practical experience with scheduling and public transport planning methods

The conference was organized for the benefit of individuals from transport operators, consulting firms and academic institutions involved in research, development or utilization of computer-aided scheduling methods in public transport. A total of 60 attendees were present for the conference in San Diego. During the conference, a total of 39 presentations were given in these subject areas, representing both research and applications. Of these, a full 35 involved formal papers. These papers were then peer-reviewed, resulting in a select number of high quality papers (22) that are represented in this volume.

The organization of this volume follows the more general structure of the conference itself. Consistent with previous volumes, the initial section is organized around the topic of vehicle and crew scheduling. These papers highlight significant advances in both areas, but also illustrate that very useful and computationally efficient methods are being developed for integrated vehicle and crew scheduling.

The second section deals more specifically with vehicle routing and timetabling. In this section, various new methods are advanced for establishing public transport timetables for railways, ferries, and school buses. For many of these cases, new vehicle routing methods must also be devised to enhance the vehicle scheduling process. Of considerable note are the advances in periodic vehicle scheduling, which is relevant to short-distance rail systems.

The third section addresses a growing topic in transport service and performance monitoring, operations management and control, and dispatching. These topics reflect a considerable growth in interest in the improvement of transport operations through the use of decision tools. The papers in this section cover applications from bus and rail vehicle tracking and travel time prediction. A number of the papers cover decision-making techniques to improve operations when there are inevitable service disruptions.

The final section includes papers dealing with more strategic-level planning of public transport services. Topics covered in these areas include network design, optimal fare and tolling policies, line planning, fleet sizing, and the level of service for demand-responsive transit services. These papers reflect a growing interest in the application of operations research tools to more strategic decisions by transit operators.

We believe that this volume captures some sense of the state of the art in this field. In this spirit, we realize that there have been significant advances since the first workshop in 1975 in the capabilities for information processing and computation, allowing us now to address and solve problems that were previously beyond reach. At the same time, we look forward to further advances, as they may be relayed in future conferences: in Leeds in 2006, and (tentatively) in Hong Kong in 2009.

Acknowledgements

Of course, organizing a conference of this caliber and publishing the proceedings relies substantially on the valuable input of many individuals and organizations. The scientific program was assembled through the international committee consisting of the following members:

- Avi Ceder, Technion - Israel Institute of Technology, Haifa, Israel
- Joachim R. Daduna, University of Applied Business Administration Berlin, Berlin, Germany
- Mark Hickman, University of Arizona, Tucson, Arizona, USA
- Raymond S.K. Kwan, University of Leeds, Leeds, United Kingdom
- Pitu Mirchandani, University of Arizona, Tucson, Arizona, USA
- Jean-Marc Rousseau, Cirano, Montreal, Quebec, Canada
- Paolo Toth, University of Bologna, Bologna, Italy
- Stefan Voß, University of Hamburg, Hamburg, Germany
- Nigel H.M. Wilson, Massachusetts Institute of Technology, Cambridge, Massachusetts, USA

We also wish to thank all of the authors and the conference participants for their contributions to making this a success. In addition, several people assisted with the peer review of the papers; these persons are listed below. Their help was of vital importance in maintaining the high quality of papers in this volume. As not all papers were included, a list of additional presentations and papers not included in this volume is also given below.

In addition, the conference was generously supported by a number of exhibitors and sponsors. Software exhibitors at the conference included:

- Trapeze Software
- GIRO Inc.
- PTV America Inc.
- VERSYSS Transit Solutions

A local tour to the San Diego Trolley was also arranged courtesy of San Diego Transit, and we appreciated a very nice presentation at the conference banquet by Thomas Larwin, who was Deputy Director of the San Diego Association of Governments at the time of the conference.

Beyond these exhibitors, the conference also received considerable financial support from the National Science Foundation, the University of Arizona Department of Civil Engineering and Engineering Mechanics, and from the Center for Advanced

Transportation and Logistics Algorithms and Systems (the ATLAS Center) at the University of Arizona.

Finally we like to thank Holger Höller for some help regarding the transfer of some papers between different word processing systems.

Referees

Hillel Bar-Gera, John Beasley, Michael Bussieck, Avi Ceder, Steven Chien, Pierluigi Coppola, Cristian Cortes, Joachim Daduna, Mauro Dell'Amico, Guy Desaulniers, Andreas Ernst, Matteo Fischetti, Charles Fleurent, Markus Friedrich, Liping Fu, Peter Furth, Vitali Gintner, Fred Glover, Sebastian de Groot, Knut Haase, Ali Haghani, Mark Hickman, Mark E.T. Horn, Dennis Huisman, Matthew Karlaftis, Isam Kaysi, Natalia Kliewer, Raymond Kwan, William H.K. Lam, C.-K. Lee, Janny Leung, Christian Liebchen, Hong K. Lo, Andreas Loebel, David Lovell, Federico Malucelli, Elise Miller-Hooks, Pitu Mirchandani, Rabi Mishalani, Rob van Nes, Dario Pacciarelli, Juaquin Pacheco, Ana Paias, Leon Peeters, Marc Peeters, Jean-Marc Rousseau, Francesco Russo, Anita Schoebel, Brian Smith, James Strathman, Leena Suhl, Sam Thangiah, Stefan Voß, and Nigel H.M. Wilson.

Presented Papers Not Included in This Volume

A. Ceder, *Network Route Design and Evaluation Methods for Passenger Ferry Service*

J. Daduna and S. Voß, *OR Applications in Public Mass Transit Processes: An Overview*

A. Dallaire, C. Fleurent, and J.-M. Rousseau, *Dynamic Constraint Generation in CrewOpt, a Column Generation Approach for Transit Crew Scheduling*

R.N. Datta, *Computer-Aided Utility Assessment of Bus Routes and Schedules Conforming to Suburban Train Schedules in Indian Urban Areas*

C. Fleurent, R. Lessard, and L. Seguin, *Transit Timetable Synchronization: Evaluation and Optimization*

M. Friedrich and K. Noekel, *Extending Transportation Planning Models: From Strategic Modeling to Operational Transit Planning*

M. Hickman, *A Method for Incorporating Reliability in Passenger Itinerary Planning*

B. Horwath, *Automated Publishing of Transit Schedules for Print & Online*

A. Kwan, M. Parker, R. Kwan, S. Fores, L. Proll, and A. Wren, *Recent Advances in TRACS*

R. Kwan, I. Laplagne, and A. Kwan, *Train Driver Scheduling With Time Windows of Relief Opportunities*

C.-K. Lee, *The Integrated Scheduling and Rostering Problem of Train Driver Using Genetic Algorithm*

S. Li and S.H. Lam, *Schedule Optimization for an Integrated Multi-operator and Multimodal Transit System*

M. Ridwan, *FiPV based Dynamic Transit Assignment*

H. Soroush, *A Bi-Attribute Shortest Path Problem with Fractional Cost Function*

S. Wegele and E. Schnieder, *Dispatching of Train Operations Using Genetic Algorithms*

R. Wong and J. Leung, *Timetable Synchronization for Mass Transit Railway*

A. Wren, *Scheduling Vehicles and Their Drivers: Forty Years´ Experience*

Mark Hickman, Tucson
Pitu Mirchandani, Tucson   May 2007
Stefan Voß, Hamburg

Contents

## Part I

Vehicle and Crew Scheduling

A Bundle Method for Integrated Multi-Depot Vehicle and Duty Scheduling in Public Transit

Ralf Borndörfer, Andreas Löbel, and Steffen Weider

Zuse Institute Berlin, Takustr. 7, 14195 Berlin, Germany, Email
{borndoerfer,loebel,weider}@zib.de

Summary. This article proposes a Lagrangean relaxation approach to solve integrated duty and vehicle scheduling problems arising in public transport. The approach is based on a version of the proximal bundle method for the solution of concave decomposable functions that is adapted for the approximate evaluation of the vehicle and duty scheduling components. The primal and dual information generated by this bundle method is used to guide a branch-and-bound type algorithm.

Computational results for large-scale real-world integrated vehicle and duty scheduling problems with up to 1,500 timetabled trips are reported. Compared with the results of a classical sequential approach and with reference solutions, integrated scheduling offers remarkable potentials in savings and drivers' satisfaction.

1 Introduction

The process of *operational planning* in public transit is traditionally organized in successive steps of timetabling, vehicle scheduling, duty scheduling, duty rostering, and crew assignment. These tasks are well investigated in the optimization and operations research literature. And enormous progress has been made in both the theoretical analysis of these problems and in the computational ability to solve them. For an overview see the proceedings of the last five CASPT conferences (Voß and Daduna (2001), Wilson (1999), Daduna *et al.* (1995), Desrochers and Rousseau (1992), and Daduna and Wren (1988)).

It is well known that the *integrated treatment* of planning steps discloses additional degrees of freedom that can lead to further efficiency gains. The first and probably best known approach in this direction is the so-called *sensitivity analysis*, a method on the interface between timetabling and vehicle scheduling that uses slight shiftings of trips in the timetable to improve the vehicle schedule. The method has been used with remarkable success in HOT and HASTUS, see Daduna and Völker (1997) and Hanisch (1990).

Vehicle and duty scheduling, the topic of this article, is another area where integration is important. The need is largest in *regional scenarios*, which often have few relief points for drivers, such that long vehicle rotations can either not be covered with legal duties at all or only at very high cost. In such scenarios the powerful optimization tools of sequential scheduling are useless. Rather, the vehicle and the duty scheduling steps must be synchronized to produce acceptable results, i.e., an *integrated vehicle and duty scheduling* method is indispensable. Urban scenarios do, of course, offer efficiency potentials as well.

The current *planning systems* provide only limited support for integrated vehicle and duty scheduling. There are frameworks for manual integrated scheduling that allow to work on vehicles and duties simultaneously, rule out infeasibilities, make suggestions for concatenations, etc. Without integrated optimization tools, however, the planner must still build vehicle schedules by hand, anticipating the effects on duty scheduling by skill and experience.

The *literature* on integrated vehicle and duty scheduling is also comparably scant. The first article on the *integrated vehicle and duty scheduling problem* (ISP) that we are aware of was published in 1983 by Ball *et al.* (1983). They describe an ISP at the Baltimore Metropolitan Transit Authority and develop a mathematical model for it. However, they propose to solve this model by decomposing it into its vehicle and duty scheduling parts, i.e., the model is integrated, but the solution method is sequential.

For the next two decades, the predominant approach to the ISP was to include duty scheduling considerations into a vehicle scheduling method or vice versa. The first approach is, e.g., presented by Scott (1985) and Darby-Dowman *et al.* (1988), who propose two-step methods that first include some duty scheduling constraints in a vehicle scheduling procedure and afterwards solve the duty scheduling problem in a second step. Examples of the opposite approach are the articles of Tosini and Vercellis (1988), Falkner and Ryan (1992), and Patrikalakis and Xerocostas (1992). They concentrate on duty scheduling and take the vehicle scheduling constraints and costs heuristically into account. A survey of integrated approaches until 1997 can be found in Gaffi and Nonato (1999).

The complete integration of vehicle and crew scheduling was first investigated in a series of publications by Freling and coauthors (Freling (1997), Freling *et al.* (2001a), Freling *et al.* (2001b), Freling *et al.* (2003)). They propose a combined vehicle and duty scheduling model and attack it by integer programming methods, especially column generation and Lagrangean relaxation is used. Computational results on several problems from the Rotterdam public transit company RET with up to 300 timetabled trips, and from Connexxion, the largest bus company in the Netherlands, with up to 653 timetabled trips are reported. A branch-and-price approach to ISP instances involving a single type of vehicles was also described by Friberg and Haase (1999) and tested on artificial data. Another approach to the single-depot ISP is presented in Haase *et al.* (2001). There a set partitioning model for the duty scheduling problem is used that ensures that also a vehicle schedule can be built. Additional constraints are introduced to count the number of vehicles. This model was tested on artificial data with up to 350 timetabled trips and up to 700 tasks on timetabled trips. It was solved by a branch and price approach using CPLEX as LP-solver.

We propose in this article an integrated vehicle and duty scheduling method similar to that of Freling et al. Our main contribution is the use of bundle techniques for the solution of the Lagrangean relaxations that come up there. The advantages of the bundle method are its high quality bounds and automatically generated primal information that can both be used to guide a branch-and-bound type algorithm. We apply this method to real-world instances from several German carriers with up to 1,500 timetabled trips. As far as we know, these are the largest and most complex instances that have been tackled in the literature using an integrated scheduling approach. Our optimization module IS-OPT has been developed in a joint research project with IVU Traffic Technologies AG (IVU), Mentz Datenverarbeitung GmbH (mdv), and the Regensburger Verkehrsbetriebe (RVB). It is incorporated in IVU's commercial scheduling system MICROBUS 2.

The article is organized as follows. Section 2 gives a formal description of the ISP and states an integer programming model that provides the basis of our approach. Section 3 describes our scheduling method. We discuss the Lagrangean relaxation that arises from a relaxation of the coupling constraints for the vehicle and the duty scheduling parts of the model, the solution of this relaxation by the proximal bundle method, in particular, the treatment of inexact evaluations of the vehicle and duty scheduling component functions, and the use of primal and dual information generated by the bundle method to guide a branch-and-bound algorithm. Section 4 reports computational results for large-scale real-world data. In particular, we apply our integrated scheduling method to mostly urban instances for the German city of Regensburg with up to 1,500 timetabled trips.

2 Integrated Vehicle and Duty Scheduling

The *integrated vehicle and duty scheduling problem* contains a vehicle and a duty scheduling part. We describe these individual parts first and conclude with the integrated scheduling problem. The exposition assumes that the reader is familiar with the terminology of vehicle and duty scheduling; suitable references are Löbel (1999) for vehicle scheduling and Borndörfer *et al.* (2003) for duty scheduling.

We use the following notation for dealing with vectors: $x \in X^A$, $X \subset \mathbb{R}$, A is some index set. For $a \in A$, $x_a \in X$ denotes the component of x corresponding to a. For $B \subset A$, x_B denotes the subvector $x_B := (x_a)_{a \in B}$. Finally, $x(B) := \sum_{a \in B} x_a$, $B \subset A$, denotes a sum over a subset of components of x.

The vehicle scheduling part of the ISP is based on an acyclic directed multigraph $G = (\mathcal{T} \cup \{s,t\}, \mathcal{D})$. The nodes of G are the set $\mathcal{T}$ of *timetabled trips* plus two additional artificial nodes s and t, which represent the beginning and the end of a vehicle rotation, respectively; s is the source of G and t the sink. The arcs $\mathcal{D}$ of G are called *deadheads*, the special deadheads that emanate from the source s are the *pull-out trips*, those entering the sink t are the *pull-in trips*. Associated with each deadhead a is a *depot* $g_a \in \mathcal{G}$ from some set $\mathcal{G}$ of *depots* (i.e., vehicle types), that indicates a valid vehicle type, and a *cost* $d_a \in \mathbb{Q}$. There may be parallel arcs in G with different depots and costs. We denote by $\mathcal{D}_g := \{a \in \mathcal{D} : g_a = g\}$ the set of deadheads that can be covered by a vehicle of type $g \in \mathcal{G}$, by $\delta_g^+(v) := \delta^+(v) \cap \mathcal{D}_g$ the outcut of node v, restricted to arcs in $\mathcal{D}_g$, and by $\delta_g^-(v) := \delta^-(v) \cap \mathcal{D}_g$ the incut of node v, restricted to arcs in $\mathcal{D}_g$.

A *vehicle rotation* or *block* of type $g \in \mathcal{G}$ is an st-path in G that uses only deadheads of type g, i.e., an st-path p such that $p \subseteq \mathcal{D}_g$ for some depot $g \in \mathcal{G}$. A vehicle schedule is a set of blocks such that each timetabled trip is contained in one and only one block. The *vehicle scheduling problem* (VSP) is to find a vehicle schedule of minimal cost. It can be stated as the following integer program:

$$
\begin{aligned}
&\text{(VSP)} & \min \quad & d^\mathsf{T} y \\
&\text{(i)} & y(\delta_g^+(v)) - y(\delta_g^-(v)) &= 0 & \forall v \in \mathcal{T}, g \in \mathcal{G} \\
&\text{(ii)} & y(\delta^+(v)) &= 1 & \forall v \in \mathcal{T} \\
&\text{(iii)} & y(\delta^-(v)) &= 1 & \forall v \in \mathcal{T} \\
&\text{(iv)} & y &\in \{0,1\}^{\mathcal{D}}
\end{aligned}
$$

The duty scheduling part of the ISP also involves an acyclic digraph $D = (\mathcal{R} \cup \{s,t\}, \mathcal{L})$. The nodes of D consist of a set of *tasks* $\mathcal{R}$ plus two artificial nodes s and t, which mark the beginning and the end of a *part of work* of a duty; again s is the source of D and t the sink. A task r can correspond either to a timetabled trip $v_r \in \mathcal{T}$ or to a deadhead trip $a_r \in \mathcal{D}$. There may also be additional tasks independent of the vehicle schedule that model sign-on and sign-off times and similar activities of drivers.

Let $\mathcal{R}_T$ and $\mathcal{R}_D$ be the sets of tasks that correspond to a timetabled trip and a deadhead trip, respectively. We assume that there is at least one task associated with every timetabled trip and every deadhead trip; these tasks correspond to units of driving work on such a trip. Several tasks for one trip indicate that this trip is subdivided by relief opportunities to exchange a driver into several units of driving work. The arcs $\mathcal{L}$ of D are called *links*; they correspond to feasible concatenations of tasks in a potential duty. A *part of work* of a duty is an st-path p in D that corresponds to certain legality rules and has some cost c_p, again determined by certain rules. A *duty* is a concatenation of one or more (usually one or two) compatible parts of work.

Denote by $\mathcal{S}$ the set of all such duties, and by $c_p, p \in \mathcal{S}$, their costs. Let further $\mathcal{S}_r := \{p \in \mathcal{S} : r \in p\}$ be the set of all duties that contain some task $r \in \mathcal{R}$ and let $\mathcal{D}_r \subset \mathcal{D}$ be the set of deadheads that contain task r. Given a vehicle schedule y, a *compatible duty schedule* is a collection of duties such that each task that corresponds to either a timetabled trip or a deadhead trip from the vehicle schedule is contained in exactly one duty, while the tasks corresponding to deadhead trips that are not contained in the vehicle schedule are not contained in any duty. The *duty scheduling problem* associated with a vehicle schedule y is to find a compatible duty schedule of minimum cost. This DSP can be stated as the following integer program:

$$
\begin{aligned}
&\text{(DSP}_y\text{)} & \min \quad & c^\mathsf{T} x \\
&\text{(i)} & x(\mathcal{S}_r) &= 1 & \forall r \in \mathcal{R}_T \\
&\text{(ii)} & x(\mathcal{S}_r) &= y_a & \forall (r,a) \in \mathcal{R} \times \mathcal{D} \text{ with } a \in \mathcal{D}_r \\
&\text{(iii)} & x &\in \{0,1\}^{\mathcal{S}}
\end{aligned}
$$

This type of model is generally solved by column generation. For duty scheduling in public transit this was first proposed by Desrochers and Soumis (1989).

The *integrated vehicle and duty scheduling problem* is to simultaneously construct a vehicle schedule and a compatible duty schedule of minimum overall cost. Introducing suitable constraint matrices and vectors, the ISP reads:

$$
\begin{array}{rll}
(\text{ISP}) & \min \ d^{\mathsf{T}}y + c^{\mathsf{T}}x & \\
(\text{i}) & Ny & = b \\
(\text{ii}) & Ax & = \mathbb{1} \\
(\text{iii}) & My - Bx & = 0 \\
(\text{iv}) & y & \in \{0,1\}^{\mathcal{D}} \\
(\text{v}) & x & \in \{0,1\}^{\mathcal{S}}
\end{array}
$$

In this model, the *multiflow constraints* (ISP) (i) correspond to the vehicle scheduling constraints (VSP) (i)–(iii); they generate a feasible vehicle schedule. The *(timetabled) trip partitioning constraints* (ISP) (ii) are exactly the duty scheduling constraints (DSP_y) (i); they make sure that each timetabled trip is covered by exactly one duty. Finally, the *coupling constraints* (ISP) (iii) correspond to the duty scheduling constraints (DSP_y) (ii); they guarantee that the vehicle and duty schedules x and y are synchronized on the deadhead trips, i.e., a deadhead trip is either assigned to both a vehicle and a duty or to none. Note that fixing variables corresponding to deadhead trips reduces the size of the subproblems as well as the number of coupling constraints by logical implications.

We remark that practical versions of the ISP include several types of additional constraints such as depot capacities, and duty scheduling base constraints (e.g., duty type capacities, average paid/working times), which we omit in this article. The inclusion of such constraints in our scheduling method is, however, straightforward.

The integrated scheduling model (ISP) consists of a multicommodity flow model for vehicle scheduling and a set partitioning model for duty scheduling on timetabled trips. These two models are joined by a set of coupling constraints for the deadhead trips, one for each task on a deadhead trip. The model (ISP) is the same as that used by Freling (1997).

3 A Bundle Method

Our general solution strategy for the ISP is a Lagrangean relaxation approach. For an introduction to this we suggest Lemarechal (2001). There also an overview of applications and variants of Lagrangean relaxation can be found.

Relaxing the coupling constraints (ISP) (iii) in a Lagrangean way decomposes the problem into a vehicle scheduling subproblem, a duty scheduling subproblem, and a Lagrangean master problem. All three of these problems are large scale, but of quite different nature. Efficient methods are available to solve vehicle scheduling problems of the sizes that come up in an integrated approach with a very good quality or even to optimality. We use the method of Löbel (1997). Duty scheduling is, in fact, the hardest part. We are not aware of methods that can produce high quality lower bounds for large-scale real-world instances. However, duty scheduling problems can be tackled in a practically satisfactory way using column generation algorithms; see Borndörfer *et al.* (2003) for the algorithm we used to "solve" our duty scheduling subproblems. In the Lagrangean master, multipliers for several tens of thousands of coupling constraints have to be determined. Here, the complexity of the vehicle and the duty scheduling subproblems demands a method that converges quickly and that can be adapted to inexact evaluation of the subproblems. The *proximal bundle method* of Kiwiel (1995) has these properties. It further produces primal information that can be used in a branch-and-bound algorithm to guide the branching decisions. Moreover, the large dimension of the Lagrangean multiplier space, a potential computational obstacle, collapses by a simple dualization.

This section discusses our Lagrangean relaxation/column generation approach to the ISP using the proximal bundle method. In a first phase, the procedure aims at the computation of an "estimation" of a global lower bound for the ISP *and* at the computation of a set of duties that is likely to contain the major parts of a good duty schedule. This procedure constitutes the core of our integrated vehicle and duty scheduling method. In a second phase, the bundle core is called repeatedly in a branch-and-bound type procedure to produce integer solutions.

3.1 Lagrangean Relaxation

We consider in this subsection a restriction (ISP_I) of the ISP to some subset of duties $I \subseteq S$ that have been generated explicitly (in some way): This set I may change (grow and shrink) from one iteration to another in our algorithm, however, for simplicity of exposition we keep it constant in the next two sections. The dynamic case will be described in Section 3.3.

$$
\begin{aligned}
(\text{ISP}_I) \quad &\min \quad d^\mathsf{T}y + c_I^\mathsf{T}x^I \\
\text{(i)} \quad & \qquad Ny \qquad\qquad = d \\
\text{(ii)} \quad & \qquad\qquad\quad A_I x^I = \mathbb{1} \\
\text{(iii)} \quad & \qquad My - B_I x^I = 0 \\
\text{(iv)} \quad & \qquad\quad y \qquad\qquad \in \{0,1\}^{\mathcal{D}} \\
\text{(v)} \quad & \qquad\qquad\qquad x^I \in \{0,1\}^I
\end{aligned}
$$

A Lagrangean relaxation with respect to the coupling constraints (ISP_I) (iii) and a relaxation of the integrality constraints (iv) and (v) results in the Lagrangean dual

$$
(\mathrm{L}_I) \quad \max_{\lambda} \left[\min_{\substack{Ny=d, \\ y\in[0,1]^{\mathcal{D}}}} (d^\mathsf{T} - \lambda^\mathsf{T}M)y + \min_{\substack{A_I x^I=\mathbb{1}, \\ x^I\in[0,1]^I}} (c_I^\mathsf{T} + \lambda^\mathsf{T}B_I)x^I \right].
$$

Define functions and associated arguments by

$$
\begin{aligned}
f_V &: \mathbb{R}^{\mathcal{R}_{\mathcal{D}}} \to \mathbb{R}, \ \lambda \mapsto \min(d^\mathsf{T} - \lambda^\mathsf{T}M)y; \quad Ny = d; \quad y \in [0,1]^{\mathcal{D}} \\
f_D^I &: \mathbb{R}^{\mathcal{R}_{\mathcal{D}}} \to \mathbb{R}, \ \lambda \mapsto \min(c^\mathsf{T} + \lambda^\mathsf{T}B_I)x^I; \ A_I x^I = \mathbb{1}; \ x^I \in [0,1]^I \\
f^I &:= f_V + f_D^I,
\end{aligned}
$$

and

$$y(\lambda) := \mathrm{argmin}_{y \in [0,1]^{\mathcal{D}}} f_V(\lambda); \ Ny = d$$
$$x^I(\lambda) := \mathrm{argmin}_{x^I \in [0,1]^I} f_D^I(\lambda); \ A_I x^I = \mathbb{1}$$

breaking ties arbitrarily. With this notation, (L_I) becomes

$$(L_I) \quad \max_{\lambda} f^I(\lambda) = \max_{\lambda} \left[f_V(\lambda) + f_D^I(\lambda) \right].$$

The functions f_V and f_D^I are concave and piecewise linear. Their sum f^I is therefore a decomposable, concave, and piecewise linear function; f^I is, in particular, nonsmooth. This is precisely the setting for the proximal bundle method.

3.2 The Proximal Bundle Method

The *proximal bundle method* (PBM) is a subgradient-type procedure to minimize concave functions. It can be adapted to handle decomposable, nonsmooth functions in a particularly efficient way.

We recall the method in this section as far as we need for our exposition. An in-depth treatment can be found in Kiwiel (1990), Kiwiel (1995).

When applied to (L_I), the PBM produces two sequences of iterates $\lambda_i, \mu_i \in \mathbb{R}^{\mathcal{R}_{\mathcal{D}}}$, $i = 0, 1, \ldots$. The points μ_i are called *stability centers*; they converge to a solution of (L_I). The points λ_i are trial points; calculations at the trial points result either in a shift of the stability center, or in some improved approximation of f^I.

More precisely, the PBM computes at each iterate λ_i linear approximations

$$\bar{f}_V(\lambda; \lambda_i) := f_V(\lambda_i) + g_V(\lambda_i)^{\mathsf{T}}(\lambda - \lambda_i)$$
$$\bar{f}_D^I(\lambda; \lambda_i) := f_D^I(\lambda_i) + g_D^I(\lambda_i)^{\mathsf{T}}(\lambda - \lambda_i)$$
$$\bar{f}^I(\lambda; \lambda_i) := \bar{f}_V(\lambda; \lambda_i) + \bar{f}_D^I(\lambda; \lambda_i)$$

of the functions f_V, f_D^I, and f^I by determining the function values $f_V(\lambda_i)$, $f_D^I(\lambda_i)$ and the subgradients $g_V(\lambda_i)$ and $g_D^I(\lambda)$. By definition, these approximations overestimate the functions f_V and f_D^I, i.e., $\bar{f}_V(\lambda; \lambda_i) \geq f_V(\lambda)$ and $\bar{f}_D^I(\lambda; \lambda_i) \geq f_D^I(\lambda)$ for all λ. Note that $\bar{f}_V$ and $\bar{f}_D^I$ are polyhedral, such that subgradients can be derived from the arguments $y(\lambda_i)$ and $x^I(\lambda_i)$ associated with the multiplier λ_i as

$$g_V(\lambda_i) := - My(\lambda_i)$$
$$g_D^I(\lambda_i) := \quad B_I x^I(\lambda_i)$$
$$g^I(\lambda_i) := - My(\lambda_i) + B_I x^I.$$

For implementation an affine function $\bar{f}$ can be stored as a tuple $(\bar{f}(0), \nabla \bar{f})$ of its function value at the origin and its gradient. We call the sets of linearizations collected until iteration i *bundles* and denote them by $J_{V,i}$ and $J_{D,i}$. The PBM uses such bundles to build piecewise linear approximations

$$\hat{f}_{V,i}(\lambda) := \min_{\bar{f}_V \in J_{V,i}} \bar{f}_V(\lambda)$$

$$\hat{f}_{D,i}(\lambda) := \min_{\bar{f}_D \in J_{D,i}} \bar{f}_D(\lambda)$$

$$\hat{f}_i := \hat{f}_{V,i} + \hat{f}_{D,i}$$

of f_V, f_D^I, and f^I. Adding a quadratic term to this model that penalizes large deviations from the current stability center μ_i, the next trial point λ_{i+1} is calculated by solving the quadratic programming problem

$$(\text{QP}_i) \quad \lambda_{i+1} := \text{argmax}_\lambda \, \hat{f}_i(\lambda) - \tfrac{u}{2} \left\| \mu_i - \lambda \right\|^2 .$$

Here, u is a positive weight that can be adjusted to increase accuracy or convergence speed. If the approximated function value $\hat{f}_i(\lambda_{i+1})$ at the new iterate λ_{i+1} is sufficiently close to the function value $f^I(\mu_i)$, the PBM stops; μ_i is the approximate solution. Otherwise a test is performed whether the predicted increase $\hat{f}_i(\lambda_{i+1}) - f^I(\mu_i)$ leads to sufficient real increase $f^I(\lambda_{i+1}) - f^I(\mu_i)$; in this case, the model is judged accurate and the stability center is moved to $\mu_{i+1} := \lambda_{i+1}$. The bundles are updated by adding the information computed in the current iteration, and, possibly, by dropping some old information. Then the next iteration starts, see Algorithm 1 for a listing (the affine functions $\tilde{f}_{V,i}$ and $\tilde{f}_{D,i}$ will be defined and explained below).

Require: Starting point $\lambda_0 \in \mathbb{R}^n$, weights $u_0, m > 0$, optimality tolerance $\epsilon \geq 0$.
1: Initialization: $i \leftarrow 0$, $J_{V,i} \leftarrow \{\lambda_i\}$, $J_{D,i} \leftarrow \{\lambda_i\}$, and $\mu_i = \lambda_i$.
2: Direction finding: Compute $\lambda_{i+1}, \tilde{g}_{V,i}, \tilde{g}_{D,i}$ by solving problem (QP_i).
3: Function evaluation: Compute $f_V(\lambda_{i+1}), g_V(\lambda_{i+1}), f_D^I(\lambda_{i+1}), g_D^I(\lambda_{i+1})$.
4: Stopping criterion: If $\hat{f}_i(\lambda_{i+1}) - f^I(\mu_i) < \epsilon(1 + |f^I(\mu_i)|)$ output μ_i, terminate.
5: Bundle update:
     Select $J_{V,i+1} \subseteq J_{V,i} \cup \{\bar{f}_V(\cdot, \lambda_{i+1}), \tilde{f}_{V,i}\}$,
     select $J_{D,i+1} \subseteq J_{D,i} \cup \{\bar{f}_D^I(\cdot, \lambda_{i+1}), \tilde{f}_{D,i}\}$.
6: Ascent test: $\mu_{i+1} \leftarrow f^I(\lambda_{i+1}) - f^I(\mu_i) > m(\hat{f}_i(\lambda_{i+1}) - f^I(\mu_i))$? λ_{i+1} : μ_i.
7: Weight update: Set u_{i+1}.
8: $i \leftarrow i + 1$, goto Step 2.

Algorithm 1: Generic PBM

Besides function and subgradient calculations, the main work in the PBM is the solution of the quadratic problem QP_i. This problem can also be stated as

$$(\text{QP}_i) \quad \max \quad v_V + v_D - \tfrac{u}{2} \left\| \mu_i - \lambda \right\|^2$$

$$(\text{i}) \qquad\qquad v_V \quad - \bar{f}_V(\lambda) \leq 0 \qquad \forall \bar{f}_V \in J_{V,i}$$

$$(\text{ii}) \qquad\qquad v_D - \bar{f}_D(\lambda) \leq 0 \qquad \forall \bar{f}_D \in J_{D,i}.$$

A dualization and some algebraic transformations using the optimality criterion $0 \in \partial \hat{f}_i(\lambda) + u(\mu_i - \lambda)$ of (QP_i) results in the equivalent formulation

$$(\text{DQP}_i) \quad \max \sum_{\bar{f}_V \in J_{V,i}} \alpha_{V,\bar{f}_V} \bar{f}_V(\mu_i) + \sum_{\bar{f}_D \in J_{D,i}} \alpha_{D,\bar{f}_D} \bar{f}_D(\mu_i)$$

$$-\frac{1}{2u} \left\| \sum_{\bar{f}_V \in J_{V,i}} \alpha_{V,\bar{f}_V} \nabla \bar{f}_V + \sum_{\bar{f}_D \in J_{D,i}} \alpha_{D,\bar{f}_D} \nabla \bar{f}_D \right\|^2,$$

$$\sum_{\bar{f}_V \in J_{V,i}} \alpha_{V,\bar{f}_V} = 1,$$

$$\sum_{\bar{f}_D \in J_{D,i}} \alpha_{D,\bar{f}_D} = 1,$$

$$\alpha_V, \alpha_D \geq 0.$$

Here, $\alpha_V \in [0,1]^{J_{V,i}}$ and $\alpha_D \in [0,1]^{J_{D,i}}$ are the dual variables associated with the constraints (QP_i) (i) and (ii), respectively. Note that (DQP_i) is again a quadratic program, the dimension of which is equal to the size of the bundles, while its codimension is only two. In our integrated scheduling method, we solve (DQP_i) using a specialized version of the *spectral bundle method* of Helmberg (2000), a variant of the PBM that can take advantage of this special structure. Given a solution (α_V, α_D) of DQP_i, the vectors

$$\tilde{g}_{V,i} := \sum_{\bar{f}_V \in J_{V,i}} \alpha_{\bar{f}_V} \nabla \bar{f}_V$$

$$\tilde{g}_{D,i} := \sum_{\bar{f}_D \in J_{D,i}} \alpha_{\bar{f}_D} \nabla \bar{f}_D$$

$$\tilde{g}_i := \tilde{g}_{V,i} + \tilde{g}_{D,i}$$

are convex combinations of subgradients; they are called *aggregated subgradients* of the functions f_V, f_D^I, and f^I, respectively. It can be shown that they are, actually, subgradients of the respective linear models of the functions at the point λ_{i+1} and, moreover, that this point can be calculated by means of the formula

$$\lambda_{i+1} = \mu_i + \frac{1}{u} \left[\sum_{\bar{f}_V \in J_{V,i}} \alpha_{V,\bar{f}_V} \nabla \bar{f}_V + \sum_{\bar{f}_D \in J_{D,i}} \alpha_{D,\bar{f}_D} \nabla \bar{f}_D \right].$$

The aggregated subgradients can be used to define linearizations of $\hat{f}_{V,i}$, $\hat{f}_{D,i}$, and $\hat{f}_i$, at λ_{i+1}:

$$\tilde{f}_{V,i}(\lambda) := \hat{f}_{V,i}(\lambda_{i+1}) + \tilde{g}_{V,i}^{\mathsf{T}}(\lambda - \lambda_{i+1})$$

$$\tilde{f}_{D,i}(\lambda) := \hat{f}_{D,i}(\lambda_{i+1}) + (\tilde{g}_{D,i})^{\mathsf{T}}(\lambda - \lambda_{i+1})$$

$$\tilde{f}_i(\lambda) := \hat{f}_i(\lambda_{i+1}) + \tilde{g}_i^{\mathsf{T}}(\lambda - \lambda_{i+1})$$

Primal approximations can be calculated using aggregated arguments as follows:

$$\tilde{x}_i := \sum_{\bar{f}_D \in J_{D,i}} \alpha_{\bar{f}_D} x(\bar{f}_D)$$

$$\tilde{y}_i := \sum_{\bar{f}_V \in J_{V,i}} \alpha_{\bar{f}_V} y(\bar{f}_V)$$

Here $x(\bar{f}_D)$ and $y(\bar{f}_V)$ are the arguments associated with the affine functions $\bar{f}_D$ and $\bar{f}_V$, respectively. The PBM (without stopping) is known to have the following properties:

- The series (μ_i) converges to an optimal solution of L_I, i.e., an optimal dual solution of the LP-relaxation of (ISP_I).
- The series $(\tilde{y}_i, \tilde{x}_i)$ converges to an optimal primal solution of the LP-relaxation of (ISP_I).
- Convergence is preserved if, at every iteration i, the bundles contain at least two affine functions, namely, the last linearizations $\bar{f}_V^I(\cdot; \lambda_i), \bar{f}_D^I(\cdot; \lambda_i)$ and the linearization of the cutting plane model $\tilde{f}_{D,i}, \tilde{f}_{V,i}$, see step 5 of Algorithm 1.

The bundle size controls the convergence speed of the PBM. If large bundles are used, less iterations are needed, however, problem (QP_i^I) becomes more difficult. We limit the bundle size for both bundles $J_{V,i}$ and J_{D_i} to 500. This is in practice no limit for our instances, since we usually perform less than 500 iterations of the bundle method. We use such large bundles because the computation time to solve problem (DQP_i) is very short in comparison to the time needed for the column generation even for this size of bundles.

3.3 Adaptations of the Bundle Method

Two obstacles prevent the straightforward application of the PBM to the ISP. First, the component problem for duty scheduling is $\mathcal{NP}$-hard, even in its LP-relaxation; the vehicle scheduling LP is computationally at least not easy. We can therefore not expect that we can compute the function values $f_V(\lambda_i)$ and $f_D^I(\lambda_i)$ and the associated subgradients $g_V(\lambda_i)$ and $g_D^I(\lambda_i)$ exactly. The algorithms of Löbel (1997) and Borndörfer et al. (2003) that we use provide in general only approximate solutions. Second, the column generation process that is carried out for the duty scheduling problem must be synchronized with the PBM. That is, the set I changes throughout the bundle algorithm.

The literature gives two versions of approximate versions of the PBM that can deal with inexact evaluations of the component functions. Kiwiel (1995) stated a version of the PBM that asymptotically produces a solution, given that ϵ-linearizations of the function f to be minimized can be found at every trial point $\mu \in \mathbb{R}^m$ for all $\epsilon > 0$, i.e., one can find an affine function $\bar{f}_\epsilon(\lambda; \mu) := f_\epsilon(\mu) + g_\epsilon(\mu)^\mathsf{T}(\lambda - \mu)$ such that $f_\epsilon(\mu) \geq f(\mu) - \epsilon$ and $f(\lambda) \geq \bar{f}_\epsilon(\lambda; \mu)$ for all $\lambda \in \mathbb{R}^m$.

Hintermüller (2001) gave another version which replaces exact subgradients of f by ϵ-subgradients. In his method it is not necessary to know or control the actual value of ϵ; his method produces solutions that are as good as the supplied ϵ-subgradients. They converge, in particular, to the optimum if the linear approximation converges to the original function.

We could use these approaches in principle in our setting, but at a high computational cost and with only limited benefit. In fact, our vehicle scheduling algorithm produces not only a primal solution, but also a lower bound and an adequate subgradient from a certain single-depot relaxation of the vehicle scheduling problem.

However, the information that can be derived from the subgradients associated with this single-depot relaxation was not very helpful in our computational experiments. Concerning the duty scheduling part, we are also able to compute a lower bound and adequate subgradients for the duty scheduling component function f_D^I for any fixed column set using exact LP-techniques. However, this is a lot of effort for a bound that is not globally valid. We remark that one can, at least in principle, also compute a lower bound for the entire duty scheduling function f_D, see Borndörfer *et al.* (2003). Such procedures are, however, extremely time consuming and do not yield high quality bounds for large-scale problems. Therefore, we use a different, much faster approach to approximate the component functions themselves by piecewise linear functions. We show below how this can be done rigorously for the vehicle scheduling part; in the duty scheduling part, the procedure is heuristic, and we simply update our approximation whenever we notice an error.

Vehicle Scheduling Function f_V. Denote by $f_V^L : \mathbb{R}^{\mathcal{D}} \mapsto \mathbb{R}$ the approximation to the value of the vehicle scheduling component function $f_V(\lambda)$ as given by some vehicle scheduling algorithm, and by $y^L(\lambda) \in [0,1]^{\mathcal{D}}$ the associated argument. We have $f_V^L(\lambda) := (d^\mathsf{T} - \lambda^\mathsf{T} M) y^L(\lambda) \geq f_V(\lambda)$, but f_V^L is in general not concave. However, we can use f_V^L to create a concave approximation $\hat{f}_{V,i}^L \geq f_V$ using a linearization at the current trial point λ_{i+1} and the linearizations stored in the bundle, namely, by setting

$$
\begin{aligned}
g_{V,i+1}^L &:= -M y^L(\lambda_{i+1}) \\
\bar{f}_V^L(\lambda; \lambda_{i+1}) &:= f_V^L(\lambda_{i+1}) + {g_{V,i+1}^L}^\mathsf{T}(\lambda - \lambda_{i+1}) \\
\hat{f}_{V,i+1}^L(\lambda) &:= \min_{\bar{f}_V \in J_{V,i} \cup \{\bar{f}_V^L(\cdot; \lambda_{i+1})\}} \bar{f}_V(\lambda).
\end{aligned}
$$

We use this approximation in the PBM Algorithm 1 by replacing f_V by $\hat{f}_{V,i}^L$. The bundle update (Step 5) is implemented as

$$
J_{V,i+1} \subset \begin{cases} J_{V,i} \cup \{\bar{f}_V^L(\cdot; \lambda_{i+1}), \tilde{f}_{V,i}\}, & \text{if } f_V^L(\lambda_{i+1}) < \hat{f}_{V,i+1}^L(\lambda_{i+1}), \\ J_{V,i}, & \text{otherwise.} \end{cases} \tag{1}
$$

Since the function $\hat{f}_{V,i+1}^L$ depends on $J_{V,i}$, we must also recalculate its value $\hat{f}_{V,i+1}^L(\mu_i)$ at the stability center in the stopping criterion and the ascent test (Steps 4 and 6) of the PBM at each iteration.

Duty Scheduling Function f_D^I. The idea is similar as in the vehicle scheduling case. Denote by I_i the duty set that is used in iteration i, by $f_D^{L,I_i} : \mathbb{R}^{\mathcal{D}} \mapsto \mathbb{R}$ a lower bound of the duty scheduling component function $f_D^{I_i}(\lambda)$ and by $x^{L,I_i}(\lambda_i)$ the argument of f_D^{L,I_i} computed again by the bundle algorithm. Here we have $f_D^{L,I_i}(\lambda) \leq f_D^{I_i}(\lambda)$, and f_D^{L,I_i} is in general not concave. Further, we know $f_D^{I_i}(\lambda) \geq f_D(\lambda)$. Thus, $f_D^{L,I_i}(\lambda)$ can be smaller or larger than $f_D(\lambda)$, the function that we actually want to maximize.

Similar, but this time heuristically, we use f_D^{L,I_i} and the current bundle to create a concave approximation $\hat{f}_{D,i}^L$ of f_D, namely,

$$g_{D,i+1}^{L} \quad := B_{I_i} x^{L,I_i}(\lambda_{i+1})$$
$$\bar{f}_D^{L,I_i}(\lambda; \lambda_{i+1}) := f_D^{L,I_i}(\lambda_i) + g_{D,i}^{L}{}^{\mathsf{T}}(\lambda - \lambda_{i+1})$$
$$\hat{f}_{D,i+1}^{L}(\lambda) \quad := \min_{\bar{f}_D \in J_{D,i} \cup \{\bar{f}_D^{L,I_i}(\lambda;\lambda_{i+1})\}} \bar{f}_D(\lambda).$$

Since each linearization is computed with respect to a subset of duties I_j, it is in general not true that $\bar{f}_D^{L,I_j} \geq f_D^{I_i}$ if $I_i \neq I_j$. It can (and does) therefore happen that we notice that the current iterate is cut off by some previously computed linearization, i.e.,

$$f_D^{L,I_i}(\lambda_{i+1}) > \bar{f}_D^{L,I_j}(\lambda_{i+1}; \lambda_j)$$

for some $j \leq i$. In this case, we have detected an error made in a previous iteration and simply remove the faulty elements from the bundle and also from the approximation. The duty scheduling bundle update in Step 5 of Algorithm 1 is implemented as

$$J_{D,i+i} \subset \begin{cases} \{\bar{f}_D \in J_{D,i} : f_D^{L,I_i}(\lambda_{i+1}) \leq \bar{f}_D(\lambda_{i+1})\} \\ \cup \{\bar{f}_D^{L,I_i}(\cdot; \lambda_{i+1}), \tilde{f}_{V,i}\}, & \text{if } f_D^{L,I_i}(\lambda_{i+1}) < \hat{f}_{D,i}^{L}(\lambda_{i+1}), \\ J_{D,i}, & \text{otherwise.} \end{cases} \quad (2)$$

This approximation must also be recomputed at the stability center in every iteration.

Combined Function f^I.   The combined approximate functions are

$$f^{L,I_i} := f_V^L + f_D^{L,I_i}$$
$$\hat{f}_i^L \quad := \hat{f}_{V,i}^L + \hat{f}_{D,i}^L.$$

Require: Starting point $\lambda_0 \in \mathbb{R}^n$, duty set I_0, weights $u_0, m > 0$, optimality tolerance $\epsilon \geq 0$.
1: Initialization: $i \leftarrow 0$, $J_{V,i} \leftarrow \{\lambda_i\}$, $J_{D,i} \leftarrow \{\lambda_i\}$, and $\mu_i = \lambda_i$.
2: Direction finding: Compute $\lambda_{i+1}, \tilde{g}_{V,i}^L, \tilde{g}_{D,i}^L$ by solving problem (QP$_i$).
3: Function evaluation: Compute $f_V^L(\lambda_{i+1}), g_V^L(\lambda_{i+1}), I_i, f_D^{L,I_i}(\lambda_{i+1}), g_D^{L,I_i}(\lambda_{i+1})$.
4: Stopping criterion: If $\hat{f}_i^L(\lambda_{i+1}) - f^{L,I_i}(\mu_i) < \epsilon(1 + |f^{L,I_i}(\mu_i)|)$ output μ_i, terminate.
5: Bundle update: Select $J_{V,i+1}, J_{D,i+1}$ as stated in (1), (2).
6: Ascent test: $\mu_{i+1} \leftarrow f^{L,I_i}(\lambda_{i+1}) - f^{L,I_i}(\mu_i) > m(\hat{f}_i^{L,I_i}(\lambda_{i+1}) - f^{L,I_i}(\mu_i))?\lambda_{i+1} : \mu_i$.
7: Weight update: Set u_{i+1}.
8: $i \leftarrow i + 1$, goto Step 2.

Algorithm 2: Inexact PBM with Column Generation

Column generation.   This is the most time consuming part of our algorithm, and we therefore enter this phase only if significant progress can be expected. Details about the column generation itself can be found in Borndörfer et al. (2003). Our strategy to generate new columns is basically to recompute the duty set when the stability center changes; we call such an iteration a *serious step*, all other iterations are called *null steps*.

The reasoning behind this strategy is as follows. The quadratic penalty term in the quadratic program QP_i ensures that the next trial value for the dual multipliers λ_{i+1} stays in the vicinity of the current stability center. When the multipliers change only little, one has reason to believe that the number and the potential effect of improving duties is also small. We therefore hope that the current duty set I_i, which has been updated when the stability center was set, does still provide a good representation of the duty space also for the new multipliers λ_{i+1}. In practice, we reduce the number of column generation phases even further by requiring a certain minimum increase ε in the objective function at the new stability center; the larger ε, the less column generation phases will occur.

Algorithm 2 gives a listing of our bundle algorithm using inexact evaluations of the component functions and column generation in the duty scheduling component.

3.4 Backtracking Procedure

The inexact proximal bundle method that we have described in this section is embedded in a backtracking procedure that aims at the generation of integer solutions. This procedure makes use of the primal information produced by the bundle method, namely, the sequence $(\tilde{y}_i, \tilde{x}_i)$. As in an LP-approach, fractional values can be interpreted as probabilities for the inclusion/exclusion of a deadhead trip or duty in an optimal integer solution.

Our computational experiments revealed that it is advantageous to fix the deadhead trips first, until the vehicle scheduling part of the problem is decided. The remaining duty scheduling problem can then be solved with the duty scheduling module of the algorithm as described in Borndörfer et al. (2003). Our strategy for fixing the deadhead variables is to fix the deadheads in the order of largest y-values. Our algorithm also examines the consequences of such fixings and, if the increase in the objective function is too large, also reverses decisions. The details on how many variables to fix at a time, up to which threshold, etc. have been determined experimentally. In general, the algorithm fixes more boldly in the beginning and more carefully towards the end.

Fig. 1 shows a typical runtime chart of our algorithm IS-OPT. The x-axis measures time in seconds, the y-axis gives statistics in two different scales, namely, on the right scale, the number of duties generated (#columns), the number of deadheads fixed to one (#fixed deadheads), and the residuum of the coupling constraints (more precisely: the norm is the square of the Euclidean norm of $\tilde{g}_i$), as well as, on the left scale, the vehicle, duty, and the integrated scheduling objective values. Here the duty scheduling value is the lower bound of the restricted DSP calculated by the PBM, and integrated scheduling objective value is simply the sum of the VSP and the DSP value.

In the first phase of the algorithm until point A a starting set of columns was generated with Lagrangean multipliers λ all at zero. In principle the DSP objective value should be strictly decreasing here, while the number of columns should grow. However, we calculated in this initial phase only rough lower bounds for the restricted DSP, which may be more or less accurate. Additionally we deleted columns with

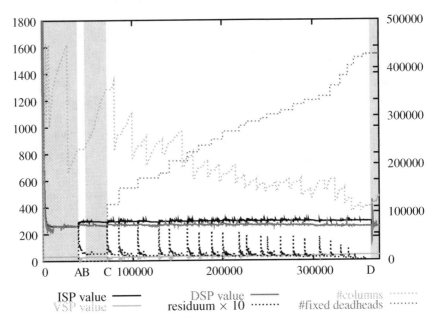

Fig. 1. IS-OPT Runtime Chart

large reduced cost if the total number of columns exceeded 450,000. Between points A and B, a series of null steps was performed, which resulted in a decreased norm and an increased ISP-value. Between points B and C, column generation phases alternated with PBM-steps, until an aggregated subgradient of small norm and thus also a "good" primal approximation of the LP-relaxation of ISP was calculated. Since the column generation process did not find enough improving columns at this point, we used the computed information to fix deadheads until (at point D) the vehicle scheduling part of the problem was completely decided. At that point, the duty scheduling component of the algorithm concluded by computing a feasible duty scheduling.

Serious steps of the PBM are marked by peaks of the norm statistic. This effect is due to the shift of the stability center in combination with the possible inclusion of additional columns in I_i. In fact, the new stability center may lie in a region where the model $\hat{f}^{L,I_i}$ of the previous iteration i is less accurate; also, new columns in I_i change the function f^{L,I_i}, which also worsens the model.

In our computational tests the algorithm rarely had to reverse a fixing decision for a deadhead and backtrack. In all our instances, the ISP objective value is very stable with respect to careful fixings of deadheads, see also Fig. 1. In fact, the gap between our estimated lower bound, i.e., the objective value prior to the first fixings, and the final objective value was never larger than 5% and only 1-2% on the average. However, we do not know the size of the gap between the estimated lower bound and the real minimum of (ISP); the mentioned behavior is therefore only a weak indicator for the quality of the final solution found by IS-OPT.

4 Computational Results

In this section, we report the results of computational studies with our integrated vehicle and duty scheduling optimizer IS-OPT for several medium- and large-scale real-world scenarios as well as for benchmark scenarios from the literature. Our code IS-OPT is implemented in C and has been compiled using gcc version 3.3.3 with switches -O4. All computations were made single-threaded on a Dell Precision 650 PC with 4 GB of main memory and a dual Intel Xeon 3.0 GHz CPU running SuSE Linux 9.0. The computation times in the following tables are in hours:minutes.

We compare our integrated scheduling method *is* with two sequential approaches. The first one, denoted by *v+d*, is a classical sequential vehicles-first duties-second approach, i.e., *v+d* first solves the vehicle scheduling part of the problem using our optimizer VS-OPT (Löbel (1997)), fixes the deadheads chosen by the vehicle schedule, and solves the resulting duty scheduling problem in a second step using our optimizer DS-OPT (Borndörfer *et al.* (2003)). The second method *d+v* uses kind of the contrary approach. A simplified integrated scheduling problem is set up that identifies drivers and vehicles, i.e., vehicle changes outside of the depot are forbidden. This "poor man's integrated scheduling model" is solved using the duty scheduling algorithm DS-OPT. The vehicle rotations resulting from this duty schedule are concatenated into daily blocks using the vehicle scheduling algorithm VS-OPT in a second step.

We calibrated the parameters of the bundle method, namely m and the series $(u_i)_{i=1,2,...}$, such that about 20% of the iterations were serious steps. We never needed more than 50 iterations of the bundle method before the first fixing of variables.

4.1 RVB Instances

The Regensburger Verkehrsbetriebe GmbH (RVB) is a medium sized public transportation company in Germany. We consider two instances that contain the entire RVB operation for a Sunday and for a workday. The structure of the RVB data is mostly urban with only four relief points. In fact, the network of the RVB is mostly star-shaped with nearly all lines meeting in a small area around the main railway station. Only there, at two stations nearby, and at the also nearby garage the drivers can change buses and begin or end duties. The RVB uses only one type of vehicle on Sundays, and three types on workdays, i.e., the Sunday scenario is fleet homogenous, while the workday scenario is a multi-depot problem. The vehicle types can only be used on trips on certain sets of (non-disjoint) lines. The Sunday scenario involves three different types of early, mid, and late duties, each with four different types of break rules. In Germany, detailed legal regulations exist about the number, the length, and the feasible positions of breaks in a duty. These regulations may also differ from one company to the other by works council agreements. We use in the RVB instances block breaks of 1×30, 2×20, and 3×15 minutes plus 1/6-quotient breaks. The most important regulations valid for all these break rules are: There is no interval without break with more than six hours working time. There is no interval without break with more than four and a half hours driving time. Between two breaks is at least half an hour of working time. A duty fulfills the 1/6-quotient break rule if every continuous segment of a duty contains at least a sixth part break time, and every break must be at least eight minutes.

The workday scenario contains in addition a type of split duties, again with the mentioned break rules per part of work. Table 1 reports further statistics on the number of timetabled trips, tasks, and deadhead trips (also equal to the number of Lagrangean multipliers). The Sunday scenario is medium-sized, while the workday scenario is, as far as we know, the largest and most complex instance that has been attacked with integrated scheduling techniques.

Table 1. Statistics on the RVB Instances

	Sunday	workday
vehicle types	1	3
timetabled trips	794	1414
tasks on tt	1248	3666
deadhead trips	47523	57646
duty types	3	4
break rules	4	4

Table 2 gives computational results for the Sunday scenario. The column 'reference' lists statistics for the solution that RVB planners had generated by hand. The next four columns give the results of two sequential $v+d$-optimizations and two integrated is-optimizations; we do not report results for the method $d+v$, because we could not produce a feasible solution for this scenario with this method. The objective function consists of a weighted sum of the number of duties, the number of pieces of work, the paid time of the duty schedule, and penalties for exceeding an average duty time. A *piece of work* is an inclusion-maximal continuous segment of a duty where a driver does not change the vehicle. Changes of vehicles should be avoided because they may lead to operational problems in case of delays of vehicles.

In the optimization runs "$v+d$ 2" and "is 2", emphasis was placed on the minimization of the number of duties, while runs "$v+d$ 1" and "is 1" tried to reproduce the average duty time of the reference solution.

Table 2. Results for the RVB Sunday Scenario

	reference	$v+d$ 1	$v+d$ 2	is 1	is 2
time on vehicles	518:33	472:12	472:12	501:42	512:55
paid time	545:25	562:58	565:28	518:03	531:31
paid break time	112:36	131:40	85:41	74:17	64:27
number of duties (slacks)	82	83	74(1)	76	66
number of vehicles	36	32	32	32	35
average duty duration	6:39	6:48	7:38	6:40	8:03
computation time	—	0:33	5:13	35:44	37:26

As expected the sequential methods reduce the number of vehicles and the time on vehicle rotations since these are the primary optimization objectives. Also they produce quite reasonable results in terms of duty scheduling. "*v+d* 1" suffers from a slight increase in duties and paid time, "*v+d* 2" yields substantial savings in duties; however, the price for this reduction is a raised average paid time. Also one task was not covered by duties in the solution (remarked by the one in brackets). Even better are the results of the integrated optimizations. "*is* 1" is perfect with respect to any statistic and produces *large* savings. These stem from the use of short duties involving less than 4:30 hours of driving time, which do not need a break; this potential improvement of the Sunday schedule is one of the most significant results of this optimization project for the RVB. Even more interesting is solution "*is* 2." This solution trades three vehicles and an increased average for another 10 duties; as longer duties must have breaks, the paid time (breaks are paid here) increases as well. Solution "*is* 2" revived a discussion at the RVB whether drivers prefer to have less, but longer duties on weekends or whether they want to stay with more, but short duties.

Table 3 lists the results of the workday optimizations. Method *d+v* could again not produce a feasible solution and is therefore omitted from the table. The objective in this scenario is far from obvious; it is given as a complicated mix of fixed and variable vehicle costs, fixed costs and paid time for duties, and various penalties for several pieces of work, split duties, etc., that can compensate each other such that one cannot really compare the solutions by means of a single statistic. Doing it nevertheless, we see that both optimization approaches clearly improve the reference solution substantially. The outcome is close. In fact, *v+d* has less paid time than *is*; in the end, however, *is* is better in terms of the composite objective function.

Table 3. Results for the RVB Workday Scenario

	reference	*v+d*	*is*
time on vehicles	1037:18	960:29	1004:27
paid time	1103:48	1032:20	1040:11
granted break time	211:53	109:11	105:23
number of duties	140	137	137
number of vehicles	91	80	82
number of pieces of work	217	290	217
number of split duties	29	39	36
average duty duration	7:56	8:03	7:55
objective value	—	302.32	291.16
computation time	—	8:02	125:55

4.2 RKH Instances

The Regionalverkehrsbetrieb Kurhessen (RKH) is a regional carrier in the middle of Germany. They provided data for the subnetworks of Marburg and Fulda which is not (yet) in industrial use; some deadheads are missing, while for some others travel times have only been estimated by means of distance calculations. In our opinion the data still captures to a large degree the structure of a regional carrier and we therefore deem it worthwhile to report the results of the conceptual study that we did with it.

Fig. 2 shows the spatial structure of the line network of Fulda, which is one part of the RKH service area. The black arcs denote the timetabled trips (drawn straight from the line's start to the end), the gray arcs indicate the potential deadhead trips. It can be seen that the trip network is hub-and-spoke-like, connecting several cities and villages among themselves and with the rural regions around them. While the deadhead network is almost complete, there are only a few relief opportunities for drivers to leave or enter a vehicle.

Table 4 gives further statistics on the RKH instances. They are similar to the RVB Sunday scenario in terms of timetabled trips and tasks, but contain much more deadhead trips. The scenarios involve three duty types, two types of split duties that differ in the maximum duty length and one type of continuous duties. Each duty type can have 1×30, 2×20, or 3×15 minutes block breaks or 1/6-quotient breaks.

Table 4. RKH Instances for the Cities of Marburg and Fulda

	Marburg	Fulda
depots	3	1
vehicle types	5	1
timetabled trips	634	413
tasks on tt	1022	705
deadhead trips	142,668	67,287

Table 5 reports the results of our optimizations. We do not report results for the method $v+d$ as we were not able to produce a feasible solution for either scenario with this method. Method $d+v$ yields useful results, but it is not able to cover all tasks/trips of the Fulda-scenario with duties and vehicles; in fact, $d+v$ left three tasks

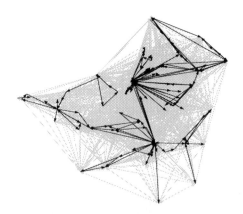

Fig. 2. The Graph of Scenario *Fulda*

and six timetabled trips uncovered (numbers in parentheses). These deficiencies are resolved in the *is*-solutions, which also look better in terms of numbers of vehicles.

Table 5. Solutions on Marburg and Fulda

	Marburg		Fulda	
	d+v	*is*	*d+v*	*is*
time on vehicles	772:02	642:41	365:41	387:37
paid time	620:27	606:30	390:08	374:53
granted break time	120:51	103:27	88:13	57:44
number of duties	73	70	41(3)	41
number of vehicles	62	50	45(6)	37
average duty duration	10:35	10:18	10:59	11:18
computation time	5:29	17:18	1:42	7:05

4.3 ECOPT Instances

Finally, we compare IS-OPT with the approach of Huisman *et al.* (2005) on the randomly generated benchmark data proposed in their article. These data consist of two sets of instances involving two and four depots, respectively. Each set contains ten instances of 80, 100, 160, 200, 320, and 400 trips; see again Huisman *et al.* (2005) for a detailed description. The duty scheduling rules associated with these examples are relatively simple. Duties are allowed to have at most one break, which must be outside of a vehicle, i.e., each break also begins a new piece of work. The only other rule is that each piece of work must be of certain minimum and maximum length. It is shown in Huisman *et al.* (2005) that in this situation one can solve the duty generation subproblem in polynomial time, i.e., exact column generation is applicable.

Tables 6 and 7 report average solution values for each of the ten instances of each problem class for the problem variant A; similar results for variant B have been omitted. All computations were done with the same set of parameters, which was optimized for speed. Row *reference* gives the sum of the numbers of vehicles and duties as published in Huisman *et al.* (2005); for the problems with 4 depots and 320 and 400 trips, no reference is given due to excessive computation time.

Table 6. Results for ECOPT-Instances with 2 Depots Variant A

trips	080	10 0	160	200	320	400
vehicles	9.4	11.2	15.0	18.6	27.0	33.3
duties	21.2	25.1	33.9	40.6	57.7	69.8
total	30.6	36.3	48.9	59.2	84.7	103.1
reference	29.8	35.6	48.3	59.1	86.8	106.1
time	00:05	00:08	00:17	00:31	01:58	03:19

Table 7. Results for ECOPT-Instances with 4 Depots Variant A

trips	080	100	160	200	320	400
vehicles	9.2	11.2	15.0	18.5	26.7	33.1
duties	20.4	24.5	32.7	40.5	56.1	68.9
total	29.6	35.7	47.7	59.0	82.8	102.0
reference	29.6	36.2	49.5	60.4	—	—
time	00:13	00:21	00:44	01:46	05:28	12:00

It can be seen that our algorithm IS-OPT performs worse than that in Huisman *et al.* (2005) for the small instances, but produces better results with increasing problem size and complexity; it can also solve the largest problem instances. We remark that IS-OPT can also produce slightly better solutions for the small instances than those reported in Huisman *et al.* (2005) by changing the optimality parameter ϵ in Algorithm 2 and by raising the threshold for deadhead fixes. This leads, of course, to longer computation times.

5 Conclusions

We have shown that it is possible to tackle large-scale, complex, real-world integrated vehicle and duty scheduling problems using a novel "bundle" algorithm for integrated vehicle and duty scheduling. The solutions produced by such an integrated approach can be decidedly better *in several respects at once* than the results of various types of sequential planning.

Acknowledgement: This research has been supported by the German ministry for research and education (BMBF), grant No 03-GRM2B4. Responsibility for the content of this article is with the authors.

References

Ball, M. O., Bodin, L., and Dial, R. (1983). A matching based heuristic for scheduling mass transit crews and vehicles. *Transportation Science*, **17**, 4–31.

Borndörfer, R., Grötschel, M., and Löbel, A. (2003). Duty scheduling in public transit. In W. Jäger and H.-J. Krebs, editors, *MATHEMATICS – Key Technology for the Future*, pages 653–674. Springer Verlag, Berlin. http://www.zib.de/PaperWeb/abstracts/ZR-01-02.

Daduna, J. R. and Völker, M. (1997). Fahrzeugumlaufbildung im ÖPNV mit unscharfen Abfahrtszeiten (in German). *Der Nahverkehr,* 11/1997, pages 39–43.

Daduna, J. R. and Wren, A., editors (1988). *Computer-Aided Transit Scheduling*, volume 308 of *Lecture Notes in Economics and Mathematical Systems*. Springer.

Daduna, J. R., Branco, I., and Paixão, J. M. P., editors (1995). *Computer-Aided Transit Scheduling*, volume 430 of *Lecture Notes in Economics and Mathematical Systems*. Springer.

Darby-Dowman, K., J. K. Jachnik, R. L. L., and Mitra, G. (1988). Integrated decision support systems for urban transport scheduling: Discussion of implementation and experience. In J. R. Daduna and A. Wren, editors, *Computer-Aided Transit Scheduling*, volume 308 of *Lecture Notes in Economics and Mathematical Systems*, pages 226–239, Berlin. Springer.

Desrochers, M. and Rousseau, J.-M., editors (1992). *Computer-Aided Transit Scheduling*, volume 386 of *Lecture Notes in Economics and Mathematical Systems*. Springer.

Desrochers, M. and Soumis, F. (1989). A column generation approach to the urban transit crew scheduling problem. *Transportation Science*, **23**(1), 1–13.

Falkner, J. C. and Ryan, D. M. (1992). Express: Set partitioning for bus crew scheduling in Christchurch. In M. Desrochers and J.-M. Rousseau, editors, *Computer-Aided Transit Scheduling*, volume 386 of *Lecture Notes in Economics and Mathematical Systems*, pages 359–378, Berlin. Springer.

Freling, R. (1997). *Models and Techniques for Integrating Vehicle and Crew Scheduling*. Ph.D. thesis, Erasmus University Rotterdam, Amsterdam.

Freling, R., Huisman, D., and Wagelmans, A. P. M. (2001a). Applying an integrated approach to vehicle and crew scheduling in practice. In S. Voß and J. R. Daduna, editors, *Computer-Aided Scheduling of Public Transport*, volume 505 of *Lecture Notes in Economics and Mathematical Systems*, pages 73–90, Berlin. Springer.

Freling, R., Wagelmans, A. P. M., and Paixao, J. M. P. (2001b). Models and algorithms for single-depot vehicle scheduling. *Transportation Science*, **35**, 165–180.

Freling, R., Huisman, D., and Wagelmans, A. P. M. (2003). Models and algorithms for integration of vehicle and crew scheduling. *Journal of Scheduling*, **6**, 63–85.

Friberg, C. and Haase, K. (1999). An exact algorithm for the vehicle and crew scheduling problem. In N. H. M. Wilson, editor, *Computer-Aided Transit Scheduling*, volume 471 of *Lecture Notes in Economics and Mathematical Systems*, pages 63–80, Berlin. Springer.

Gaffi, A. and Nonato, M. (1999). An integrated approach to extra-urban crew and vehicle scheduling. In N. H. M. Wilson, editor, *Computer-Aided Transit Scheduling*, volume 471 of *Lecture Notes in Economics and Mathematical Systems*, pages 103–128, Berlin. Springer.

Haase, K., Desaulniers, G., and Desrosiers, J. (2001). Simultaneous vehicle and crew scheduling in urban mass transit systems. *Transportation Science*, **35**(3), 286–303.

Hanisch, J. (1990). Die Regionalverkehr Köln GmbH und HASTUS (in German). http://www.giro.ca/Deutsch/Publications/publications.htm.

Helmberg, C. (2000). Semidefinite programming for combinatorial optimization. Technical report ZR00-34. Zuse Institute Berlin.

Hintermüller, M. (2001). A proximal bundle method based on approximate subgradients. *Computational Optimization and Applications*, (20), 245–266.

Huisman, D., Freling, R., and Wagelmans, A. P. M. (2005). Multiple-depot integrated vehicle and crew scheduling. *Transportation Science*, **39**, 491–502.

Kiwiel, K. C. (1990). Proximal bundle methods. *Mathematical Programming*, **46**(123), 105–122.

Kiwiel, K. C. (1995). Approximation in proximal bundle methods and decomposition of convex programs. *Journal of Optimization Theory and Applications*, **84**(3), 529–548.

Lemarechal, C. (2001). Lagrangian relaxation. In M. Jünger and D. Naddef, editors, *Computational Combinatorial Optimization*, volume 2241 of *Lecture Notes in Computer Science*, pages 112–156, Berlin. Springer.

Löbel, A. (1997). *Optimal Vehicle Scheduling in Public Transit*. Ph.D. thesis, TU Berlin. http://www.zib.de/bib/diss/index.en.html.

Löbel, A. (1999). Solving large-scale multi-depot vehicle scheduling problems. In N. H. M. Wilson, editor, *Computer-Aided Transit Scheduling*, volume 471 of *Lecture Notes in Economics and Mathematical Systems*, pages 195–222, Berlin. Springer.

Patrikalakis, I. and Xerocostas, D. (1992). A new decomposition scheme of the urban public transport scheduling problem. In M. Desrochers and J.-M. Rousseau, editors, *Computer-Aided Transit Scheduling*, volume 386 of *Lecture Notes in Economics and Mathematical Systems*, pages 407–425, Berlin. Springer.

Scott, D. (1985). A large scale linear programming approach to the public transport scheduling and costing problem. In J.-M. Rousseau, editor, *Computer Scheduling of Public Transport 2*. Amsterdam, Elsevier.

Tosini, E. and Vercellis, C. (1988). An interactive system for extra-urban vehicle and crew scheduling problems. In J. R. Daduna and A. Wren, editors, *Computer-Aided Transit Scheduling*, pages 41–53, Berlin. Springer.

Voß, S. and Daduna, J. R., editors (2001). *Computer-Aided Scheduling of Public Transport*, volume 505 of *Lecture Notes in Economics and Mathematical Systems*. Berlin, Springer.

Wilson, N. H. M., editor (1999). *Computer-Aided Transit Scheduling*, volume 471 of *Lecture Notes in Economics and Mathematical Systems*. Berlin, Springer.

A Crew Scheduling Approach for Public Transit Enhanced with Aspects from Vehicle Scheduling

Vitali Gintner[1], Natalia Kliewer[2], and Leena Suhl[2]

[1] Decision Support & Operations Research Lab and International Graduate School for Dynamic Intelligent Systems, University of Paderborn, Warburger Str. 100, D-33100 Paderborn, Germany, Email: gintner@dsor.de

[2] Decision Support & Operations Research Lab, University of Paderborn, Warburger Str. 100, D-33100 Paderborn, Germany, Email: {kliewer, suhl}@dsor.de

Summary. This paper presents a new approach for solving the crew scheduling problem in public transit. The approach is based on interaction with the corresponding vehicle scheduling problem. We use a model of the vehicle scheduling problem which is based on a time-space network formulation. An advantage of this procedure is that it produces a bundle of optimal vehicle schedules, implicitly given by the solution flow. In our approach, we give this degree of freedom to the crew scheduling phase, where a vehicle schedule is selected that is most consistent with the objectives of crew scheduling.

1 Introduction

Scheduling of vehicles and of crews are two main problems arising in public transport scheduling, because there are the main resources necessary to service passengers. The main objective of vehicle and crew scheduling is to use a minimum amount of resources per required service. Traditionally, vehicle and crew scheduling problems have been approached in a sequential manner, so that vehicles are first assigned to trips, and in a second phase, crews are assigned to the vehicle blocks generated before. However, this procedure implies that the crew duties are based on a fixed underlying vehicle schedule. The crews' schedule flexibility is thereby restricted, which sometimes leads to an infeasible or inefficient crew schedule.

The fact of possibly losing efficiency or feasibility has motivated several researchers to work on simultaneous vehicle and crew scheduling. In the last years, different ways have been proposed to combine bus and driver scheduling. These approaches can be divided into two main groups, namely partial and complete integration.

Most of the techniques of the first category schedule vehicles during a heuristic approach to crew scheduling. Many of these heuristics are based on the procedure proposed by Ball *et al.* (1983). Similar procedures were proposed by Tosini and

Vercellis (1988), Falkner and Ryan (1992), and Patrikalakis and Xerocostas (1992). Another technique for a partial integration is to include crew considerations in the vehicle scheduling process. Approaches of this sub-category include Darby-Dowman *et al.* (1988) – an interactive part of a decision support system – and Scott (1985), who heuristically determines vehicle schedules while taking crew costs into account. For a detailed overview of these papers, we refer to Freling (1997).

Approaches of the second category (complete integration of vehicle and crew scheduling) have only appeared very recently. The first mathematical formulation was by Patrikalakis and Xerocostas (1992), followed and slightly changed by Freling *et al.* (1995). An exact algorithm for the single-depot vehicle and crew scheduling problem was proposed by Friberg and Haase (1999). Both the vehicle and crew scheduling aspects are modeled by using a set partitioning formulation of the problem. The solution approach combines column generation and cut generation in a branch-and-bound (B&B) algorithm. Haase *et al.* (2001) propose an approach which solves the crew scheduling problem (CSP) while incorporating side constraints for the vehicles. This is done in such a way that the solution of this problem guarantees that an overall optimal solution is found after constructing a compatible vehicle schedule.

A complete integration of vehicle and crew scheduling for the multiple-depot case is treated by Desaulniers *et al.* (2001), and, very recently, by Huisman *et al.* (2005), Huisman (2004). Their approaches are based on Lagrangian relaxation combined with column generation. However, these methods are hardly applicable to huge real-world problems, with multiple depots and heterogeneous fleet. As a result, algorithms incorporated in commercially successful computer packages keep using the sequential approach or, sometimes, offer integration on the user level.

The solution approach presented in this paper can be assigned to the first category, namely to the partial integration of vehicle and crew scheduling. It solves the vehicle scheduling problem first and the crew scheduling problem afterwards. In contrast to the traditional sequential approach, in our method scheduling of crews is based not only on one given optimal vehicle schedule but on a set of optimal vehicle schedules with minimum fleet size and minimal operational costs. This is possible due to the specific model used for solving the Multiple-Depot Vehicle Scheduling Problem (MDVSP), known to be $\mathcal{NP}$-hard (see Bertossi *et al.* (1987)). We use a multi-commodity flow formulation to solve the MDVSP, which is based on the time-space network as described in Kliewer *et al.* (2002), Kliewer *et al.* (2005), Gintner *et al.* (2005). The model guarantees a minimal fleet size and minimal operational costs for vehicles (deadhead cost and idle time outside of a depot).

An optimal solution of our formulation for MDVSP is a flow in the underlying network. Due to our time-space formulation each flow can be decomposed into many different sets of paths, because this decomposition is not unique. Each path represents a day route (vehicle block) for a vehicle, while each path set builds an optimal vehicle schedule. In our approach, we give this freedom over the choice of decomposition to the crew scheduling phase in order to select a vehicle schedule that harmonizes with the objectives of the crew scheduling. Note that we only select one of the optimal solutions, so that optimality of vehicle schedules is preserved.

To solve the crew scheduling problem, we use a column generation approach applied to a set partitioning formulation. However, in our method, the vehicle schedule is not given explicitly. Possible crew duties are generated not only based on a single optimal vehicle schedule but on the optimal flow in the time-space network used for the MDVSP. Since all decompositions of the optimal flow produce an optimal vehicle schedule (with respect to the fleet size and the operational costs for a vehicle), the final vehicle schedule can be created afterwards, depending on the final crew schedule.

We have tested our proposed approach on randomly generated and real-world data instances and compared it to the traditional method. Due to the additional flexibility in duty generation, a better crew schedule was produced.

The paper is organized as follows. In Section 2, we briefly define the MDVSP and the time-space network based model which we use for it. In Section 3, we discuss the traditional crew scheduling problem based on a given vehicle schedule. Section 4 provides the proposed crew scheduling approach, which is based on an interaction with the corresponding vehicle scheduling problem. In Section 5, we show some computational results on randomly generated and real-world data instances. Finally, a summary is given in Section 6.

2 Vehicle Scheduling

The *Vehicle scheduling problem (VSP)* deals with assigning vehicles to trips so that the total vehicle costs are minimal. The total vehicle costs usually consist of a fixed component for using each vehicle and variable costs as a function of travel distance and time. A vehicle schedule is feasible if all trips are assigned to a vehicle and if each vehicle starts in a depot, performs a sequence of trips and ends in the same depot.

In the one depot case with a homogeneous fleet (all vehicles are identical) we have the standard *Single-Depot Vehicle Scheduling Problem (SDVSP)*. It is well known that the SDVSP can be solved in polynomial time (see, e.g., Freling (1997)). The problem with more than one depot and/or heterogeneous fleet (more than one vehicle type) is defined as the *MDVSP*. In this case all vehicles have to be assigned to a depot (*home depot*). Furthermore, some trips may be assigned only to vehicles from a certain subset of depots and/or vehicle types. In some practical cases there are also other types of constraints, such as depot capacity constraints, which specify a maximum number of vehicles for every depot. The MDVSP is shown to be $\mathcal{NP}$-hard by Bertossi *et al.* (1987) if there are at least two depots. Moreover, Löbel (1997) shows that even ϵ-approximation of the MDVSP is $\mathcal{NP}$-hard.

For the last decades, a lot of attention has been given to the MDVSP in the literature. Most approaches base on a multi-commodity flow formulation (see, e.g., Forbes *et al.* (1994), Mesquita and Paixão (1999), Löbel (1997)). The most popular network model for the MDVSP is a so-called *connection based network*, where each possible connection between compatible trips is presented by an arc. A drawback of such a network is the number of possible connections which increases quadratically with the number of trips. Thus, models with several thousand trips become too large to be solved directly by standard optimization tools in a reasonable time. There are different techniques to reduce the number of possible connections. Some approaches discard arcs with too long waiting time; other approaches generate arcs applying the column generation idea to the network flow representation. Further special solution techniques, such as column generation or branch-and-price with Lagrangian relaxation, have been introduced in order to solve problems of practical size (see, e.g., Löbel (1999)).

Very recently, Kliewer et al. (2002), Kliewer et al. (2005) proposed a new way to model the MDVSP. They use a *time-space network* which is known from the airline scheduling background (see Hane et al. (1995)). The main contribution of this network is that connections between compatible trips are presented implicitly by the flow. Thus, the number of arcs in such a network is only a fraction of this in equivalent connection-based network.

In this paper, we exploit another property of the time-space network, namely that a solution flow can be decomposed into a multitude of different optimal vehicle schedules. Therefore, we give some details of this modeling approach next. For a full description, we refer the reader to Kliewer et al. (2002), Kliewer et al. (2005), Gintner et al. (2005).

MDVSP Formulation Based on the Time-Space Network

Nodes in a time-space network correspond to points in time and space. Each trip is represented by two nodes (one for the departure and one for the arrival event; each event referring to the corresponding station) and a *trip arc* in-between. Two additional arcs (*depot arcs*) for each trip represent possible pull-out and pull-in trips from and to the depot, respectively. The from-depot arc (to-depot arc) connects the corresponding departing node (arriving node) of the trip with a depot node which represents the start point of the pull-out trip (end point of the pull-in trip).

All nodes are grouped by corresponding stations and sorted by ascending time. We create a *waiting arc* between two consecutive nodes at the same station if there is not enough time to perform a round-trip to the depot. Waiting arcs represent vehicles waiting at a station. Thus, a trip arriving on its end station can be implicitly connected with each trip departing later from the same station through a flow using waiting and/or depot arcs.

One special requirement in bus traffic is that empty movements (*deadheading*) are basically possible between all stations, i.e., after each trip, a bus may move to any of the other stations to take over a trip starting elsewhere. Thus, we have to provide a connection between all compatible trips. But instead of doing it explicitly by creating an arc for each connection as in the connection-based network, we take advantage of the special structure of a time-space network and its ability to forward the flow through the waiting and/or depot arcs. For each trip i, we consider a *deadhead (dh) arc* from its arriving node to the first available departing nodes on every other station. Note that for each trip, there is at most one dh-arc to each station. All later trips are connected with i through the dh-arc and a sequence of waiting and/or depot arcs. Moreover, not all such dh-arcs are needed. Some of them can be omitted due to forwarding the flow on the source station as well. We refer to Kliewer *et al.* (2002), Kliewer *et al.* (2005), Gintner *et al.* (2005) for the detailed description of this and further aggregation techniques. Thus, all possible connections between compatible trips are implicitly included. Let n and m be the number of trips and stations, respectively. Then the number of arcs is $\mathcal{O}(nm)$ instead of $\mathcal{O}(n^2)$ for the connection-network model, while usually $n \gg m$ holds. Kliewer *et al.* (2005) report that the number of arcs in the time-space network amounts only 1-3% of the arcs in an equivalent connection-based network. Thus, the problem size could be reduced significantly without reducing the solution space because all compatible trips are implicitly connected.

Finally, we create a *circulation arc* from the last to the first depot node. The network is a directed acyclic graph. A path from the first to the last depot node represents a day schedule for one vehicle. The capacity of the arcs is set to one for trip and depot arcs and to C for all remaining arcs, where C is the maximum number of vehicles available at the corresponding depot. Fig. 1 shows an example of a time-space network for an instance with six trips and one depot.

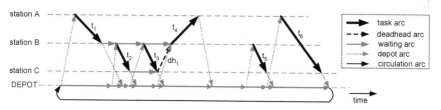

Fig. 1. Basic Structure of the Time-Space Network Model

For the case with more than one depot and/or vehicle type, such a network is built for each combination of depot and vehicle type. As some trips can only be performed by a certain vehicle type or from a certain subset of depots, each network contains only trips allowed for the corresponding depot and vehicle type.

The MDVSP is formulated as a min-cost multi-commodity network flow problem. We associate vehicle costs with each arc in the network according to the corresponding driving/waiting activity. Fixed costs for using a vehicle are associated with each circulation arc. An integer variable for each arc indicates the flow value through the arc. We minimize the total vehicle costs. For each node in the network, there is a flow conservation constraint in the MIP model. An additional set of constraints ensures that each trip is covered by a vehicle. The proposed MIP formulation is solved using the all-purpose solver CPLEX 9.0. Note that this time-space-network based model has very good MIP behavior. The IP-gap is infinitesimally small or non-existent and almost all variables have integer values in the optimal (basis) solution of the LP-relaxation.

Flow Decomposition
The solution vector describes the solution flow (a set of selected arcs) in each net-

work. Each flow unit represents a vehicle starting in the first depot node, flowing through the network arcs and returning back through the circulation arc into the first depot node. In order to create a feasible vehicle schedule, the solution flow has to be decomposed into paths. However, such flow decomposition is usually not unique since there are many possibilities to determine an optimal schedule.

Fig. 2a shows an example of this situation. Consider three arrivals t_1, t_2, t_3 and three departures t_4, t_5, t_6. If the flow value on the dh-arc equals three units in the optimal solution, there is still a degree of freedom with respect to connecting these trip arcs. It is obvious that there are six possible ways to connect the arrivals with the departures in the optimal solution. Multiple decompositions do not only occur in aggregated arcs, but also in nodes, see Fig. 2b.

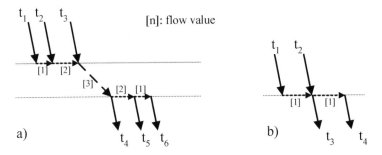

Fig. 2. Multiplicity of Decompositions

A possible approach to constructing a set of feasible paths is to decompose locally in each node. Due to the flow conservation constraints, the number of flow units entering a node equals the number of units leaving that node. Therefore, it is sufficient to connect each entering flow unit with a leaving flow unit for each node (e.g., using *Last-In-First-Out* order).

In our approach, we postpone the decomposition decision to the crew scheduling phase in order to select a vehicle schedule that is consistent with the objectives of crew scheduling. Note that we only select among optimal solutions, so that optimality of the vehicle schedules is always preserved. Further details of the proposed approach are given in Section 4.

3 Traditional Crew Scheduling

In the following, we assume that the VSP has already been solved and a set of *vehicle blocks* defining the vehicle schedule is known. For each block, a set of *relief points*, i.e., locations where a driver in the vehicle can be replaced by a new driver, is given. A *task* is defined by two consecutive relief points and represents the minimum portion of work that can be assigned to a crew. A *piece of work* is one or more consecutive tasks performed by a driver on one vehicle block without a break. The feasibility of the pieces of work is restricted by a minimum and a maximum duration. A *duty* consists of one or more pieces of work executed by the same driver.

The Crew Scheduling Problem (CSP) deals with assigning tasks to duties such that each task is performed, each duty is feasible, and the total cost of the duties is minimized. A duty is feasible if it satisfies several constraints corresponding to work regulations for crews. Typical examples of such constraints are maximum working time without a break, minimum break duration, maximum duty duration and so on. These constraints can vary between different types of duties, e.g., tripper, early, late and split duties.

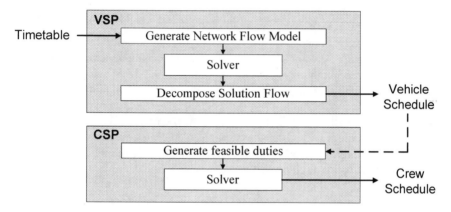

Fig. 3. Traditional Crew Scheduling Approach

Fig. 3 shows the main phases of the solution approach for vehicle and crew scheduling problems and the interactions between them. Again, an optimal vehicle schedule serves as input for the traditional CSP.

The CSP is $\mathcal{NP}$-hard even in the case of very simple feasibility requirements for duties (e.g., only spread time or working time constraints), see, e.g., Fischetti *et al.* (1987) and Fischetti *et al.* (1989). Since the beginning of the 70s, several researchers have worked on approaches to computerize crew scheduling. The most common approaches formulate CSP as a set partitioning/covering problem (SPP/SCP). Because of the large number of variables involved, column generation techniques are often applied in order to solve the LP-relaxation, and the process is embedded in a B&B framework to produce integer solutions (see, e.g., Desrochers and Soumis (1989) and Falkner and Ryan (1992)). Other authors apply a dual heuristic based on Lagrangian relaxation for solving the master problem (see Carraresi *et al.* (1995) and Freling (1997)). For a good literature overview of existing approaches for the CSP, we refer to Huisman (2004).

Our traditional crew scheduling approach is based on a SCP formulation. The objective is to minimize the total duty costs which are usually a combination of fixed costs such as wages and variable costs such as overtime payment. A set of cover constraints guarantees that each task is included in at least one of these duties.

We chose a set covering formulation instead of a set partitioning one because it is easier to solve from the computational point of view. The over-covers of the tasks can always be deleted in order to convert a set covering solution into a set partitioning solution. From the practical point of view, the over-covers of the tasks mean that the person who is assigned to such a duty will make a trip as a passenger.

The solution algorithm is a combination of column generation and Lagrangian relaxation. The Lagrangian dual problem is solved using a subgradient method. Since a piece of work is a feasible sequence of consecutive tasks on the same vehicle block restricted only by its duration, we can easily enumerate all feasible pieces at the beginning. Then, in the column generation pricing problem, we generate new duties with negative reduced cost by enumerating all possible combinations of pieces of work and checking if such a combination is feasible.

Feasible integer solutions are found by applying the default B&B algorithm of CPLEX for the set of columns generated during the column generation. Note that we apply column generation only for the root node of the B&B-tree. Thus, there is no guarantee that the integer solution is optimal, unless the gap between LP and IP solutions is zero.

We assume that all crews have their own depot. Therefore, a duty of a single crew member contains only tasks on vehicles from that depot. However, it is not necessary that every duty starts and ends in this depot. Thus, in the case of multiple depots, we solve a separate CSP for each depot.

4 Crew Scheduling Enhanced with Aspects from Vehicle Scheduling

The traditional CSP described in the previous section is a common method used in most commercial optimization tools. However, this procedure has the drawback that crew duties are based on a fixed underlying vehicle schedule. Often, several optimal vehicle schedules exist but the traditional crew scheduling considers only one of them. Yet a vehicle scheduling solution that is not considered may in fact lead to a better crew schedule.

The reasons for the propagation of the traditional approach may be found in the methods of solving the VSP in the previous phase. Most of these methods provide only one optimal vehicle schedule. We use an alternative approach based on the time-space network formulation for solving the VSP. An additional advantage of this procedure is that it produces a bundle of optimal vehicle schedules, implicitly given by the solution flow.

Fig. 4 shows an example of how the multiplicity of the optimal vehicle schedules affects crew scheduling. We consider the solution flow of a problem with five trips $t_1, \ldots, t_5$. The result of the vehicle scheduling is the optimality graph presented in the figure. Dotted arrows represent selected depot trips $d_1, \ldots, d_4$ and deadhead dh_1. In order to obtain a certain vehicle schedule, the presented solution has to be decomposed in the node in the middle into two paths. There are two possibilities of combining two inflowing arcs dh_1 and t_2 with two outgoing arcs t_3 and t_4. This results in two equivalent vehicle schedules with two vehicle blocks each. The first contains vehicle blocks $B_1 = \{d_1, t_1, dh_1, t_4, t_5, d_4\}$ and $B_2 = \{d_2, t_2, t_3, d_3\}$ while the second includes blocks $B'_1 = \{d_1, t_1, dh_1, t_3, d_3\}$ and $B'_2 = \{d_2, t_2, t_4, t_5, d_4\}$.

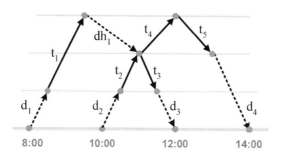

Fig. 4. Multiplicity of Optimal Vehicle Schedules

Following the example in Fig. 4, we apply the traditional crew scheduling approach separately to the schedules obtained. Furthermore, we assume that the maximum duration of a piece of work is limited between two and four hours. Thus, performing crew scheduling based on the first vehicle schedule produces a solution with three drivers, namely two performing B_1 (B_1 is too large to be covered by a single driver and has to be divided into two duties) and one performing B_2. However, the crew scheduling solution based on the second vehicle schedule needs only two drivers, namely one for each vehicle block.

In our approach, instead of decomposing the solution flow during vehicle scheduling, we give this degree of freedom to the crew scheduling phase, where a vehicle schedule is selected that is most consistent with the objectives of crew scheduling. Note that we only select among optimal solutions, so that optimality of vehicle schedules is preserved. We denote the proposed crew scheduling problem by CSP2.

Fig. 5 shows the interaction between vehicle and crew scheduling. The vehicle scheduling is interrupted one step before the last. Instead of decomposing the optimal flow into a vehicle schedule at this point, we leave it until the crew scheduling phase. The set of tasks and corresponding pieces of work cannot be generated directly because the vehicle blocks are not present. They are generated with an alternative method, which will be described in the next subsection.

After a crew schedule is found, a compatible vehicle schedule can always be created afterwards (because all decompositions of the solution flow of the MDVSP-network produce vehicle schedules which are equivalent with respect to fleet size and to operational vehicle costs). In fact, from the bundle of the optimal vehicle schedules, we select one that is most consistent with the objectives of crew scheduling.

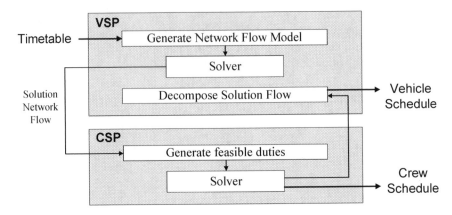

Fig. 5. Vehicle and Crew Scheduling: The Proposed Approach

4.1 Mathematical Formulation

At first, we define a *crew scheduling network (CSN)* which is an extension of the vehicle scheduling network (see Section 2) containing only arcs selected in the solution of the VSP (in the following we consider only the single depot case since CSP2 is solved for each depot separately). For each arc in the CSN, we associate a flow value resulting from the VSP-solution. Furthermore, we delete the circulation arc and all waiting arcs representing waiting in the depot. Thus, all remaining arcs in the CSN represent vehicle activities and have to be covered by duties. Moreover, due to the arc aggregation, some dh- and waiting arcs can be used by several vehicles (indicated by flow value greater than one). Such arcs have to be covered by an appropriate number of duties. Note that the CSN is not necessarily strongly connected.

A piece of work is represented by a path in the CSN. However, not each path represents a feasible piece of work because piece feasibility is restricted by duration and not each node represents a relief point. We distinguish between nodes which are relief points and nodes which are not (each node represents either start or end of a trip). If there are relief points during a trip (between start and end of the trip), we divide the arc which represents such a trip into several arcs by inserting additional nodes for each relief point (according to its corresponding location and time).

Let d_k be the cost of duty $k \in K$, where K is the set of all feasible duties. Define $K(e) \in K$ as the set of duties covering arc $e \in E$, where E is the set of arcs building the CSN. Let z_e be the flow value of arc $e \in E$. Consider binary decision variable x_k indicating whether duty k is selected in the solution or not. We formulate CSP2 as a generalized SPP.

$$min \quad \sum_{k \in K} d_k x_k \qquad (1)$$

$$\sum_{k \in K(e)} x_k = z_e \qquad \forall \, e \in E \qquad (2)$$

$$x_k \in \{0, 1\} \qquad \forall \, k \in K \qquad (3)$$

The objective is to minimize total duty costs. In fact, the constraint set (2) guarantees that the number of duties, which contain a task indicated by arc e is equal to the number of vehicles using arc e. This ensures that there always is a driver for a bus movement.

4.2 Column Generation Approach

For solving CSP2, we propose a solution algorithm which is based on a combination of column generation and Lagrangian relaxation. At first, we compute a feasible solution by using the traditional crew scheduling approach, which means that we solve the MDVSP, decompose the solution flow into an optimal vehicle schedule, and afterwards solve the CSP for each depot. We take the resulting duties from this solution as an initial set of columns for the column generation approach. Thereby, we save the intermediate solution of the MDVSP (*Solution Network Flow* in Fig. 5), which is the basis for the crew scheduling network described above.

Master problem
The main part of the approach is used to compute a lower bound. As in the previous section we solve the *master problem* with Lagrangian relaxation. After relaxing the partitioning constraints (2) in a Lagrangian way, the remaining subproblem can be solved by pricing out columns with negative reduced costs. The Lagrangian dual problem is solved by applying *subgradient optimization*.

Pricing problem
Furthermore, in each iteration of column generation, we generate and add new duties with negative reduced costs (*pricing problem*) to the current set of columns. Recall that vehicle blocks are not known in CSP2. Instead, we implicitly consider a bundle of vehicle schedules given by the flow in the CSN. This leads to a greater freedom in the generation of pieces of work and consequently to many more feasible duties compared to traditional CSP.

The column generation pricing subproblem is solved using a two phase procedure which is similar to method proposed by Freling (1997): in the first phase, the CSN is used to generate a set of pieces of work which serves as input for the second phase where duties are generated. We modify costs of arcs in the CSN according to duals provided from solving the master problem in such a way that the costs of a path are equal to the reduced costs of the corresponding piece of work. Then, the set of pieces is defined by using an all-pair shortest path algorithm involving all nodes which represent a relief point. However, only paths that satisfy the duration are generated.

In the second phase duties are generated. Duties have to satisfy certain feasibility conditions. In particular, they consist of a maximum number of pieces. In our case, the maximum number is equal to 2. This is the reason why we simply enumerate all possible combinations of pieces and check whether each combination is feasible. The reduced cost of a duty can be easily computed when those of the pieces are already known. In the pricing problem, we search only for duties with negative reduced costs in order to satisfy the column generation optimality condition. New duties are added to the current master problem which is then resolved. The master and pricing phases are repeated alternately until no column with negative reduced cost can be found or another termination criterion is satisfied.

Feasible solution

After the column generation process is terminated, a feasible crew schedule has to be constructed. We investigate two methods for finding a feasible IP-solution: one is the *Branch-and-Bound* procedure of CPLEX for columns which are generated during column generation; the second approach is a local search heuristic based on a *Simulated Annealing* algorithm. The B&B-method may produce an optimal solution for some problems (especially small problems) very fast. However, the drawback of this method is that its solution time is unpredictable and may even be exponential in the worst case. In fact, we can limit the solution time of the B&B-process, but then the solution quality of problems which are hard to solve is very poor.

The second approach which we have investigated for finding a feasible IP-solution is a local search heuristic based on the Simulated Annealing algorithm combined with the *Volume algorithm* of Barahona and Anbil (2000). The Volume algorithm is an extension of the subgradient method which also provides an approximate primal solution in addition to a dual solution. We apply the Volume algorithm to columns generated during column generation and use the primal information to construct an initial feasible solution which is improved using a Simulated Annealing approach afterwards. Moreover, we create a pool of primal solutions from different iterations of the Volume method. In each replication of the Simulated Annealing heuristic one solution is chosen as a starting point for the local search. The second IP-approach provides very good solutions in a given time frame. Thus, this method outperforms the default B&B of CPLEX for problems which are hard to solve. However, the local search heuristic always consumes a predefined time.

In our IP-procedure, we combine the advantages of both methods. The overall approach starts with the B&B-procedure of CPLEX with a time limit (in our case 10 minutes). If the problem is hard to solve with B&B, i.e., the procedure does not terminate within that time, then B&B is stopped with the incumbent solution and the second approach is started afterwards. Finally, we choose the best solution of both methods.

Note, we do not generate new columns during the IP-procedure. Thus, there is no guarantee that the integer solution is optimal, unless the gap between the LP and IP solution is zero. However, due to the good initial set of columns in column generation, the solution of CSP2 is at least as good as the solution of the traditional crew scheduling approach.

4.3 Vehicle Schedule

After the CSP is solved, the feasible vehicle schedule must be built, depending on the final crew schedule. Since all decompositions of the optimal flow of the MDVSP-network correspond to an optimal vehicle schedule and the crew scheduling problem was solved based on this bundle of optimal solutions, an optimal vehicle schedule can always be built afterwards.

Once a crew schedule is given, all pieces of work are set. Recall that the pieces of work are sequences of consecutive tasks performed by a single driver on one vehicle. Thus, a vehicle block can be presented as a sequence of pieces of works without additional vehicle movements in-between. Since all tasks (trip tasks and deadhead tasks) are covered by a set of pieces of work, the final vehicle schedule can be defined as a minimal set of vehicle blocks which cover all pieces of work from the final crew schedule. Due to the flow conservation constraints, the number of pieces of work arriving at a node is equal to the number of pieces of work leaving it except for the first and the last depot nodes, which represent the source and the sink, respectively.

5 Computational Results

We tested our approach on some randomly generated and real-world instances. All tests were performed on an Intel P4 3.4GHz/2GB personal computer running Windows XP.

We consider five different duty types, namely *tripper, early normal, day normal, late normal and split*, where tripper consists of a single piece of work and the remaining types consist of two pieces. We use the same duty rules and cost functions as described in Huisman (2004), Huisman *et al.* (2005).

We denote results of the traditional crew scheduling approach with the label *CSP*. Results of our crew scheduling approach enhanced with aspects from vehicle scheduling are labeled as *CSP2*.

Results for Random Instances

We use randomly generated instances published in Huisman (2004), Huisman *et al.* (2005) and available at *http://www.few.eur.nl/few/people/huisman/instances.htm*. There are six sets of instances with two depots and six sets of instances with four depots. The sets differ in the number of trips and contain 80, 100, 160, 200, 320 and 400 trips, respectively. Thus, there are twelve sets together. Each set consists of ten data instances. The detailed description, characteristics, and the way of generating these data instances can be found by Huisman (2004).

Tables 1 and 2 give an overview of the accumulated number of drivers and computational time for all ten instances for each data set. The number of drivers saved by using our approach (CSP2) compared to traditional crew scheduling is shown in the row 'GAP'. As one can see, the number of drivers provided by our approach is always less than this number provided by traditional crew scheduling. The difference achieves up to eight drivers for the largest instance. However, the execution time for CSP2 increases as well because there are many more possible duties which have to be considered. Note that the computational times for CSP2 do not contain times needed for computing the first initial solution because it was given by CSP.

Note that the results presented in Tables 1 and 2 can not be directly compared with results published in Huisman (2004) because we use another vehicle schedul-

Table 1. Results for Random Instances – 2 Depots

	#trips	80	100	160	200	320	400
CSP	#drivers	249	305	370	445	603	742
	cpu (sec)	13	13	58	58	1250	1770
CSP2	#drivers	245	304	363	439	597	733
	cpu (sec)	26	16	120	306	3910	4050
	GAP	4	1	7	6	6	7

Table 2. Results for Random Instances – 4 Depots

	#trips	80	100	160	200	320	400
CSP	#drivers	274	319	394	466	630	782
	cpu (sec)	9	10	25	38	250	1460
CSP2	#drivers	27.3	318	389	464	623	774
	cpu (sec)	13	12	53	89	2210	3065
	GAP	1	1	5	2	7	8

ing approach. In fact, vehicle schedules we computed are optimal and have the same objective values as published in Huisman (2004). However, the optimal assigning of trips to depots may not be unique. Therefore, two vehicle schedules with the same objective value may consist of different assignment to depots and different vehicle blocks. As mentioned in Section 3, drivers can perform only tasks which are assigned to the same depot. Therefore, two equivalent vehicle schedules with different partitioning of trips to depots state different bases to form duties and consequently may provide different crew schedules.

Results for Real-World Instances

We also tested our algorithm on two large real-world instances with 2047 and 2633 trips, respectively. Further properties of the data instances are provided in the Table 3. Since the crew scheduling is solved separately for each depot, we split the first instance into two independent problems A1 and A2 with respect to the partitioning of tasks to depots in the corresponding vehicle schedule. Similarly, the second problem is split into B1, B2 and B3. The second part of the Table 3 shows results of traditional and our crew scheduling. Except for the instance B3, the number of drivers can be improved by using our approach.

Table 4 presents detailed results for both methods of crew scheduling. Rows '#pows' and '#duties' provide the number of pieces of work and the number of possible duties, respectively. We can conclude that the CSP2 considers more pieces of work which results in many more possible duties. However, due to applying the column generation approach, only most promising of them are iteratively selected and finally passed to the B&B procedure. The presented computational time is separated into time spent in column generation (cpu CG) and time for computing an integer

Table 3. Results for Real-World Instances

name	A		B		
data properties					
#trips	2047		2633		
#tasks	2545		3075		
#depots	2		3		
#vehicles	114		126		
subprob.	A1	A2	B1	B2	B3
#tasks	793	1752	1320	1446	309
results(#drivers)					
CSP	118	302	124	144	29
CSP2	117	300	123	141	29
GAP	1	2	1	3	0

solution (cpu IP). However, we limit the time spent in B&B to 60 minutes. The total execution time is denoted by 'cpu total'.

Table 4. Detailed Results for Real-World Instances

	name	A1	A2	B1	B2	B3
	# pows	5,030	10,715	27,773	33,044	6,183
	# duties	483,130	1,341,340	5,182,764	9,590,720	730,778
CSP	cpu CG (sec)	21	60	413	496	48
	cpu IP (sec)	8	143	678	652	672
	cpu total (sec)	29	203	1091	1148	720
	# pows	8,445	23,655	40,881	49,791	6,400
	# duties	1,200,464	6,837,831	13,603,777	17,248,574	832,141
CSP2	cpu CG (sec)	35	215	602	643	51
	cpu IP (sec)	10	521	3750	3915	1265
	cpu total (sec)	45	736	4352	4558	1316

For the instances B1 and B2, the B&B is stopped after the given time limit with an integrality gap of four and eight drivers, respectively. These incumbent solutions could not be improved by applying the local search heuristic presented in Section 4.2. Therefore, the number of drivers for the method CSP2 (see Table 3) may theoretically be improved by increasing the time limit for the B&B procedure or by applying a sophisticated IP-heuristic.

6 Conclusion

The results reported in the previous section indicate that the quality of the crew scheduling solution can be improved by using the proposed crew scheduling method.

We have considered two methods of solving the crew scheduling problem. The first method is the traditional one, vehicles are first assigned to trips, and in the second phase crews are assigned to the vehicle blocks. However, this procedure implies that the crew duties are based on a fixed underlying vehicle schedule. The crews' schedule flexibility is thereby restricted, which sometimes leads to an infeasible or inefficient crew schedule. Our method couples the vehicle and the crew scheduling phases. We use a specific model of the vehicle scheduling problem which is based on the time-space network formulation. An advantage of this procedure is that it produces a bundle of optimal vehicle schedules, implicitly given by the solution flow. In our approach, we give this degree of freedom to the crew scheduling phase, where a vehicle schedule is selected that is most consistent with the objectives of crew scheduling.

We have tested the proposed method on some medium-size randomly generated and large real-world instances.

References

Ball, M. O., Bodin, L., and Dial, R. (1983). A matching based heuristic for scheduling mass transit crews and vehicles. *Transportation Science*, **17**, 4–31.

Barahona, F. and Anbil, R. (2000). The volume algorithm: Producing primal solutions with a subgradient method. *Mathematical Programming*, **87**(3), 385–399.

Bertossi, A. A., Carraresi, P., and Gallo, G. (1987). On some matching problems arising in vehicle scheduling models. *Networks*, **17**, 271–281.

Carraresi, P., Girardi, L., and Nonato, M. (1995). Network models, lagrangean relaxation and subgradient bundle approach in crew scheduling problems. In J. Daduna, I. Branco, and J.M.P. Paixão, editors, *Computer-Aided Transit Scheduling*, pages 188–212. Springer, Berlin.

Darby-Dowman, K., Jachnik, J. K., Lewis, R. L., and Mitra, G. (1988). Integrated decision support systems for urban transport scheduling: Discussion of implementation and experience. In J. Daduna and A. Wren, editors, *Computer-Aided Transit Scheduling*, pages 226–239. Springer, Berlin.

Desaulniers, G., Cordeau, J.-F., Desrosiers, J., and Villeneuve, D. (2001). Simultaneous multi-depot bus and driver scheduling. Technical report, TRISTAN IV preprints.

Desrochers, M. and Soumis, F. (1989). A column generation approach to the urban transit crew scheduling problem. *Transportation Science*, **23**, 1–13.

Falkner, J. C. and Ryan, D. M. (1992). Express: Set partitioning for bus crew scheduling in Christchurch. In M. Desrochers and D. Rousseau, editors, *Computer-Aided Transit Scheduling*, pages 359–378. Springer, Berlin.

Fischetti, M., Martello, S., and Toth, P. (1987). The fixed job schedule problem with spread-time constraints. *Operations Research*, **35**, 849–858.

Fischetti, M., Martello, S., and Toth, P. (1989). The fixed job schedule problem with working-time constraints. *Operations Research*, **37**, 395–403.

Forbes, M. A., Hotts, J. N., and Watts, A. M. (1994). An exact algorithm for multiple depot vehicle scheduling. *European Journal of Operations Research*, **72**, 115–124.

Freling, R. (1997). *Models and Techniques for Integrating Vehicle and Crew Scheduling*. Ph.D. thesis, Tinbergen Institute, Erasmus University Rotterdam.

Freling, R., Boender, C. G. E., and ao, Paixão, J. M. P. (1995). An integrated approach to vehicle and crew scheduling. Technical report 9503/a, Econometric Institute, Erasmus University Rotterdam.

Friberg, C. and Haase, K. (1999). An exact branch and cut algorithm for the vehicle and crew scheduling problem. In N. Wilson, editor, *Computer-Aided Transit Scheduling*, pages 63–80. Springer, Berlin.

Gintner, V., Kliewer, N., and Suhl, L. (2005). Solving large multiple-depot multiple-vehicle-type bus scheduling problems in practice. *OR Spectrum*, **27**(4), 507–523.

Haase, K., Desaulniers, G., and Desrosiers, J. (2001). Simultaneous vehicle and crew scheduling in urban mass transit systems. *Transportation Science*, **35**, 286–303.

Hane, C., Barnhart, C., Johnson, E. L., Marsten, R. E., Nemhauser, G. L., and Sigismondi, G. (1995). The fleet assignment problem: Solving a large integer program. *Mathematical Programming*, **70**(2), 211–232.

Huisman, D. (2004). *Integrated and Dynamic Vehicle and Crew Scheduling*. Ph.D. thesis, Tinbergen Institute, Erasmus University Rotterdam.

Huisman, D., Freling, R., and Wagelmans, A. P. M. (2005). Multiple-depot integrated vehicle and crew scheduling. *Transportation Science*, **39**, 491–502.

Kliewer, N., Mellouli, T., and Suhl, L. (2002). A new solution model for multi-depot multi-vehicle-type vehicle scheduling in (sub)urban public transport. In *Proceedings of the 13th Mini-EURO Conference and the 9th meeting of the EURO working group on transportation, Politechnic of Bari*.

Kliewer, N., Mellouli, T., and Suhl, L. (2005). A time-space network based exact optimization model for multi-depot bus scheduling. *European Journal of Operations Research, in press (online available)*.

Löbel, A. (1997). *Optimal Vehicle Scheduling in Public Transit*. Ph.D. thesis, Technische Universität Berlin.

Löbel, A. (1999). Solving large-scale multiple-depot vehicle scheduling problems. In N. Wilson, editor, *Computer-Aided Transit Scheduling*, pages 193–220. Springer, Berlin.

Mesquita, M. and Paixão, J. (1999). Exact algorithms for the multiple-depot vehicle scheduling problem based on multicommodity network flow type formulations. In N. Wilson, editor, *Computer-Aided Transit Scheduling*, pages 223–246. Springer, Berlin.

Patrikalakis, I. and Xerocostas, D. (1992). A new decomposition scheme of the urban public transport scheduling problem. In M. Desrochers and J. Rousseau, editors, *Computer-Aided Transit Scheduling*, pages 407–425. Springer, Berlin.

Scott, D. (1985). A large linear programming approach to the public transport scheduling and cost model. In J. Rousseau, editor, *Computer Scheduling of Public Transport 2*, pages 473–491. North Holland, Amsterdam.

Tosini, E. and Vercellis, C. (1988). An interactive system for the extra-urban vehicle and crew scheduling problems. In J. Daduna and A. Wren, editors, *Computer-Aided Transit Scheduling*, pages 41–53. Springer, Berlin.

Vehicle and Crew Scheduling: Solving Large Real-World Instances with an Integrated Approach

Sebastiaan W. de Groot[1] and Dennis Huisman[2]

[1] ORTEC bv, Gouda, the Netherlands sgroot@ortec.nl
[2] Erasmus Center for Optimization in Public Transport (ECOPT) & Econometric Institute, Erasmus University Rotterdam, P.O. Box 1738, NL-3000 DR Rotterdam, the Netherlands huisman@few.eur.nl

Summary. In this paper we discuss several methods to solve large real-world instances of the vehicle and crew scheduling problem. Although there has been an increased attention to integrated approaches for solving such problems in the literature, currently only small or medium-sized instances can be solved by such approaches. Therefore, large instances should be split into several smaller ones, which can be solved by an integrated approach, or the sequential approach, i.e., first vehicle scheduling and afterwards crew scheduling, is applied.

In this paper we compare both approaches, where we consider different ways of splitting an instance varying from very simple rules to more sophisticated ones. Those ways are extensively tested by computational experiments on real-world data provided by the largest Dutch bus company.

1 Introduction

In the literature on vehicle and crew scheduling, not much attention has been paid to the problem of splitting up large instances into several smaller ones such that a good overall solution is obtained. Algorithms are developed to solve a certain problem, either optimally or heuristically, and they are tested on self made problem instances, or on (small) instances from practice which the algorithm can still solve. If a real-world instance has to be solved and it seems to be too large for the algorithm to solve it, the problem is just split up into several smaller instances, the algorithm is used to solve those smaller instances and the results are combined such that there is an overall solution. This solution is then feasible, but of course, even if the algorithm itself provides an optimal solution, optimality for the overall problem is likely to be lost. The way the instance has been divided up is almost never an issue in the literature. However, different divisions can result in completely different final outcomes; one splitting can result in a much better solution than another one. Therefore, the instances are mostly divided according to some logical rules.

For example, in the field of crew scheduling, Fores *et al.* (2001) describe this problem. In 1998, they subdivided a large instance of ScotRail into two smaller instances according to a geographic division. Since this resulted in some strange outcomes, several tasks were exchanged between the different divisions. After several days of trial and error, they found a reasonable splitting of the instance such that the optimal solutions of both smaller instances seemed to give a reasonable overall solution. In 2000, they were able to solve the large instance optimally. They checked the performance of the splitting and indeed the optimal solution of the complete instance was the same as the solution which they obtained by splitting up the instance several years before.

Haghani *et al.* (2003) describe a comparative analysis of different approaches to solve large-scale vehicle scheduling problems with route time constraints. This can be seen as a special case of the integrated vehicle and crew scheduling problem, namely where a duty exactly coincides with a vehicle and the only constraint is a maximum duty length. They compared several approaches on a large real-world instance in Baltimore which consists of multiple depots. Since they could not solve this problem exactly, they considered three approaches. The first approach (see also Haghani and Banihashemi (2002)) used CPLEX to solve a reduced problem instance, i.e., several variables in the large IP were just omitted. In the second and third approach, they solved several smaller, single-depot instances with an exact algorithm. The difference between both approaches is the way in which the problem is split up. One is based on the current solution of the public transport company, the other on the outcome of the first approach. They showed that this last approach outperformed the first one.

For the integrated vehicle and crew scheduling problem only small and medium-sized instances have been solved (see, e.g., Huisman *et al.* (2005)). Therefore, we try to answer the following questions in this paper.

1. How can large instances be split up into several smaller ones such that applying an integrated approach on those instances can be done in a reasonable computation time?
2. Does such a splitting approach outperform the sequential approach when the latter is used to solve the large instance at once?
3. Does it outperform the integrated approach when this is terminated after a certain computation time?

Furthermore, we compare different ways of splitting the problem and we give some results on several real-world instances from Connexxion. Finally, we use these ideas to find a solution for large problem instances which we could not solve before with an integrated approach.

The paper is organized as follows. In Section 2, we describe the integrated vehicle and crew scheduling problem and summarize a mathematical formulation and algorithm for this problem, which we introduced in an earlier paper (Huisman *et al.* (2005)). We discuss several splitting approaches in Section 3. Finally, a computational study is provided in Section 4.

2 Multiple-Depot Integrated Vehicle and Crew Scheduling

Several approaches to tackle the integrated variant of the vehicle and crew scheduling problem are recently proposed in the literature (see, e.g., Freling (1997), Haase
and Friberg (1999), Haase *et al.* (2001) and Freling *et al.* (2003) for the single-depot
case, and Gaffi and Nonato (1999), Huisman *et al.* (2005) and Huisman (2004) for
the multiple-depot case). In Huisman *et al.* (2005), two different algorithms are proposed. Both are based on different mathematical formulations, which are themselves
extensions of the single-depot case formulations proposed by Freling *et al.* (2003)
and Haase *et al.* (2001), respectively. Because the first algorithm performed slightly
better, we will only consider this one in the remainder of the paper. Before we discuss
that algorithm, we will first provide a formal problem definition and a mathematical
formulation.

2.1 Problem Definition

The *multiple-depot vehicle and crew scheduling problem* (MD-VCSP) combines the
multiple-depot vehicle scheduling problem (MDVSP) and the *crew scheduling problem* (CSP). Given a set of *trips* within a fixed planning horizon, it minimizes the
total sum of vehicle and crew costs such that both the vehicle and the crew schedule
are feasible and mutually compatible. Each trip has fixed starting and ending times,
and can be assigned to a vehicle and a crew member from a certain set of depots.
Furthermore, the travelling times between all pairs of locations are known. A vehicle
schedule is feasible if (1) all trips are assigned to exactly one vehicle, and (2) each
trip is assigned to a vehicle from a depot that is allowed to drive this trip. From a vehicle schedule it follows which trips have to be performed by the same vehicle and this
defines so-called vehicle *blocks*. The blocks are subdivided at *relief points*, defined
by location and time, where and when a change of driver may occur and drivers can
enjoy their break. A *task* is defined by two consecutive relief points and represents
the minimum portion of work that can be assigned to a crew. These tasks have to
be assigned to crew members. The tasks that are assigned to the same crew member
define a crew *duty*. Together the duties constitute a crew schedule. Such a schedule
is feasible if (1) each task is assigned to one duty, and (2) each duty is a sequence of
tasks that can be performed by a single crew, both from a physical and a legal point
of view. In particular, each duty must satisfy several complicating constraints corresponding to work load regulations for crews. Typical examples of such constraints
are maximum working time without a break, minimum break duration, maximum
total working time, and maximum duration. Finally, a *piece (of work)* is defined as a
sequence of tasks on one vehicle block without a break that can be performed by a
single crew member without interruption.

We distinguish between two types of tasks, viz., *trip tasks* corresponding to trips,
and *dh-tasks* corresponding to deadheading. A *deadhead* is a period that a vehicle is
moving to or from the depot, or a period between two trips that a vehicle is outside
of the depot (possibly moving without passengers).

2.2 Mathematical Formulation

Let $N = \{1, 2, ..., n\}$ be the set of trips, numbered according to increasing starting time. Define D as the set of depots and let s^d and t^d both represent depot d. Moreover, define E as the set of *compatible trips*, where two trips i and j are compatible if a vehicle can perform trip j directly after trip i. We define the vehicle scheduling network $G^d = (V^d, A^d)$, which is an acyclic directed network with nodes $V^d = N^d \cup \{s^d, t^d\}$, and arcs $A^d = E^d \cup (s^d \times N^d) \cup (N^d \times t^d)$. Note that N^d and E^d are the parts of N and E corresponding to depot d, since it is not necessary that all trips can be served from every depot. Let c_{ij}^d be the vehicle cost of arc $(i, j) \in A^d$.

To reduce the number of constraints, we assume that a vehicle returns to the depot if it has an idle time between two consecutive trips which is long enough to let it return. In that case the arc between the trips is called a *long arc*; the other arcs between trips are called *short arcs*. Denote A^{sd} (A^{ld}) as the set of short (long) arcs.

Furthermore, K^d denotes the set of duties corresponding to depot d and f_k^d denote the crew cost of duty $k \in K^d$, respectively. The subset of duties covering the trip task corresponding to trip $i \in N^d$ is denoted by $K^d(i)$, where we assume that a trip corresponds to exactly one task. $K^d(i, j)$, $K^d(s^d, j)$ and $K^d(i, t^d)$ denote the set of duties covering dh-tasks corresponding to deadhead (i, j), (s^d, j) and $(i, t^d) \in A^d$, respectively. Decision variables y_{ij}^d indicate whether an arc (i, j) is used and assigned to depot d or not, while x_k^d indicates whether duty k corresponding to depot d is selected in the solution or not. The MD-VCSP can then be formulated as follows.

$$\min \sum_{d \in D} \sum_{(i,j) \in A^d} c_{ij}^d y_{ij}^d + \sum_{d \in D} \sum_{k \in K^d} f_k^d x_k^d \tag{1}$$

$$\sum_{d \in D} \sum_{j:(i,j) \in A^d} y_{ij}^d = 1 \quad \forall i \in N \tag{2}$$

$$\sum_{d \in D} \sum_{i:(i,j) \in A^d} y_{ij}^d = 1 \quad \forall j \in N \tag{3}$$

$$\sum_{i:(i,j) \in A^d} y_{ij}^d - \sum_{i:(j,i) \in A^d} y_{ji}^d = 0 \quad \forall d \in D, \forall j \in N^d \tag{4}$$

$$\sum_{k \in K^d(i)} x_k^d - \sum_{j:(i,j) \in A^d} y_{ij}^d = 0 \quad \forall d \in D, \forall i \in N^d \tag{5}$$

$$\sum_{k \in K^d(i,j)} x_k^d - y_{ij}^d = 0 \quad \forall d \in D, \forall(i, j) \in A^{sd} \tag{6}$$

$$\sum_{k \in K^d(i,t^d)} x_k^d - y_{it^d}^d - \sum_{j:(i,j) \in A^{ld}} y_{ij}^d = 0 \quad \forall d \in D, \forall i \in N^d \tag{7}$$

$$\sum_{k \in K^d(s^d,j)} x_k^d - y_{s^d j}^d - \sum_{i:(i,j) \in A^{ld}} y_{ij}^d = 0 \quad \forall d \in D, \forall j \in N^d \tag{8}$$

$$x_k^d, y_{ij}^d \in \{0, 1\} \quad \forall d \in D, \forall k \in K^d, \forall(i, j) \in A^d \tag{9}$$

The objective is to minimize the sum of total vehicle and crew costs. The first three sets of constraints, (2)-(4), correspond to the formulation of the MDVSP. Constraints (5) assure that each trip task will be covered by a duty from a depot if and only if the corresponding trip is assigned to this depot. Furthermore, constraints (6), (7) and (8) guarantee the link between dh-tasks and deadheads in the solution, where deadheads corresponding to short and long arcs in A^d are considered separately.

2.3 Algorithm

An outline of the algorithm is shown in Fig. 1.

Step 0: Initialization
Solve MDVSP and CSP for every depot and take as initial set of columns the duties in the CSP-solution.
Step 1: Computation of dual multipliers
Solve a Lagrangian dual problem with the current set of columns. This gives a lower bound for the current set of columns.
Step 2: Deletion of columns
If there are more columns than a certain minimum amount, then delete columns with positive reduced cost greater than a certain threshold value.
Step 3: Generation of columns
Generate columns with negative reduced cost.
Compute an estimate of a lower bound for the overall problem. If the gap between this estimate and the lower bound found in Step 1 is small enough (or another termination criterion is satisfied), go to Step 4;
otherwise, return to Step 1.
Step 4: Construction of feasible solution
Solve a second Lagrangian dual problem with the set of columns generated in Step 3, where the optimal solution of the subproblem gives feasible vehicle schedules. Solve for each depot the crew scheduling problem corresponding to the feasible vehicle schedules.

Fig. 1. Solution Method for MD-VCSP

First, we compute a feasible solution by using the sequential approach, which means we compute the optimal solution of the MDVSP and afterwards, we solve for each depot a CSP given the vehicle schedule for that depot. To solve the MDVSP, we use the model described in Huisman *et al.* (2004) and the all-purpose solver CPLEX. The approach we used to solve the CSP is described in Freling *et al.* (2003).

The main part of the algorithm is used to compute a lower bound and we use therefore a column generation algorithm. The *master problem* is solved with Lagrangian Relaxation. Furthermore, we generate the duties in the column generation subproblem (*pricing problem*). For details about the master and pricing problem, we refer to Huisman *et al.* (2005). Since we do not want to get a very large master problem, columns with high positive reduced costs will be removed. This only happens if there are more columns than a certain minimum number. Finally, in Step 4 we compute feasible solutions.

3 Different Ways of Splitting

In this section we describe several approaches of splitting a large instance of the MD-VCSP into several smaller ones. The different approaches can be divided into two categories:

1. splitting the problem into several single-depot vehicle and crew scheduling problems (SD-VCSPs), i.e., assign each trip to a depot;
2. splitting an instance into a predetermined number of smaller ones.

We will start the discussion with the first category. The most simple way is a random assignment of the trips to the depots. Although this is not interesting in itself, a more sophisticated rule should always beat this trivial one. The more interesting assignments of trips to depots are the following:

- assign each trip to the depot closest to its start location;
- assign each trip to the depot closest to its end location;
- assign each trip to the depot closest to a combination of its start and end location;
- solve the MDVSP and assign each trip to the depot where it is assigned to in the MDVSP.

The first three rules are based on the geographical structure of the problem and can be based on distances or travel times. However, the last rule requires solving of another, much simpler, optimization problem, namely the multiple-depot vehicle scheduling problem, and uses that solution. Note that even the MDVSP is a $\mathcal{NP}$-hard problem. Moreover, recall that the solution approach on the MD-VCSP starts with solving the MDVSP to obtain an initial feasible solution. Therefore, the extra effort is very low. Of course, it is possible to recombine certain smaller SD-VCSPs again to larger MD-VCSPs. This is especially attractive if certain subproblems are so small that recombining does not result in a too large problem again. Another possibility is to use this assignment only as a splitting of the instance and to consider more depots again during the optimization.

The second category is dividing the trips instead of the depot(s) into several small subproblems. We assume here that we have given a maximum number of trips per subproblem. This leads to a certain minimum number of subproblems. Below, we give an overview of such divisions.

- Assign each trip arbitrarily to a subproblem such that the maximum number of trips in a subproblem is not exceeded.
- Solve the MDVSP and assign all trips executed by the same vehicle to the same subproblem. However, the vehicles themselves are assigned arbitrarily to a subproblem.
- Solve the MDVSP and assign all trips executed by the same vehicle to the same subproblem. Moreover, assign the vehicles in consecutive order to the subproblems.

- Solve the MDVSP and assign all trips executed by the same vehicle to the same subproblem. Moreover, assign the vehicles with the highest correlation to the same subproblem.

The first three ways of dividing speak for themselves. The fourth one needs some further explanation. We calculate the correlation w_{ij} between two vehicle blocks with the algorithm suggested in Fig. 2.

$w_{ij} := 0$.
For each different line number l in vehicle block i:
     $\delta_i :=$ number of trips in block i with line number l;
     $\delta_j :=$ number of trips in block j with line number l;
     if $\delta_j > 0$, then $w_{ij} := w_{ij} + \delta_i + \delta_j - 1$;
     otherwise, $w_{ij} := w_{ij}$.

Fig. 2. Algorithm to Compute w_{ij}

It can be easily seen that the weight is only positive if both vehicle blocks have at least one trip in common of the same bus line.

We define a weighted graph $G = (V, E)$ with V as the set of nodes, where a node corresponds to a vehicle block and E as the set of edges. There is an edge (i, j) between each pair of nodes with its weight equal to w_{ij}. The assignment of the vehicle blocks to different subproblems corresponds now to the partitioning of the graph in certain subgraphs such that the total weight of the cuts is minimal and the different parts have an (almost) equal size, where the size of a part is defined as the sum of the number of trips executed by each vehicle block in that part. A well-known algorithm for bipartition is the one of Kernighan and Lin (1970). Hendrickson and Leland (1993) have generalized this algorithm for partitioning in more than two parts. We use this algorithm to partition our graph.

After the problem has been divided into several subproblems and they have been solved with an integrated approach, we can still recombine some parts of the problem such that the solution can be improved. Since the last step of the algorithm consists of solving a CSP for a certain vehicle schedule, we can recombine all vehicle schedules for each depot and solve one large CSP. Notice that this is possible, since the bottleneck of solving an integrated approach is not the CSP. We will see in the next section that this recombining significantly improves the solutions.

4 Computational Results

In this section we test our algorithms on two large data sets from Connexxion, which is the largest bus company in the Netherlands. The first set consists of 1104 trips and four depots in the area between Rotterdam, Utrecht and Dordrecht, three large cities in the Netherlands. The second set contains 1372 trips and six depots in the triangle

Rotterdam, Hoek van Holland, Leiden. We use eight subsets of the first set to test the splitting methods described in the previous section. Then, we choose the best one and perform that approach on the total set. This approach is also used to tackle the second set. The eight subsets are called instance 1 until 8, the complete set 1 is called instance 9 and set 2 is instance 10. In Subsection 4.1 we describe some other properties of these data instances.

All tests in this subsection are executed on a Pentium IV 1.8GHz personal computer (512MB RAM) with the following parameter settings. Notice that all computation times are denoted in minutes.

1. The objective is to minimize the total sum of vehicles and duties, i.e., we only consider fixed costs and the cost of a vehicle is equal to the cost of a duty. For solving the MDVSP in the sequential approach and in the initial step for the integrated approach we use an additional fictitious cost in the variable vehicle costs, viz., for every minute a vehicle is empty outside the depot a cost equal to 1 is incurred.
2. The pricing problems are solved independently for each depot and each type of duty. Moreover, we generate at most 1500 duties for each combination of a depot and type of duty.
3. The maximum number of iterations in the subgradient algorithm to solve the master problem (Step 1) is $500 + 3k$ in the k-th iteration of the column generation algorithm. However, for constructing the feasible solutions in Step 4, the number of iterations is only 10, since in that case the subproblem is $\mathcal{NP}$-hard. Such a small number of iterations is sufficient, since we already start with good multipliers, namely the best ones of the last iteration in the previous step. We construct 10 feasible solutions from which the best one will be selected.
4. The column generation algorithm is stopped if the difference between the current and estimated lower bound is smaller than 0.1% or if the computation time of the lower bound phase is more than 4 hours (2 hours for cases where the problem is divided). Notice that in the latter case we do not have a proven lower bound.

4.1 Properties of the Real-World Data Instances

The restrictions that we have taken into account are as follows. A driver can only be relieved by another driver at the start or end of a trip at certain specified locations or at the depot. If a driver starts/ends his duty at the depot, there is a sign-on/sign-off time of 10 and 5 minutes, respectively. If a driver starts/ends his duty at another relief location, an extra time of 15 minutes plus the deadhead time between this location and the depot is added to the length of the duty. There are five different types of duties, one tripper type consisting of one piece with a length between 30 minutes and 5 hours, and four normal types consisting of two pieces with the properties described in Table 1.

Table 1. Properties of the Different Duty Types

type	1 (early)		2 (day)		3 (late)		4 (split)	
	min	max	min	max	min	max	min	max
start time		8:00			13:15			
end time		16:30		18:14				19:30
piece length	0:30	5:00	0:30	5:00	0:30	5:00	0:30	5:00
break length	0:45		0:45		0:45		1:30	
duty length		9:45		9:45		9:45		12:00
work time		9:00		9:00		9:00		9:00

4.2 Sequential and Integrated Approach

In Table 2, an overview of the results of the sequential and the integrated approach is provided. For each instance, we give the number of trips and the average number of depots to which a trip may be assigned. Furthermore, we give the number of vehicles, duties and the sum of these two as well as the computation time for the sequential and the integrated approach. Finally, we report the best lower bound given by the integrated approach. As can be seen from this table the integrated approach gives much better results than the sequential one. We were only able to compute lower bounds for five of the eight instances, given the maximum computation time of 4 hours for the lower bound phase.

Table 2. Results Without Splitting

	instance	1	2	3	4	5	6	7	8
	number of trips	194	210	220	237	304	386	451	653
	av. depots/trip	1.60	2.47	1.52	2.38	2.48	1.27	1.67	1.74
seq.	vehicles	19	33	27	34	40	32	47	67
	duties	35	56	49	62	75	61	86	125
	V+D	54	89	76	96	115	93	133	192
	cpu (min.)	1	0	0	0	1	2	2	3
int.	vehicles	19	33	27	34	40	32	47	67
	duties	29	52	40	55	66	59	75	117
	V+D	48	85	67	89	106	91	122	184
	cpu (min.)	155	32	94	43	244	260	254	275
	lower	44	77	64	81	95	-	-	-

4.3 Assigning Trips to Depots

In Section 3 we suggested four different methods to assign a trip to a depot. These approaches have been tested to split real-world Instance 2 (see Subsection 4.1), containing four depots, into two subproblems. Notice that this can be done in seven different ways (four with a single-depot and a 3-depot instance and three with two

2-depot instances). Table 3 provides the results of these divisions where the trips are assigned to a depot at random (average results over three runs), or using one of the four methods, i.e., closest to the start location, closest to the end location, closest to a combination of start and end location or according to the solution of the MDVSP. Notice that, e.g., 12-34 means that Depots 1 and 2 are in one subdivision, while 3 and 4 are in the other one.

Table 3. Sum of Vehicles and Crew Duties with Splitting Depots – Instance 2

	123-4	124-3	134-2	234-1	12-34	13-24	14-23	av.
random	95	99	93.7	93	91.7	101.7	95.3	95.6
start	104	104	89	88	89	110	102	98.0
end	96	101	90	86	91	101	97	94.6
start-end	94	98	90	83	88	99	92	92.0
MDVSP	86	87	85	83	84	87	86	85.4

From Table 3 we can immediately conclude that dividing based on the MDVSP is much better than on one of the geographical rules. Some of these do not even outperform a random assignment. We refer to De Groot (2003) for similar results on other instances. Therefore, we will only consider these types of divisions of the depots in the remainder of this section.

4.4 Splitting of the Trips

The different methods for the second category introduced in Section 3 have been tested on the eight real-world problem instances discussed in Subsection 4.1. We refer to De Groot (2003) for a detailed overview of the results of these tests. Here, we only provide an overview of those methods that performed well. These are the following methods.

- Solve the MDVSP and assign each trip to the depot where it is assigned to in the MDVSP. Afterwards divide the trips into two sets: one set with the trips assigned to the largest depot, i.e., the one with most trips assigned to it, and the other set with the remainder of trips. Divide those sets again into sets of at most 200 trips such that the trips executed by the same vehicle (resulting from the earlier solved MDVSP) should be in the same subproblem and the vehicles are assigned to the different subproblems in consecutive order (**Method A**).
- Same as Method A. However, the vehicles are now divided such that the ones with high correlation are as much as possible in the same subproblem (**Method B**).
- Same as Method A. However, the depots are not split first (**Method C**).
- Same as Method B. However, the depots are not split first (**Method D**).
- Same as Method C. However, the subproblems consists of at most 150 trips instead of 200 (**Method E**).

- Same as Method D. However, the subproblems consists of at most 150 trips instead of 200 (**Method F**).

Before we continue our discussion on methods of the second category, we first look at the effect of recombining the different crew scheduling problems per depot at the end. Since the effect on the computation time of this step can be neglected, we only compare the solution values. In Table 4 we provide this comparison for Method C.

Table 4. Sum of Vehicles and Crew Duties With/Without Recombining CSPs – Method C

instance	1	2	3	4	5	6	7	8
with	49	86	70	89	105	91	122	182
without	49	87	71	91	108	91	126	188

As can be seen from Table 4 the saving of recombining can be quite large (up to six duties). Therefore, we recommend to use this option always and thus we take this option into account for the other methods as well.

In Table 5, we report the total number of duties and the maximum computation time for one subproblem (cpu) in minutes for the methods A until F. The number of vehicles is not mentioned since it is independent of the method and the same as in Table 2. The total computation time is also not mentioned, since one of the advantages of splitting is that the algorithm can run on parallel machines.

Table 5. Results Splitting on Instances 1 - 8

instance	1	2	3	4	5	6	7	8
trips	194	210	220	237	304	386	451	653
depots/trip	1.60	2.47	1.52	2.38	2.48	1.27	1.67	1.74
A duties	31	51	43	57	66	59	75	117
cpu	17	7	5	7	20	72	44	30
B duties	31	51	43	57	66	58	77	117
cpu	17	7	5	7	20	56	47	36
C duties	29	53	43	55	65	59	75	115
cpu	155	3	9	2	27	59	34	22
D duties	29	53	43	56	66	58	74	114
cpu	155	3	7	3	32	127	42	41
E duties	30	53	43	55	67	57	75	118
cpu	10	3	9	2	13	9	12	12
F duties	31	53	43	56	66	58	76	118
cpu	18	3	8	3	6	19	17	12

If we look at the results we need to make a distinction between Instance 1, Instances 2-5, Instance 6, and Instances 7 and 8. For Instance 1, Methods C and D provide the same results as the standard integrated approach, since there is no splitting at all. Furthermore, Methods A and B are the same. That is, the problem is divided into two subproblems, which reduces the computation time significantly but needs two duties more. For Instances 2-5 Methods A and B are the same. Here, we can see that the solutions are mostly slightly worse if we split the problems. However, the computation times reduce significantly. As mentioned earlier, for the largest three instances, the lower bound phase of the integrated approach was terminated after a maximum computation time and then feasible solutions were constructed. Here, we can already see an important benefit of the splitting idea. The solutions of some of the methods are better, while the others are equal. Moreover, the computation times are reduced dramatically. For the Instances 7 and 8, we can even see that most of the splitting methods provide better results. Moreover, the computation times become reasonably small. If we would run the subproblems on parallel machines the computation time would be less than one hour on each machine. For all instances, we can see that splitting the problem leads to much better results than the fast and simple sequential approach. If we compare the different methods with each other, we can conclude that Methods A and B perform worse than the others. If we compare C with D and E with F, i.e., using a more advanced approach to divide the vehicle blocks over the subproblems, then we can conclude that they are quite similar. Therefore, it does not make much sense to use this more complicated division. Moreover, if we compare E with C or F with D, then we see that the impact of smaller subproblems (at most 150 or 200 trips), is significant on the computation time, which could be expected of course, but small on the quality of the solutions. Altogether, we conclude that Method E performs well and has a low computation time. Therefore, we will use this one in the next subsection to solve the large instances.

4.5 Large Instances

Since we have shown that these methods to split an instance perform well, we consider the two large data sets introduced in the beginning of this section. Recall that those sets consist of 1104 and 1372 trips, and are called Instance 9 and 10, respectively. Furthermore, notice that the Instances 1 until 8 were derived from Instance 9 and that Instance 10 is completely independent. Although Instances 9 and 10 have four and six depots, on average each trip can only be assigned to 1.71 and 3.64 depots, respectively. Since Method E performed as the best one in the previous subsection, we use this method here. Moreover, we compared it with the sequential approach and the integrated approach with a maximum computation time. The results are shown in Table 6.

As can be seen from this table, the computation time of the integrated approach can far exceed the time limit of 4 hours for computing a lower bound. This can be explained by the fact that other steps take more time. For instance, the computation time of the MDVSP is about 9 and 35 minutes for Instances 9 and 10, respectively, while this was negligible before. Moreover, it can take some time before an iteration in the lower bound phase is finished. Since an iteration is always finished, the final computation time of the lower bound phase can exceed the time limit. Finally, the computation of the CSPs in Step 4 takes longer and this is done 10 times for

Table 6. Results Splitting on Instances 9 & 10

	instance	9	10
	vehicles	109	117
seq	duties	185	224
	cpu	10	46
int	duties	179	219
	cpu	336	474
E	duties	178	210
	cpu	35	62

each subproblem. We can also see that the computation time of one subproblem in Method E can rise over one hour, while it was at most 13 minutes before. This can be explained by the larger sizes of the subproblems. Although the maximum size of a subproblem is 150 trips, this was never reached before. For these larger instances the number of trips in a subproblem comes closer to this maximum.

If we look at the results, we can see that the splitting method saves 7 and 14 duties compared to the sequential approach, and 1 and 9 duties compared to the integrated one. This is a reduction in labor force of 0.6% and 4.1%, respectively, which is quite significant. Moreover, the computation times are reduced drastically. Therefore, we can conclude that these splitting methods clearly outperform the sequential approach as well as the integrated one with a time limit.

5 Conclusions

In this paper we discussed several methods to split large problem instances of the integrated vehicle and crew scheduling problem into several smaller instances. We first applied these approaches to small instances, where we were able to calculate lower bounds on the optimal solutions and a feasible solution with the integrated approach on the complete instance. We showed that the effect of dividing these instances did not deteriorate the quality of the solutions a lot. Later on, we applied these ideas to large instances and showed that those could be solved now, which was not possible before. Furthermore, we showed that the saving compared with the simple, sequential approach is large. Finally, we recommend the use of such splitting methods to solve practical instances instead of dividing the problem in a 'logical' way.

References

De Groot, S. W. (2003). Een geïntegreerde aanpak van voertuig- en personeelsplanning toegepast op grote probleeminstanties, master's thesis (in Dutch). School of Economics, Erasmus University Rotterdam.

Fores, S., Proll, L., and Wren, A. (2001). Experiences with a flexible driver scheduler. In S. Voß and J. R. Daduna, editors, *Computer-Aided Scheduling of Public Transport*, pages 137–152. Springer, Berlin.

Freling, R. (1997). *Models and Techniques for Integrating Vehicle and Crew Scheduling*. Ph.D. thesis, Tinbergen Institute, Erasmus University Rotterdam.

Freling, R., Huisman, D., and Wagelmans, A. P. M. (2003). Models and algorithms for integration of vehicle and crew scheduling. *Journal of Scheduling*, **6**, 63–85.

Gaffi, A. and Nonato, M. (1999). An integrated approach to extra-urban crew and vehicle scheduling. In N. H. M. Wilson, editor, *Computer-Aided Transit Scheduling*, pages 103–128. Springer, Berlin.

Haase, K. and Friberg, C. (1999). An exact branch and cut algorithm for the vehicle and crew scheduling problem. In N. H. M. Wilson, editor, *Computer-Aided Transit Scheduling*, pages 63–80. Springer, Berlin.

Haase, K., Desaulniers, G., and Desrosiers, J. (2001). Simultaneous vehicle and crew scheduling in urban mass transit systems. *Transportation Science*, **35**, 286–303.

Haghani, A. and Banihashemi, M. (2002). Heuristic approaches for solving large-scale bus transit vehicle scheduling problem with route-time constraints. *Transportation Research Part A*, **36**, 309–333.

Haghani, A., Banihashemi, M., and Chiang, K.-H. (2003). A comparative analysis of bus transit vehicle scheduling models. *Transportation Research Part B*, **37**, 301–322.

Hendrickson, B. and Leland, R. (1993). An improved spectral load balancing method. In R. F. Sincovec, D. E. Keyes, M. R. Leuze, L. R. Petzold, and D. A. Reed, editors, *Proceedings of the Sixth SIAM Conference on Parallel Processing for Scientific Computing*, pages 953–961. SIAM.

Huisman, D. (2004). *Integrated and Dynamic Vehicle and Crew Scheduling*. Ph.D. thesis, Tinbergen Institute, Erasmus University Rotterdam.

Huisman, D., Freling, R., and Wagelmans, A. P. M. (2004). A robust solution approach to the dynamic vehicle scheduling problem. *Transportation Science*, **38**, 447–458.

Huisman, D., Freling, R., and Wagelmans, A. P. M. (2005). Multiple-depot integrated vehicle and crew scheduling. *Transportation Science*, **39**, 491–502.

Kernighan, B. and Lin, S. (1970). An efficient heuristic procedure for partitioning graphs. *Bell Systems Technical Journal*, **29**, 291–307.

Line Change Considerations Within a Time-Space Network Based Multi-Depot Bus Scheduling Model

Natalia Kliewer[1], Vitali Gintner[2], and Leena Suhl[1]

[1] Decision Support & Operations Research Lab, University of Paderborn, Warburger Str. 100, D-33100 Paderborn, Germany, Email: kliewer@uni-paderborn.de; suhl@uni-paderborn.de

[2] Decision Support & Operations Research Lab and International Graduate School for Dynamic Intelligent Systems, University of Paderborn, Warburger Str. 100, D-33100 Paderborn, Germany, Email: vitali@uni-paderborn.de

Summary. The vehicle scheduling problem, arising in public transport bus companies, addresses the task of assigning buses to cover a given set of timetabled trips. It considers additional requirements, such as multiple depots for vehicles and vehicle type groups for timetabled trips as well as depot capacities. An optimal schedule is characterized by minimal fleet size and minimal operational costs including costs for unloaded trips and idle time spent outside the depot. This paper discusses the multi-depot, multi-vehicle-type bus scheduling problem for timetabled trips organized in bus lines. We use time-space-based networks for problem modeling. The cost-optimal vehicle schedule may involve several line changes for a given bus within a working day which might not be desirable from the practical point of view. Some bus companies prefer to pose a restriction for bus line changes as well. Because the network flow based model works with trips and not lines, it does not explicitly take into account line changes. In this contribution, we discuss several methods to find schedules with an acceptable number of line changes.

1 Planning of Vehicle Schedules in Public Transport

This paper discusses the vehicle scheduling problem in public transport companies, with the goal of assigning buses to cover a given set of timetabled trips, organized in bus lines with well-defined start and end stations as well as intermediate stops. One trip with fixed departure and arrival times as well as start and end locations cannot be shared by several buses but has to be taken over by exactly one bus. The task is to build a set of rotations (vehicle schedule), such that each trip of a given timetable is covered by exactly one rotation.

We consider the scheduling of vehicles under constraints and objectives arising in urban and suburban public transport. Thus, each timetabled trip can be served by a vehicle belonging to a given set of vehicle types – vehicle type group. The intersection of allowable vehicle type groups for all trips served by one bus rotation must be not empty. Each vehicle has to start and end its work day in the same depot.

After serving one timetabled (loaded) trip, each bus can serve one of the trips starting later from the station where the vehicle is standing, or it can change its location by moving unloaded to any another station (deadhead trip – unloaded trip between two end stations) in order to serve the next loaded trip starting there. This unconstrained deadheading is the main difference compared to an analogue problem in airline scheduling described in Hane *et al.* (1995). Within a bus rotation consisting of several (loaded) service trips chained with each other, the use of deadhead trips often provides an improvement in order to serve all trips of a given timetable by a minimum number of buses.

With respect to the typical "camel-shaped" timetable structure, it can be favorable to return to the depot in the middle of the day between the morning and the afternoon peaks, because waiting time in the depot implies smaller costs compared to idle time at other end stations outside the depot.

Thus a working day for one bus is defined as a sequence of trips, deadheads, waiting times at stations and pull-out/pull-in trips from/to the assigned depot. Since deadhead trips mean an additional cost factor, they should only be used if they imply a benefit for the total schedule. Waiting time costs should be avoided as well. Section 2 describes how this decision situation can be modeled as a time-space network based optimization problem.

Being obliged to save total schedule operation costs, more and more public transport companies plan mixed-line instead of pure-line vehicle schedules. However, within schedules that are cost-minimal, the planners strive for a low number of different lines per bus rotation. Each bus company has its own constraints on the number of lines, which at most can be served by one driver or one bus. In our practical experience this number varies from one to eight different lines per working day.

Section 3 compares total costs of mixed-line and pure-line schedules. Since the proposed time-space network model leads to non-negative integer variables instead of single flow variables, the optimal flows have to be split into single flows in order to define a vehicle schedule. The decomposition method may take into account a secondary objective function, in this case - the line purity of each single bus rotation. In Section 4 we describe different flow decomposition strategies with the goal to reduce the number of line changes while maintaining the optimal costs.

The next section briefly describes a time-space network based modeling approach, proposed for multi-depot vehicle scheduling in Kliewer *et al.* (2006).

2 Solving the MDVSP with a Time-Space Network Based Approach

The task of vehicle planning in public transport is known in literature as the vehicle scheduling problem. We consider here a bus network with multiple depots and multiple vehicle types, thus dealing with the Multiple Depot Vehicle Scheduling Problem (MDVSP in the following). MDVSP means in the sense of this paper the MDMVTBSP - the multi-depot, multi-vehicle-type bus scheduling problem. It is well-known that the MDVSP with heterogeneous fleet is $\mathcal{NP}$-hard (see Bertossi *et al.* (1987)). The combinatorial complexity of the multi-depot bus scheduling problem is determined by numerous possibilities to assign vehicle types to each trip, to build sequences of trips for particular buses, and to assign buses to certain depots. To represent these sequences of trips, exact modeling approaches known in the literature consider explicitly all possible connections - pairs of trips that can be served successively.

In Kliewer *et al.* (2002) and Kliewer *et al.* (2006) we introduced a time-space network based exact optimization model which guarantees minimal fleet size and minimal operational costs. Our solution approach consists in building a network structure for each depot-vehicle type combination. The arcs of such a network represent possible activities which can be carried out by one vehicle of corresponding vehicle type, assigned to a corresponding depot. The arc costs are computed using travel distance rate and time spent outside the depot rate, both user-defined.

First we define a time line for each station connecting the arriving and departing events with waiting arcs at one station to represent standing vehicles. Timetabled trips are represented by arcs, connecting corresponding events - departure in the start station to arrival in the end station. Compatible trips in different stations are connected by arcs for possible deadheads. Unlike well-known network flow models (compare, e.g., Forbes *et al.* (1994), Daduna and Paixão (1995), Löbel (1999)) or set partitioning models (see Ribeiro and Soumis (1994)) from the literature we only insert non-redundant deadhead arcs. A deadhead arc for a certain connection of two compatible trips is redundant if the same connection can be achieved using other deadhead arcs and waiting arcs in connected time lines. It leads to a crucial size reduction of the corresponding mathematical models compared to well-known network flow models.

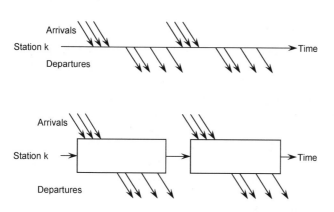

Fig. 1. Nodes as Aggregated Series of Immediate Arrivals and Following Departures

In analogy to stations we build a time line for each depot, although there may not be scheduled trips starting or ending directly in a depot. In the next step we insert arcs for possible depot trips. From the depot time line we insert arcs to start points of scheduled trip arcs and from end points of scheduled trip arcs to the depot time line with associated deadhead costs. Because it is more favorable for buses to stand at a depot than at other stations, we place a higher cost for waiting arcs outside the depots, therefore avoiding long waiting times outside the depots.

We build the nodes of the time-space network by aggregating an arrivals series with the immediately following departures series as shown in Fig. 1. In this way all stations, including depots, are represented as ordered sets of connection nodes, linked together by waiting arcs. Finally a circulation flow arc connects the last node in the depot time line to the first node in this time line.

The cost components include fixed costs for required vehicles as well as variable operational costs. On each layer, there is one circulation flow arc. This arc is provided with fixed cost for the corresponding vehicle type and represents vehicles parking over night in the depot. Waiting arcs and deadhead arcs are provided with corresponding operational costs. The variable costs consist of distance-dependent travel costs and time-dependent costs for time spent outside the depot – the case where a driver is obliged to stay with the bus. All cost components depend on vehicle type. Since the fixed vehicle cost components are usually orders of magnitude higher than the operational costs, the optimal solution always involves the minimal number of vehicles. If required, each circulation flow arc gets an upper (and/or lower) bound for the number of available vehicles. Upper bounds on the loaded trip-arcs are equal to one.

The resulting network flow model contains one network layer for each depot (as defined above), where 0/1-variables on trip arcs and integer flow variables on other arcs are defined. The solution vector describes the flow solution in each network layer with minimal total costs. Each flow unit represents a vehicle starting in the first depot node, flowing through the network arcs and returning back through the circulation arc into the first depot node. In the following we describe the mathematical formulation for the MDVSP based on the time-space network.

Mathematical Formulation   Let $N = \{1, 2, \ldots, n\}$ be the set of trips, and let D be the set of depots (in the following, we define the depot as a combination of a depot and a vehicle type). We define the vehicle scheduling network $G^d = (V^d, A^d)$ corresponding to depot d, which is an acyclic directed network described above with nodes V^d and arcs A^d.

Let c_{ij}^d be the vehicle cost of arc $(i, j) \in A^d$, which is usually some function of travel and idle time. The vehicle cost of arcs representing idle time activity in the depot is 0. Furthermore, a fixed cost for using a vehicle is set on the circulation arc. Let $N^d(n) \in A^d$ be the arc corresponding to the trip n in the vehicle scheduling network G^d.

Decision variable x_{ij}^d indicates whether an arc (i, j) is used and assigned to the depot d or not. For each decision variable an upper bound is defined as follows:

$$u_{ij}^d = \begin{cases} 1 \text{ , if } x_{ij}^d \text{ corresponds to a timetable trip} \\ u^d \text{ , if } x_{ij}^d \text{ corresponds to a circulation arc,} \\ \quad \text{(where } u^d \text{ is the capacity for depot } d) \\ M \text{ , otherwise,} \\ \quad \text{(where } M \text{ is the maximum number of available vehicles)} \end{cases}$$

The MDVSP can be formulated as follows.

$$min \quad \sum_{d \in D} \sum_{(i,j) \in A^d} c_{ij}^d x_{ij}^d \tag{1}$$

$$\sum_{\{j:(i,j) \in A^d\}} x_{ij}^d - \sum_{\{j:(j,i) \in A^d\}} x_{ji}^d = 0 \quad \forall\, i \in V^d, \forall\, d \in D \tag{2}$$

$$\sum_{d \in D, (i,j) \in N^d(n)} x_{ij}^d = 1 \quad \forall\, n \in N \tag{3}$$

$$0 \leq x_{ij}^d \leq u_{ij}^d \quad \forall\, (i,j) \in A^d, \forall\, d \in D \tag{4}$$

$$x_{ij}^d \ integer \quad \forall\, (i,j) \in A^d, \forall\, d \in D \tag{5}$$

The objective (1) is to minimize the sum of total vehicle costs. Constraints (2) are the typical flow conservation constraints, indicating that the flow into each node equals the flow out of each node, while constraints (3) assure that each trip must be covered by exactly one vehicle. In this way we obtain a time-space network based multi-commodity flow formulation.

Thus we solve the mathematical model with branch-and-cut, obtaining lower bounds for the minimization problem by LP-relaxations of the original MIP-formulation. Our modeling approach enables us to solve real-world problem instances with thousands of scheduled trips by direct application of standard optimization software such as MOPS (Suhl (2000)) or ILOG CPLEX (ILOG (2003)).

In order to create a feasible vehicle schedule, the flow solution has to be decomposed in paths. It is an important characteristic of the time-space network formulation that due to the aggregation of possible connections, any feasible flow, including also an optimal flow, represents a bundle or a class of vehicle schedules. All of them have minimal total costs but different other characteristics. With the help of a suitable flow decomposition procedure, we extract a vehicle schedule with an optimal flow and desired characteristics (see Section 4).

3 Mixed-Line Versus Pure-Line Vehicle Scheduling

We have tested our approach on several data sets from real life cases. Three different instances from the public transport companies of Halle and Munich are used here in order to illustrate the cost savings caused through mixed-line bus scheduling. The first instance - city_H, has 2047 scheduled trips from 19 lines, 2 depots for stationing of buses, belonging to 3 vehicle types. The second instance - city_Mun14, has 2452 scheduled trips from 23 lines, 2 depots and homogeneous bus fleet. The largest instance - city_Mun, has over 11 thousand scheduled trips with 55 allowed depot-vehicle type combinations.

Interesting is the relationship in the size of the mathematical models, corresponding to the conventional explicit-connection based modeling approaches from the literature and to the time-space based approach, that we applied to the bus scheduling problem. While connection based approaches would contain over 5 million variables for explicit deadhead connections, our mathematical model for city_Mun14 instance has only 75.000 of such variables and can be solved by branch-and-cut to optimality using dual simplex of ILOG CPLEX 9.0 for LP-relaxations on 2,1 GHz processor in 22 seconds (see Table 1). Due to confidentiality reasons we do not show here the original but only scaled total and operational cost values.

Table 2 illustrates the cost difference between pure-line and mixed-line schedules for three instances. Mixed-line scheduling leads to reductions of both operational costs and number of vehicles. Over 5% less busses are needed to serve city_Mun14 timetable with mixed-line bus rotations instead of pure-line rotations. Due to confidentiality reasons we do not show here the original cost values for city_Mun instance but only the savings.

Mixed-line bus schedules may involve trips of several different lines per bus rotation. Thus it makes sense to schedule mixed-line bus rotations due to cost savings, but we need some strategies how to reduce or to limit the number of different lines per bus rotation. How we can maintain such objectives?

The computing of an optimal bus schedule consists of two stages: at first we compute the minimum cost flow in the constructed network by solving the IP-formulation of the multi-commodity flow problem, then we decompose this flow into a set of paths – these are the required bus rotations.

The optimal flow solution of the mixed-line formulation describes several vehicle schedules, with different statistics of line changes. Each extracted bus schedule may involve several line changes for a given bus within a working day which might be more or less desirable from the practical point of view. The line consideration can be a part of a flow decomposition strategy; in this case we are not forced to lose the cost optimality. The disadvantage of such methods is the impossibility to guarantee a strict upper bound for the number of different lines per bus rotation.

Although it probably is more important to reduce the number of line changes for drivers, some bus companies prefer to pose a restriction for bus line changes as well. Because the time-space network based flow model works with trips and not lines, it does not explicitly take into account line changes. For this case, the consideration of line changes as a cost component in the network model can be unavoidable. Thus the mathematical model receives a cost trade-off between schedule operating cost and line-considering cost component.

In the following we discuss several methods to find bus schedules with an acceptable number of line changes.

Table 1. Properties of Data Instances, Model Size and Optimization Time

instance	stop points	layers	trips	matches	explicit connections in TSN model		rows columns nonzeros	IP opt. time
city_Mun14	60	2	2452	5014262	75215	(1.5%)	12981 100354 205614	22s
city_Mun	160	55	11063	51108336	1083311	(1.25%)	280854 1504171 3315811	10h
city_H	21	6	2047	2115896	26412	(1.25%)	15000 56543 119660	143s

Table 2. Cost Savings Through Mixed-line Instead of Pure-line Schedules

instance	# of vehicles	operational cost	total cost
city_Mun14 (2452 trips of 23 lines, 2 depots, 1 vehicle types)			
pure-line schedule	113	2409887	192814887
mixed-line schedule	107	2387027	182682027
savings	6	22860	10132860
savings in %	**5.31%**	**0.95%**	**5.26%**
city_Mun (11063 trips of 165 lines, 18 depots, 12 vehicle types)			
pure-line schedule	553		
mixed-line schedule	417		
savings	136	2866	
savings in %	**24.59%**	**9.96%**	**24.84%**
city_H (2047 trips of 19 lines, 2 depots, 3 vehicle types)			
pure-line schedule	117	134005	337005
mixed-line schedule	115	13138	332138
savings	2	2866	4866
savings in %	**1.71%**	**2.14%**	**1.14%**

4 Flow Decomposition with Lines Consideration

A large number of possible flow decomposition algorithms may be constructed to decompose a given flow. Line-considering approaches use the fact that the described optimization model usually has not only one, but many optimal solutions with varying number of line changes. We present a heuristic method with the goal to reduce the number of line changes. Furthermore, we discuss an exact model based on the set partitioning problem (SPP) to find a solution with least line changes among all optimal schedules. Because there are many ways to measure the solution quality, we provide several objective functions, such as minimizing the total number of line changes within the schedule or minimizing the maximum number of line changes within one given rotation.

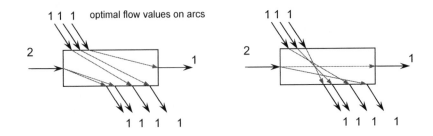

Fig. 2. FIFO- vs. LIFO-decomposition for Given Flow Solution

Fig. 2 shows two different possible decompositions of flow through one node of the time-space network. Flow feasibility, especially the feasibility of the optimal flow, assures the balance of incoming and outgoing flow units. Now we have to assign each incoming flow unit to one outgoing flow unit. With given optimal flow values on arcs as shown in Fig. 2, different assignments are possible to build an optimal vehicle schedule. For example, the left rectangle contains FIFO-decomposition - first departure will be taken by a bus which arrived first. LIFO-decomposition in the right rectangle means the bus with latest arrival has to serve the first departure.

4.1 MinAlt and XMinAlt Flow Decomposition

FIFO- and LIFO-decompositions do not consider line changes explicitly. For the case where homogeneous bus rotations are required, we developed and tested new decomposition strategies.

Table 3. Improvements for City_Mun14 Instance by New Decomposition Strategies Compared to LIFO and FIFO

# of lines	LIFO	FIFO	MinAlt	XMinAlt	LineArcs
1	12	5	5	11	47
2	16	21	20	28	36
3	18	15	16	18	18
sum	46	41	41	62	101
4	16	23	22	10	6
5	22	21	23	19	0
6	12	15	16	12	0
7	8	7	4	4	0
8	3	0	1	0	0
sum	61	66	66	53	6
≤3 lines	**42.99%**	**38.32%**	**38.32%**	**57.94%**	**94.39%**

Table 4. Improvements for City_Mun Instance by New Decomposition Strategies Compared to LIFO and FIFO

# of lines	LIFO	FIFO	MinAlt	XMinAlt	LineArcs
1	81	69	72	73	198
2	75	76	73	86	108
3	69	64	70	75	72
sum	225	209	215	234	378
4	49	61	57	57	26
5	45	53	48	48	9
6	37	40	43	32	3
7	29	26	25	26	0
8	17	13	12	9	1
9	8	8	9	7	0
10	6	3	4	2	0
11	1	3	2	2	0
12	0	1	2	0	0
sum	192	208	202	183	39
≤3 lines	**53.96%**	**50.12%**	**51.56%**	**56.12%**	**90.65%**

Table 5. Improvements for City_H Instance by New Decomposition Strategies Compared to LIFO and FIFO

# of lines	LIFO	FIFO	MinAlt	XMinAlt	LineArcs
1	3	0	3	69	90
2	30	21	35	28	21
3	34	38	36	6	4
sum	67	59	74	103	115
4	21	29	22	2	0
5	14	16	8	3	0
6	6	5	5	4	0
7	6	3	3	1	0
8	0	1	2	1	0
9	0	1	0	0	0
10	0	1	1	0	0
11	1	0	0	1	0
sum	48	56	41	12	0
≤3 lines	**58.26%**	**51.30%**	**64.35%**	**89.57%**	**100.00%**

The first strategy is a "straight forward" one. It is obvious to link at first the scheduled trips belonging to the same line, and then the remaining arcs. The results of this algorithm are shown in Tables 3, 4 and 5 in MinAlt (Minimal Alternation) columns. We count the number of "good" bus rotations, containing trips of at most three different lines. Public transport companies usually consider a rotation with no more than three different lines as being "good". The MinAlt-strategy supplies an improvement of 6% and 12% for city_H compared to LIFO- and FIFO-strategy, respectively. But it does not supply any improvement for both the city_Mun14 and the city_Mun problem instances.

MinAlt is a greedy strategy, acting only locally. A further improvement could be achieved by considering for each decision the decisions made before. Every activity (flow unit on certain arc) gets a list with the line IDs of all service trips which are already chained in one bus rotation containing this arc. We provide each possible match with costs, showing how well both lists fit to each other. We then solve an assignment problem in each node. This strategy, called XMinAlt (for eXtended Minimal Alternation), leads to further improvement for the city_H instance. We gain 25% more "good" bus rotations compared to local MinAlt strategy and 31-38% compared to LIFO or FIFO. This strategy produces also better results for the city_Mun14 instance - there are 15-19% more "good" bus rotations.

4.2 SPP-Decomposition

We observe in Section 4.1 an improvement in line consideration, which is, however, not necessarily satisfying in reality. The next step in handling the problem of line changes is an exact set partitioning model to find a solution with least line changes among all optimal schedules. After the mathematical model is solved to optimality, the set of activities to be served by buses is finally fixed. Now we have to decompose the optimal flow into a set of paths leading from source node of each network layer to sink node of this layer. Each path from the first node in the depot time line to the last node of this time line is one possible bus rotation. The columns of the SPP are binary decision variables of flow units for each possible path, which can be extracted from the optimal flow solution. They indicate whether the bus rotation is selected in the solution schedule or not. The rows are bus activities, such as trips, deadheads, waiting times at stations and in depot and pull-out/pull-in trips from/to the assigned depot.

The objective is to select a minimum cost set of columns such that each row is contained exactly once in one of these columns. In other words, each activity must be served by exactly one bus.

The objective function minimizes the sum of the number of different lines in selected bus rotations and/or the number of line changes. In the case of a given strict upper bound for the number of different lines per bus rotation, the objective is minimization of the maximum number of different lines within one given rotation. These two objectives correspond to requirements which we met in practice.

As different ways to measure the solution quality are conceivable, we provide several objective functions, such as minimizing the total number of line changes within the schedule or minimizing the maximum number of line changes within one given rotation.

In the operational practice we suggest to use the SPP-decomposition as an add-on strategy, which re-optimizes only the "bad" vehicle blocks with too many different lines.

5 Additional Line Arcs in the Network Model

The total SPP-decomposition can take a long time because we should enumerate all possible paths in the bus activities network. Furthermore, depending on the data, it is not always possible to find an optimal solution with at most the allowed number of line changes. Thus, we furthermore present an optimization model which combines both objectives, minimizing cost and minimizing the number of line changes. The model is embedded in a decision support system which allows the user to set priorities and to experiment with different approaches, objective functions, and parameters. For this purpose we extend the network model by inserting a new kind of arc: line arcs. These arcs are provided with a bonus for "line-purity" as negative costs and can be used by flow units connecting trip arcs belonging to the same line (see Fig. 3).

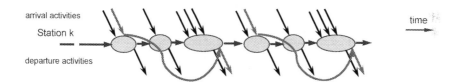

Fig. 3. Inserting Line Arcs in the Network

The IP-formulation gets additional flow constraints, allowing the usage of line arcs, only if both connected service arcs are used. The user can now manage the trade-off between cost minimization and line purity by modifying the bonus value for using the line arcs. Fig. 4 shows the computational results for each strategy on all instances. Concerning different lines, Minimal Alternation strategy provides a bus schedule with similar quality as FIFO and LIFO. Extended Minimal Alternation significantly improves line-purity of the vehicle blocks. Applying the SPP-decomposition for re-optimization of all "bad" vehicle blocks, having four or more different lines, leads to further improvement compared to the Extended Minimal Alternation results (see Fig. 5 for city_H statistics). After inserting line arcs we obtain nearly the pure-line schedule with the same fleet size (115 buses - compare to pure-line scheduling, which needs 117 buses!) and a marginal operational cost increase.

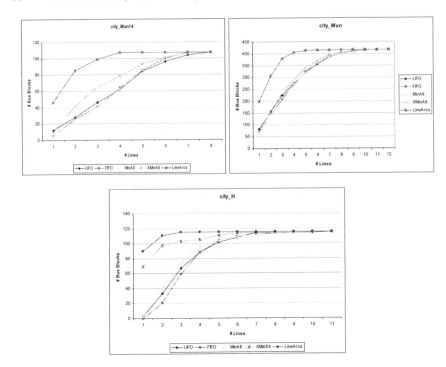

Fig. 4. Dominance of Line Arcs and XMinAlt Strategies for All Instances

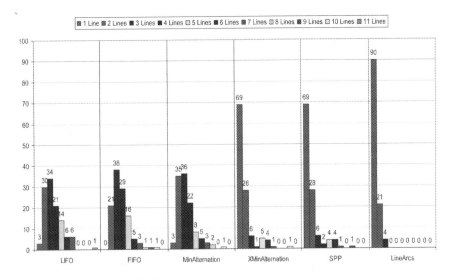

Fig. 5. Line Statistics for City_H Instance

6 Conclusion and Outlook

This contribution discusses the vehicle scheduling problem in public transport companies and particularly the consideration of lines in the mixed-lines bus schedules.

We implemented the time-space network based modeling approach as a software component which has been integrated in commercial software packages to support planning processes in public transport. This software component generates mathematical models for given instances and solves them to optimality. We have carried out tests on real-life timetables of several public transport companies in Germany, such as Halle and Munich.

Thus, we used two ways to consider the line information:

1. The line consideration as a part of flow decomposition strategy. In this case we are not forced to lose the cost optimality.
2. The consideration of line changes as cost component in the network of possible bus activities. Thus, the mathematical model receives a cost trade-off between schedule operating cost and line-considering cost component.

The first two approaches for the line consideration are based on the fact that the optimal solution of the optimization model based on proposed time-space network usually describes many optimal vehicle schedules with varying number of line changes. We present heuristic algorithms which search among possible optimal schedules, with the goal to reduce the number of line changes. Furthermore, we discuss an exact set partitioning model to find a solution with the smallest number of line changes among all optimal schedules. An appropriate modification of the network model makes possible to trade between cost optimality and line purity by modifying the bonus values for using additional line arcs connecting trips of the same line.

The cumulative number of bus rotations with not more than a given number of lines is shown in Fig. 4. The presented methods are integrated in a commercial system for scheduling in bus companies (*ptv interplan*) of the software development company PTV AG and are already used in the planning of the vehicle schedules of several public transport companies.

References

Bertossi, A., Carraresi, P., and Gallo, G. (1987). On some matching problems arising in vehicle scheduling models. *Networks*, **17**, 271–281.

Daduna, J. R. and Paixão, J. M. P. (1995). Vehicle scheduling for public mass transit – an overview. In J. R. Daduna, I. Branco and J.M.P. Paixão, editors, *Computer-Aided Transit Scheduling*, Lecture Notes in Economics and Mathematical Systems 430, pages 76–90. Springer, Berlin.

Forbes, M., Hotts, J., and Watts, A. (1994). An exact algorithm for multiple depot vehicle scheduling. *European Journal of Operational Research*, **72**, 115–124.

Hane, C., Barnhart, C., Johnson, E., Marsten, R., Nemhauser, G., and Sigismondi, G. (1995). The fleet assignment problem: Solving a large integer program. *Mathematical Programming*, **70**(2), 211–232.

ILOG (2003). *Cplex v8.0 User's Manual*. ILOG, Gentilly, France.

Kliewer, N., Mellouli, T., and Suhl, L. (2002). A new solution model for multi-depot multi-vehicle-type vehicle scheduling in (sub)urban public transport. In *Proceedings of the 13th Mini-EURO Conference and the 9th meeting of the EURO working group on transportation, Politechnic of Bari*.

Kliewer, N., Mellouli, T., and Suhl, L. (2006). A time-space network based exact optimization model for multi-depot bus scheduling. *European Journal of Operational Research*, **175**, 1616–1627.

Löbel, A. (1999). Solving large-scale multiple-depot vehicle scheduling problems. In N. Wilson, editor, *Computer-Aided Transit Scheduling*, pages 193–220. Springer, Berlin.

Ribeiro, C. and Soumis, F. (1994). A column generation approach to the multiple-depot vehicle scheduling problem. *Operations Research*, **42**, 41–52.

Suhl, U. (2000). Mops - mathematical optimization system. *OR News*, **8**, 11–16.

Scheduling Models for Short-Term Railway Traffic Optimisation

Alessandro Mascis[1], Dario Pacciarelli[2], and Marco Pranzo[2]

[1] Bombardier Transportation Italy S.p.A., Via Cerchiara 125, 00131 Roma, Italy.
alessandro.mascis@it.transport.bombardier.com
[2] Dipartimento di Informatica e Automazione, Università degli Studi Roma Tre, Via della vasca navale 79, 00146 Roma, Italy. {pacciarelli, mpranzo}@dia.uniroma3.it

Summary. In this paper we report on the results of a research project on train traffic control systems, supported by the European Commission. The results of the project include the development of new optimisation models and algorithms for traffic management, and a general architecture for train traffic control, capable of managing both fixed block and moving block signaling safety concepts. This paper focuses in particular on models and algorithms for real time conflict resolution. Computational results are reported, based on a portion of the Dutch railway network, on the high-speed line Paris-Brussels-Amsterdam.

1 Introduction

This paper deals with the results of a research project on train traffic control systems supported by the European Commission, entitled Project No. TR4004 IV FP - DG XIII Telematics, acronym COMBINE. The project involves suppliers and users of rail traffic systems, software houses and universities from different European Countries. Its goal is to analyze opportunities and problems for traffic management related to the introduction of the moving block signaling standard ERTMS. The results of the project include the development of a general architecture for a train traffic control system and new optimization models and algorithms for traffic management.

Due to its inherent complexity, the management and control of rail operations is usually organized in a hierarchically structured planning process to generate and maintain train schedules. The strategy consists of developing off-line a detailed timetable for each train, often called the *master schedule*, and by operating in real time with strict adherence to these timetables (Hallowell and Harker (1996)). When unforeseen events occur, such as the temporary unavailability of some resources, which make infeasible the planned timetables, it is necessary to partially modify in real time the master schedule in order to restore feasibility. Modifications may include changing precedence between trains and/or their planned speed. This on-line process is called train dispatching or *conflict resolution* (CR) in the first case, and *speed regulation* in the second case.

Even if the resolution of conflicts is presently performed by human dispatchers all over the world, several computerized *Traffic Management Systems* (TMS) have been designed and implemented to support them to re-schedule the train movements and to prevent them from making wrong decisions, such as causing a deadlock situation. Among the published results, we cite the papers by Dorfman and Medanic (2004), Adenso-Díaz *et al.* (1999), Cai *et al.* (1998), Higgins *et al.* (1997), Sahin (1999) and the papers of Kraay and Harker (1995), Hallowell and Harker (1996), Hallowell and Harker (1998). In any case, models at the on-line control and planning level are not designed to replace the human decision maker, who is always in charge to take the decision of implementing a solution.

One aim of the COMBINE project is to move a step further in the direction of automating the train traffic control process, by enabling the TMS to implement some traffic control actions without the authorization of the human dispatcher. A significant difference between a decision support system and a partially automated system, like the COMBINE TMS, is that while the former one can provide a solution which is not feasible in reality, a partially automated TMS must either provide a solution which can be really implemented, or ask for the help of a human decision maker. To this aim, detailed optimization models are necessary, in order to guarantee that a solution, which is feasible for the optimization model, is always also physically feasible.

It is worth noting that the TMS is not in charge of the safety of the rail network. In fact, there exist underlying safety systems that, when necessary, can take the control of the trains by imposing emergency braking in order to avoid collisions between trains.

The paper is organized as follows. Section 2 introduces the train scheduling problem or conflict resolution problem. Section 3 introduces and describes the architecture of the COMBINE TMS. In Section 4 we first introduce the notation and the alternative graph formulation, then we formulate the conflict resolution problem by means of an alternative graph. Finally, we describe the solution procedure adopted to solve the conflicts. Section 5 deals with the solution procedures for the Speed Regulation System. In Section 6 we illustrate the computational experiences, which are based on the so-called Breda triangle, in the Dutch part of the high-speed line Paris-Brussels-Amsterdam. Finally some conclusions follow in Section 7.

2 Problem Description

In this section we introduce the conflict resolution problem. There are two different technologies to ensure safety in the railway networks: the *fixed block* technology and the *moving block* technology. Since there are many different national standards, in this paper we refer to the Dutch NS54 fixed block signaling and to the European standard ERTMS for the moving block technology.

In its basic form a fixed block railway network is composed by *track segments* and *signals*. Signals allow to control the traffic on the network, and to avoid any potential collision among trains. There are signals before every station, passing loop, junction, etc., as well as along the lines. A block section is a track segment between two signals. Signaling systems vary quite a lot from country to country. However, the basic mechanism is as follows. A signal may turn into three or more colors, say red, yellow, or green. A red signal means that the subsequent block section is either out of service or occupied by another train, a yellow signal means that the subsequent block section is empty, but the following block section is occupied by another train, and a green signal means that the next two block sections are empty. A train is allowed to enter a block section depending both on its speed and on the signal color. Slow trains can enter a block section only if the signal is either green or yellow, fast trains can enter a block section at high speed only if the signal is green. Hence, each block section can host at most one train at a time. A block section takes a minimum time to be traversed, which is known in advance for each train, depending on the train and infrastructure characteristics. Besides the traversing time, a delay may occur at the end of a block section if the signal is red or yellow. The combinatorial structure of the train scheduling problem is therefore similar to that of the blocking job shop scheduling problem, a block section corresponding to a blocking machine, and a train corresponding to a job.

With the moving block technology, at any time the exact position and speed for each train are known. Signals are not necessary in this case, since the safety of the trains is ensured by regulating and controlling their respective speeds. Safety standards impose a maximum speed for each train, depending on the distance from the preceding train, necessary to grant the space for completely blocking the train in case of emergency. Hence, track segments in this case are multiple capacity resources.

In both cases, i.e., fixed and moving blocks, stopping or slowing a train causes a remarkable loss of time and energy, due to the long braking distances, followed by acceleration of large masses. More important, if a railway line slopes up over a certain gradient, then there are some freight trains that should not decrease their speed under a certain limit, otherwise they would not be able to reach the top, due to horsepower reasons. Therefore, in a feasible schedule, there are some freight trains that must not decelerate too much. However, in a good schedule, fast trains should always have a good speed profile, i.e., a speed profile that permits low energy consumption. This means that in a fixed block railway network some trains should always find green signals, whereas slow trains should always find green or yellow signals. On the other hand, in a moving block railway network fast and freight trains should not suffer too many speed variations.

The real-time management of rail operations requires checking if the off-line timetables are coherent with the current train positions and speeds. If unforeseen events cause a train not to follow exactly its planned timetable, then an action is required in order to restore the feasibility in the schedule. In this paper we deal with this short term planning process, which is often called conflict resolution. More precisely, a *conflict* is any unforeseen event which makes the planned timetables infeasible (see, e.g., Kraay and Harker (1995)). A conflict occurs, e.g., when two trains require the same resource, i.e., the same segment of track, at the same time. The conflict resolution problem requires determining a new feasible plan of meets and overtakes as close as possible to the master schedule, i.e., such that the delay at all the stations is minimized. In particular, in this paper we address the problem of minimizing the maximum delay.

3 Traffic Management System Architecture

In this section we describe the architecture of the TMS developed in the COMBINE project, as far as the modules for automated train control are concerned. The architecture of the TMS is shown in Fig. 1, where two different layers inside the TMS can be distinguished: the *conflict resolution system* (CRS) and the *speed regulator* (SR).

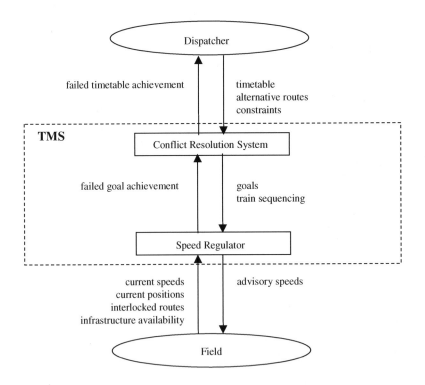

Fig. 1. Train Management System Architecture

At the highest hierarchical level there is the human dispatcher in charge for controlling the rail network. The dispatcher evaluates the rail network status and controls the traffic flows in the network. The human dispatcher, in the COMBINE TMS, focuses on important planning decisions only and leaves to the TMS all other minor decisions. In other words, while the human dispatcher is able to make major decisions, such as canceling a connection or changing the route of a train, the computerized dispatcher can only re-schedule train movements, thus maintaining in real time a conflict-free schedule for each train, compatible with the real time situation. Three different operating possibilities can be identified:

- (Manual Mode) The dispatcher decides to manually solve the conflicts.
- (Mixed Mode) The dispatcher can interact with the TMS modifying the planned timetable or imposing precedence relations between trains.
- (Supervision Mode) The dispatcher supervises the work of the automatic TMS.

In the manual mode the dispatcher manually solves every conflict arising in the rail network.

In the mixed mode the dispatcher can impose to the TMS some constraints in order to guide the solution process. A typical constraint is a fixed precedence relation among two trains or a given route for a train. By constraining the TMS the dispatcher can influence the behavior of the system guiding the algorithm towards good quality solutions.

In the supervision mode, the TMS is in charge of solving the conflicts, and the main role of the dispatcher is to control the work of the TMS. In any case and at any time, in the supervision mode, the dispatcher can switch to the manual mode to assure a better circulation. Moreover, in some critical situations the TMS might not be able to find a feasible solution, thus requiring the dispatcher's help. In these situations the dispatcher has to take the control of the network by solving the arising conflicts manually. Usually in these situations major changes in the timetable are required in order to restore a feasible situation.

The CR layer takes as input the position, the speed and the planned timetable, usually obtained by some off-line algorithm, for each train circulating in the rail network. Moreover, as mentioned before, in the mixed mode, a set of precedence relations could be directly added to the problem by the dispatcher. In other words, given the current network status, the aim of the CRS is to obtain in real time a conflict-free schedule, as close as possible to the planned timetable.

The output of the CRS is a set of precedence constraints among trains and a set of *goals* for each train. A goal specifies a relevant point along the line to be met by the train, such as a station, a junction, or the end of the current resource, an interval [earliest, latest] possible time to reach the position, and an interval [minimum, maximum] speed for the train at the goal position.

The SR module is in charge of regulating the speed profile of each train in the network with the aim of respecting all goals and saving energy. In other words, the SR module generates a speed profile for each train, such that the train is able to reach the position specified by all goals within the given margins of time and speed. Speed regulation is expected to become a significant aspect of traffic control under the moving block technology, whereas it is usually managed with simple static rules under the traditional fixed block technology. The SR layer takes the feasible plan produced by the CRS as input, and for each train decides the train speed needed to reach the goal while reducing the energy consumption.

Finally, the output of the SR is sent to the field level. In our experiments, the field has been modeled using a detailed rail simulator compliant to the NS54 signaling system and the ERTMS standard.

In the COMBINE TMS the SR procedure is executed every time the rail network status is updated, whereas the CRS is invoked, and a new feasible plan is obtained, only if the SR is not able to reach all the goals. Note that, as long as the SR is able to reach all the goals the CRS algorithm is not executed. In this way the CRS is executed only a small number of times, and the solution of the TMS is "stable," i.e., it changes rarely over time. If the CRS is not able to respect all the planned timetable constraints then the help of the dispatcher is requested.

4 Conflict Resolution

In this section we describe in details the CR system developed in the COMBINE project. First we introduce the mathematical notation used to model the train scheduling problem, then we show how the alternative graph formulation (Mascis (1997), Mascis and Pacciarelli (2002)) is able to represent in details the train scheduling problem. Finally, we describe the algorithm developed for the CRS, based on the alternative graph formulation. As already observed in Section 2, the combinatorial structure of the train scheduling problem is similar to that of the blocking job shop scheduling problem, a block section corresponding to a blocking machine, and a train corresponding to a job. In what follows, we describe the alternative graph formulation for the blocking job shop problem, we then extend the model to the CR context.

4.1 Models

Following the traditional terminology used in scheduling theory, we refer to a train as a job, whereas we refer to a track segment as a machine (i.e., a resource that is used by a job). In the usual definition of the job shop problem a job must be processed on a set of machines (i.e., a train must pass through a given set of track segments). The sequence of machines for each job is prescribed; the processing of a job on a machine is called an *operation* and it cannot be interrupted. We have therefore a set of operations $\{o_0, o_1, \ldots, o_n\}$ which have to be performed on m machines $\{m_1, m_2, \ldots, m_m\}$. Each operation o_i requires a specified amount of processing p_i on a specified machine m_i (or $M(i)$), and cannot be interrupted from its starting time t_i to its completion time $c_i = t_i + p_i$. o_0 and o_n are dummy operations, with zero processing time, that we call "start" and "finish," respectively. Each machine can process only one operation at a time.

There is a set of precedence relations among operations. A *precedence relation* (i, j) is a constraint on the starting time of operation o_j, with respect to t_i. More precisely, the starting time of the successor o_j must be greater or equal to the starting time of the predecessor o_i plus a given *time lag* f_{ij}, which in this model can be either positive, null or negative. A positive time lag may represent, e.g., the fact that operation o_j may start processing only after the completion of o_i, plus a possible setup time. A time lag smaller or equal to zero represents a synchronization between the starting times of the two operations. Finally, we assume that o_0 precedes $o_1, \ldots, o_n$, and o_n follows $o_0, \ldots, o_{n-1}$. Precedence relations are divided into two sets: *fixed* and *alternative*. Alternative precedence relations are partitioned into pairs.

A *schedule* is an assignment of starting times $t_0, t_1, \ldots, t_n$ to the respective operations $o_0, o_1, \ldots, o_n$, such that all fixed precedence relations, and exactly one for each pair of the alternative precedence relations, are satisfied. Without loss of generality we assume $t_0 = 0$. The goal is to minimize the starting time of operation o_n. This problem can be formulated as a particular *disjunctive program*, i.e., a linear program with logical conditions involving operations "and" ($\wedge$, conjunction) and "or" ($\vee$, disjunction), as in Balas (1979).

$$
\begin{aligned}
&\min\ t_n - t_0 \\
&\text{s.t.}\ \ t_j - t_i \geq f_{ij} && (i, j) \in F \\
&\qquad (t_j - t_i \geq a_{ij}) \vee (t_k - t_h \geq a_{hk})\ ((i, j), (h, k)) \in A
\end{aligned}
\tag{1}
$$

Associating a node to each operation, Problem (1) can be usefully represented by the triple $\mathcal{G} = (N, F, A)$ that we call *alternative graph* (Mascis and Pacciarelli (2002)). The alternative graph is as follows. There is a set of nodes N, a set of directed arcs F and a set of pairs of directed arcs A. Arcs in the set F are *fixed* and f_{ij} is the length of arc $(i, j) \in F$. Arcs in the set A are *alternative*. If $((i, j), (h, k)) \in A$, we say that (i, j) and (h, k) are *paired* and that (i, j) is *the alternative* of (h, k). Finally, a_{ij} is the length of the alternative arc (i, j).

A *selection S* is a set of arcs obtained from A by choosing at most one arc from each pair. The selection is *complete* if exactly one arc from each pair is chosen. Given a pair of alternative arcs $((i, j), (h, k)) \in A$, we say that (i, j) is *selected* in S if $(i, j) \in S$, whereas we say that (i, j) is *forbidden* in S if $(h, k) \in S$. Finally, the pair is *unselected* if neither (i, j) nor (h, k) is selected in S. Given a selection S, let $\mathcal{G}(S)$ indicate the graph $(N, F \cup S)$. A selection S is *consistent* if the graph $\mathcal{G}(S)$ has no positive length cycles. With this notation each schedule is associated with a complete consistent selection on the corresponding alternative graph. The *makespan* of a consistent selection S is the length of a longest path from node 0 to node n in $\mathcal{G}(S)$. Given a selection S, we denote the value of a longest path from i to j in $\mathcal{G}(S)$ by $l^S(i, j)$.

4.2 Train Scheduling Formulation

In this section a description of the alternative graph model for the conflict resolution problem is given. We first address the case of a fixed block signaling system. Then, at the end of this section, we extend the results to deal with the moving block case and with mixed situations.

A railway network can be modeled as a set of track lines and signals, as described in Section 2, and a block section is a track segment between two signals. In the alternative graph model of the conflict resolution problem a node in the alternative graph corresponds to the time at which a given train enters a given block section. In this model fast trains require two or more empty block sections at a time, in order to travel at their maximum speed, and this can be easily modeled by suitably choosing the alternative pairs. Fig. 2 shows an example for the case of two trains moving in the same direction: train A is a slow train and train B is a fast train, nodes i and j refer to the same block section k. Here, p_{hk} is the travel time for train h and block section k. If train B precedes A on block section k, train A must wait until the section is empty, i.e., until train B enters section $k + 1$. On the contrary, if train A enters block section k before B, then train B must wait until the next two sections are empty, i.e., until train A reaches block section $k + 2$.

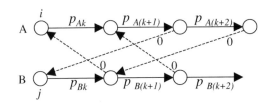

Fig. 2. The Graph Representation for a Slow and a Fast Train

We observe that different trains have different further requirements. For energy saving and horsepower reasons, fast trains and freight trains should not decrease their speed under a certain limit. These constraints can be easily modeled by specifying a maximum time for moving from one point to another of the network. The requirement that a passenger train should not be too late at the stop stations can also be easily modeled as a due date constraint.

Fig. 3 shows a small railway network with four block sections (denoted as 1, 7, 9, and 10), a simple station with two platforms (denoted as 3 and 4), and four special resources, called routes (denoted as 2, 5, 6 and 8), each of them including all the track segments in a junction. These resources have capacity one. At time t there are three slow trains in the network. Train A is a freight train, going from block section 1 to block section 10, and passing through Platform 3 without stopping. Here, α is the time needed for train A to pass through all block sections at the lowest speed allowed. Train B is a passenger train going from block section 9 to block section 1, and passing through Platform 4. Train C is a passenger train going from block section 7 to block section 1, and stopping on Platform 4. Its departure time from the station is β. Finally, the planned times for trains A, B and C to leave the network are γ, δ and χ, respectively.

In Fig. 4 the alternative graph for this example is reported. For the sake of clarity we make use of a different notation here. Each node of the alternative graph is de-

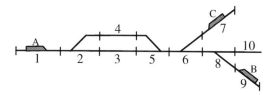

Fig. 3. A Small Rail Network

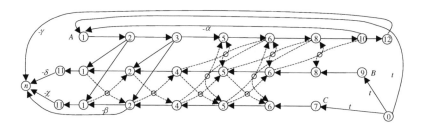

Fig. 4. The Alternative Graph for the Example with Three Trains

noted by the pair (train, block section). A pair of alternative arcs is represented by connecting the two arcs with a small circle in Fig. 4. Each alternative pair of arcs is associated to the usage of a common resource. In particular, trains A and B share resources 1, 2, 5, 6, and 8. Trains A and C share resources 1, 2, 5, and 6. Trains B and C share resources 1, 2, 4, 5, and 6. Note that the initial position of train A implies that B and C are not allowed to precede A on block sections 1 and 2, and therefore we have the selected alternative arcs $(A2, B1)$, $(A2, C1)$, $(A3, B2)$ and $(A3, C2)$. The respective forbidden alternative arcs are not depicted. On all the alternative arcs there is an arbitrarily small weight $\epsilon > 0$.

The fixed arcs with negative weight represent the minimum speed constraint for train A and the delays of the three trains at some relevant points of the network. In particular, arc $(A10, A1)$, with weight $-\alpha$, corresponds to requiring a maximum time α for train A to travel from block section 1 to 10. Due to minimum and maximum travel time constraints, in a feasible schedule the train speed is always kept within the feasible interval.

The planned departure time β of train C from the station (resource 4) is modeled with arc $(C2, n)$ with weight $-\beta$. Similarly, arcs $(A12, n)$, $(B11, n)$ and $(C11, n)$ with weight $-\gamma$, $-\delta$ and $-\chi$, respectively, model the planned exit time of each train from the network. With this model, given a complete consistent selection S, the length of the longest path from 0 to n in $\mathcal{G}(S)$ equals the maximum delay of the three trains in the associated schedule. In fact, $l^S(0, C2)$ is the departure time of Train C from the station, and therefore $l^S(0, C2) - \beta$ is the delay of Train C at the station. Similarly, $l^S(0, C11)$, $l^S(0, A12)$, and $l^S(0, B11)$ are the exit times of the three trains from the network, and therefore $l^S(0, C11) - \chi$, $l^S(0, A12) - \gamma$, and $l^S(0, B11) - \delta$ are their respective exit delays.

The case of a moving block signaling system is now addressed. This case is slightly more complicated to model than the fixed block case. A moving block section can be represented as a resource with multiple capacity in which two consecutive trains cannot enter simultaneously, but rather with a minimum time lag depending on train speed. Since the overtaking is not allowed within a resource, the model must represent this fact.

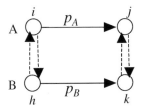

Fig. 5. The Alternative Graph Model for a Moving Block Signaling System

Fig. 5 shows an example for a moving block section with two trains (A and B). There are two pairs of alternative arcs $((i, h), (k, j))$ and $((h, i), (j, k))$. The minimum separation at the beginning [at the end] of the block section equals the length of arcs (i, h) and (h, i) [(j, k) and (k, j)]. The non-overtaking constraint follows from the fact that, if an arc from any of the two pairs is selected, then an arc from the other pair is forbidden. For example, if (i, h) is selected from the first pair, then (h, i) must be forbidden in the second in order to avoid positive length cycles in the graph.

It is worth noting that this representation is not able to limit the number of trains simultaneously using the same moving block section, thus resulting in an infinite capacity resource. However, in practical applications, the capacity of a moving block section is rarely reached, and the number of trains simultaneously using the same moving block section can be easily checked in a post-processing phase.

Fig. 6 shows an example of a mixed situation. In this case the junction in bold, labeled with number 3, is equipped with fixed block technology, while the following block section, numbered with 4, is equipped with the moving block technology.

The alternative graph for the Train A and the Train B is shown in Fig. 7, where the shaded nodes represent the actual position of the two trains. In this example there are three pairs of arcs, the pair $((j, k), (l, i))$ representing the conflict arising in the block section (resource 3), and the pairs $((j, l), (m, h))$ and $((l, j), (h, m))$ representing the conflict arising in the multiple capacity resource 4.

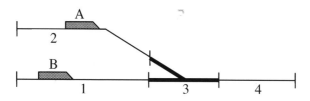

Fig. 6. Example of a Mixed Situation

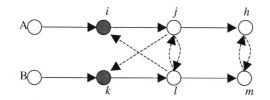

Fig. 7. The Alternative Graph Model for a Mixed Situation

4.3 Conflict Resolution Procedure

The CRS is responsible for train scheduling, and it is the critical system from the computational perspective. In fact, finding the optimal solution to a problem formulated by means of the alternative graph is an $\mathcal{NP}$-hard problem. More generally, the problem of deciding whether a deadlock-free schedule exists or not, being fixed the initial positions and routes of the trains is an $\mathcal{NP}$-complete problem (Mascis and Pacciarelli (2002)). Unfortunately, within a real time environment it is necessary to solve the problem under severe time requirements. Hence, the COMBINE CRS uses a fast heuristic algorithm to find a feasible solution to the Problem (1). If the algorithm fails in finding a feasible solution, it means either that there is no feasible solution respecting all the constraints, or that the heuristic is unable to find one. In both cases the system requires the help of the human dispatcher to restore feasibility.

In order to respect the strict time bound the CRS only considers those trains that are or will be present in the network within a given time window, called the *planning horizon*, thus obtaining a significant reduction in the size of the problem. With a short planning horizon only few trains, and few conflicts, are considered, whereas a longer planning horizon leads to a larger number of circulating trains and a larger number of possible conflicts. There is a trade-off between the size of the planning horizon time window and the quality of the solution found by the CRS. In fact the solutions found with few circulating trains could be myopic, since the CRS does not take into account conflicting trains not in the planning horizon. On the other hand a conflict arising far in the future is not important as a closer conflict, since other unforeseen events could still affect the far conflict. In other words there is a priority in the conflicts; conflicts arising in near future are more important than others that could arise far in the future. Moreover, the size of the resulting alternative graph is strictly dependent on the number of circulating trains, i.e., the smaller the planning horizon the smaller the alternative graph is.

The CR algorithm can be considered basically as a sequence of three independent phases: *pre-processing*, *plan creation* and *post-processing*. Every time a sequence is completed the output of the algorithm is given as input to the SR. In what follows we describe in details the three phases composing the algorithm.

Pre-processing: The pre-processing phase can be divided in two basic subtasks: the update scenario phase and the graph building phase.

The update scenario phase is responsible for filling the internal data structures of the CRS with the current route status and train position and speed and, when available, with a new plan received by the dispatcher. The current position and the speed of a train influence the minimum travel time needed for moving through the subsequent track segments.

The second task of the pre-processing operations is the graph building phase. In the graph building phase the alternative graph representing the rail network is built. Every train is represented in the alternative graph by a chain of nodes and fixed arcs, representing the sequence of actions to be performed by the train: e.g., perform route x, enter track y, enter track z, etc. A travel time is associated with each action; this time is evaluated in the update scenario task, assuming the train is running at a constant speed and without taking into account any conflict. In order to reduce computational times we update the alternative graph instead of rebuilding it completely. New trains are added to the alternative graph model as they enter the planning horizon. The duration of each operation is updated according to the new position and speed of the train and the length of the arc is modified accordingly. If the train route is modified by the dispatcher, the train is removed and added again as a new train entering in the planning horizon.

As mentioned before, the dispatcher has the chance of imposing some precedence constraints between trains, i.e., imposing that a train should enter a conflicting resource before another train. The set of constraints received by the dispatcher is represented with a set of fixed arcs that is added to the alternative graph during the building graph task. A check is performed to verify if the graph is feasible, i.e., with no positive length cycles. If the resulting graph is infeasible then a new plan is required from the dispatcher, and the TMS switches to the manual mode.

In order to reduce the computing time, the build graph subtask does not generate in the alternative graph all the pairs needed to represent the problem. The alternative pairs are added to the graph only when needed. More precisely, in the preprocessing step a plan of earliest/latest possible arrival and departure times for the trains at a set of key points is computed. Then, for each resource in the network, a conflict can arise only for those pairs of trains that are allowed to pass through the resource at the same time, i.e., such that the respective intervals of earliest/latest possible arrival/departure times for the trains overlap. Hence, we add a pair of alternative arcs only for these trains and resources. A time window, and consequently the number of alternative pairs, is increased whenever a train violates it. Computational experience shows that even a large network with high traffic conditions can be modeled with a reasonable number of pairs of alternative arcs, thus allowing us to solve it within a very short time.

Plan Creation: Our scheduling procedure, shown in Fig. 8, is a constructive greedy algorithm that repeatedly enlarges a feasible partial solution. If an infeasible selection is reached, the algorithm performs a backtrack and explores another branch of the enumeration tree. The aim of the search is to find a feasible solution such that the maximum delay of a train at each stop is never larger than a given quantity.

Procedure Conflict Resolution

1. **while** a conflict is found
2. **begin**
3.   Add to the graph the alternative pair representing the conflict.
4.   Solve the conflict by selecting the pair.
5.   **if** the graph is infeasible **then**
6.   **begin**
7.   Perform backtrack and choose the alternative arc.
8.   **if** no backtrack is possible **then exit** (found an infeasible solution).
9.   **end**
10. **end**
11. **exit** (feasible solution found).

Fig. 8. The Conflict Resolution Procedure

A conflict arises when a train asks for a resource already in use by another train in case of fixed blocks or when a train overtakes another train in the moving block case. More precisely in the fixed block case it arises when a Train A enters a resource R_x before Train B leaves the resource R_x. Whereas in the moving block case a conflict occurs if Train A enters resource R_x before Train B and Train B exits from R_x before Train A.

The conflicts are detected by means of a topological visit of the alternative graph, and the algorithm solves the conflicts with higher priority first. The CR algorithm solves the conflicts giving the precedence to the conflicting train that minimizes the increase in the delay. More formally let $((i, j), (h, k))$ be the alternative pair detected by the topological visit. The pair is selected according to the following expression

$$\min\{l^S(0, i) + a_{ij} + l^S(j, n), l^S(0, h) + a_{hk} + l^S(k, n)\} \qquad (2)$$

where $l^S(x, y)$ denotes the length of the longest path in $\mathcal{G}(S)$ from node x to node y. In other words, the criterion adopted to solve the conflicts can be considered as giving the precedence to the *a posteriori* more delayed train.

Note that in some situations there is no choice on how to select an alternative pair. For example, let us consider an alternative pair $((i, j), (h, k))$ such that there exists a path in $\mathcal{G}(S)$ from node j to node i, and let $l^S(j, i)$ be its length. Then, if

$$l^S(j, i) + a_{ij} > 0 \qquad (3)$$

selecting the arc (i, j) would cause a positive length cycle in the graph. Hence, that arc has to be forbidden and its alternative selected. For some resources the planned timetable defines intervals earliest/latest on the earliest and latest entry time allowed on that resource. If selecting an alternative pair causes a train not respecting those constraints, then the Condition (3) permits to identify positive length cycles in the graph and thus immediately to select the pair in the other direction.

Post-processing: When a satisfactory solution has been found by the CR algorithm, a post-processing is applied to it. The main task of the post-processing phase is to specify a set of *goals* for each train and each relevant point visited by the train. A goal contains the following information:

- a relevant point along the line to be met by the train, such as a station, a junction, or the end of the current resource,
- an interval [earliest, latest] possible time to reach the position,
- an interval [minimum, maximum] speed for the train at the goal position.

In other words, each train has to reach its next goal within given margins of time and speed. The definition of goals starts from the output of the plan creation phase, in which trains are scheduled to travel at maximum speed through all block sections. If a train reaches a station early with respect to the timetable, or if a train has to wait for another train at a junction, then in the post-processing phase the earliness or the waiting time is distributed backwards along the train path whenever this does not cause a delay to the previous trains. In doing so, the train is allowed to travel at a lower speed, thus saving energy, while reaching on time all the relevant points.

After the post-processing phase, the resulting goals and precedence relationships between booking actions (the plan graph) are sent to the SR.

5 Speed Regulation

In this section we briefly describe the SR. A more detailed description of the speed regulation procedures is given in Mascis *et al.* (2002), we provide only a brief summary here. The SR is responsible for controlling the train speed. Different SR procedures are necessary when dealing with fixed block and moving block technologies. In fact, in the moving block technology the advisory speed is given by the minimum between possible local speed restrictions and the maximum speed related to the distance from the next train, whereas in the fixed block case the speed depends on signals. In other words, the advisory speed depends on the status of the next block sections (available/not available). Note that this difference does not affect the architecture of the system, but only the computation of the SR.

As shown in the previous section the CRS process sends a plan to the SR. A CRS plan contains an ordered set R of resources and the associated goals and routes for each train and a set of precedence relations between the routes to be booked. Recall that a goal specifies a position, such as the end of the current resource, and the [earliest,latest] possible time to reach the position, and the [minimum,maximum] possible speed at the goal position. Reaching the goal at the minimum time and with the maximum speed typically allows to reduce the delays but causes an increase in the energy consumption. The opposite holds when reaching the goal at the maximum time and minimum speed.

In both fixed block technology and moving block technology, the SR performs a sequence of three independent phases: *update scenario*, *safety check*, and *speed evaluation*. Every time a sequence is completed a new sequence can start.

- In the *update scenario* phase the SR updates the status of the network and the train positions and speeds. Also the plan of precedences and goals is updated when the CRS provides a new plan.
- In the *safety check* phase a simple and very fast check on the train speed is performed in order to avoid that the underlying safety system takes the control of the train with undesired safety braking. Two different limitations on the maximum speed allowed to a train can be distinguished: a "static" limitation due to the status of the network, and a "dynamic" limitation due to a preceding train having smaller speed. In the fixed block case the maximum speed allowed to a train is always dependent on static limitations, in particular it depends on the maximum speed allowed by the infrastructure, and on the distance between the current train position and the position of the next red signal. In the moving block case, the maximum speed allowed for a train also depends on dynamic limitations, i.e., on the distance between the current train position and the position of the first train ahead.
- The *speed evaluation* phase verifies if the train can reach the goal. The speed evaluation calculates the speed profile for all the trains, while respecting the precedence constraints imposed by the CRS. If one is interested in optimizing the punctuality, then the SR looks for a solution in which the trains reach their respective goals at the center of the time window and with the maximum speed. If one is interested in minimizing the energy consumption, then the SR looks for a solution in which the trains reach their respective goals at the latest value of the time window and with the minimum speed. If the computation proves that no feasible solution exists, the SR sends a warning message to the CRS, requesting a new plan and a new set of goals.

6 Computational Experiences

The COMBINE system has been tested with a detailed simulator of a portion of the Dutch railway network, more precisely the Breda triangle, in the Dutch part of the high-speed line Paris-Brussels-Amsterdam (hereafter called Breda junction). The test site is depicted in Fig. 9. A mini-station with a loop enables passing and recovery of required train orders on the area boundary. A junction of two train tracks enables crossing movements of trains, and it is assumed that there are no power supply limitations. The maximum speed on the bold lines is 300 km/h, and in the tunnel is 280 km/h. On the medium tracks the maximum speed is 170 km/h, whereas on the thin lines the maximum speed is 140 km/h in the first 400 meters from the main line, 110 km/h otherwise. TGV's run on the main line from Amsterdam to Brussels. Shuttle trains run from Rotterdam to Breda and from Brussels to Breda, where merging and exiting is done via fly-overs. The high speed line will be used by TGV's. Some of its sections will be jointly used by national high speed Shuttle trains.

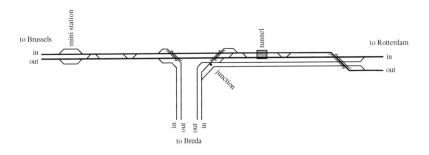

Fig. 9. The Test Site (Breda Junction)

This site has been chosen as a case study for the COMBINE project since at the time of writing this was one of the first sites undergoing real world implementation of the ERTMS Level 3 system. The approach has been tested by using a detailed rail simulator fully compliant with ERTMS Level 3 specifications. The rail simulator takes into account the characteristics of the rolling stock, rail tracks, radio transmissions, driver reaction times, etc. In particular, we call *control loop delay* the minimum time between two consecutive updates of the rail network status. The control loop delay is dependent on a number of technological variables, such as radio transmission delays and others. In any case the TMS should be able to obtain a new solution within the control loop delay time, otherwise the safety layer could take control of the trains and impose undesired emergency braking. In the computational experiments the control loop delay is fixed, for all tests, at 20 seconds.

In all the tests the TMS optimization algorithm is compared with a simple dispatching rule (First In - First Out, FIFO), which is the most commonly used rule for train dispatching. The comparison between TMS and FIFO is carried out showing a set of information presented in graphical form. The description of such information, as well as definitions necessary to avoid any misunderstanding for the reader, is presented in the following. Let us define the "entry delay" as the difference between the actual entry time and the planned entry time, i.e., the difference between the instant when the observed train enters the control area and the instant when the observed train is scheduled to enter the control area according to the timetable. We call "exit delay" the difference between the actual exit time and the planned exit time, i.e., the difference between the instant when the observed train leaves the control area and the instant when the observed train is scheduled to leave the control area according to the timetable. The "total tardiness normalized to entry delay" shows the sum of the exit delays, as a percentage of the sum of the entry delays. The "normalized energy consumption" shows the energy consumption as a percentage of the energy consumption for the reference case, i.e., the FIFO case.

We describe here two representative test situations in this section, called AT1 and AT2, and we analyze the influence and the benefits of the TMS versus the FIFO control strategy. A broader analysis of the TMS performance is reported in Mascis *et al.* (2002). In these tests we considered the planned traffic over the high speed line for year 2015 and no priority distinction among trains. Some perturbations (entry delay) have been added to the planned traffic in order to generate conflicts among the circulating trains. Since each test involves stochastic disturbances, and in order to collect sufficient data for a statistically sound analysis, each test consisted of four replications of five consecutive hours.

These tests address the behavior of the TMS in order to assess the effectiveness of optimization algorithms, in conditions where a delay recovery margin is available. These tests are characterized by the fact that timetables are defined taking into account suitable delay recovery margins. In other words, planned travel times, for each train, are higher than their minimum values.

6.1 Hindering Conflict Test

In the first test case, hereafter called AT1, the Shuttle 138604 from Belgium to Breda enters the control area with large delays (between 780 and 840 seconds), so that a conflict arises with the TGV 104 from Belgium to Rotterdam.

With the FIFO rule, the Shuttle 138604 passes through the mini-station on the secondary line and joins the high speed line preceding TGV 104. The TGV is hindered by the Shuttle until the latter leaves the high speed line. This turns out into significant delays for TGV 104, whereas Shuttle 138604 recovers most of its initial delay. The delay collected by TGV 104 causes a convergence/hindering conflict with Shuttle 138601 from Breda to Rotterdam. In this case Shuttle 138605 joins the high speed line preceding TGV 104, which leaves the control area with a large delay.

Whereas TMS uses the secondary line inside the mini-station in order to allow TGV 104 to overtake Shuttle 138604, that is slowed down below the maximum speed allowed inside the station, so that it is no more hindered by the latter and leaves the control area on schedule. No other conflict arises.

As shown in Fig. 10, with the FIFO rule, the Shuttle 138604 is able to drastically reduce its delay from 805 seconds to 233 seconds, but the TGV 104 exits with 307 seconds of delay, and all the other trains exit before their scheduled time, since they all drive at maximum speed. When the TMS is running, the exit delay of Shuttle 138604 is doubled in comparison with the FIFO case, but it is halved in comparison with the entry delay. All the other trains respect the timetable.

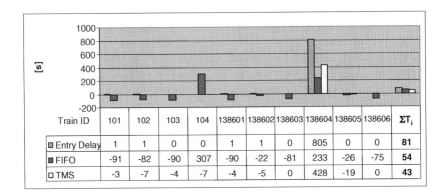

Fig. 10. AT1. Entry and Exit Delay

6.2 Convergence Conflict Test

Now we address the second test case (AT2). Trains coming from Rotterdam enter the control area with large delays (between 800 and 900 seconds for TGVs, between 300 and 360 seconds for Shuttles), so that convergence/hindering conflicts are likely to arise between the TGV 101 from Rotterdam to Belgium and the Shuttle 138602 from Breda to Belgium, when joining the high speed line.

With the FIFO case Shuttle 138602 runs with the speed scheduled by the original plan and approaches the convergence point before the delayed TGV 101, joining the high speed line preceding it. The TGV is hindered by the Shuttle up to the control area border and its exit delay is larger then the entry one. Shuttle 138602 leaves the control area on schedule.

When TMS is active the algorithm slows down Shuttle 138602 before the convergence point so that it joins the high speed line just behind the delayed TGV 101. This has some consequences on Shuttle punctuality, but allows the TGV 101 to recover a significant part of its initial delay, running at maximum speed throughout the control area.

With the FIFO rule, as shown in Fig. 11, the Shuttle 138603 and Shuttle 183606 are able to recover partially their entry delay from 325 seconds to 51 seconds, and from 332 to 58 seconds. The TGV 103 exits with 654 seconds of delay, thus reducing the entry delay, whereas the delay of TGV 101 increases from 838 to 1051 seconds. All the other trains exit before their scheduled time, since they all drive at maximum speed. On the other hand when the TMS is running, the exit delay of Shuttles 138603 and 183603 are completely recovered, but Shuttle 138602 exits the Breda junction with 186 seconds of exit delay. Both the TGV 101 and 103 are capable of reducing their delays from 838 to 605 and from 861 to 627 seconds, respectively.

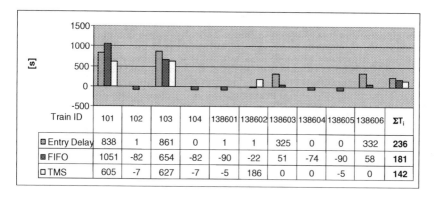

Train ID	101	102	103	104	138601	138602	138603	138604	138605	138606	ΣT_i
Entry Delay	838	1	861	0	1	1	325	0	0	332	**236**
FIFO	1051	-82	654	-82	-90	-22	51	-74	-90	58	**181**
TMS	605	-7	627	-7	-5	186	0	0	-5	0	**142**

Fig. 11. AT2. Entry and Exit Delay

6.3 Discussion

Table 1 summarizes the performance of the TMS solutions with respect to those provided by the FIFO rule, as far as both punctuality and energy saving are concerned. In particular, such tests demonstrated the benefits deriving from the implementation of optimization algorithms which make decisions based on the knowledge of the global traffic status, with respect to a system where simple control rules are used.

Table 1. AT1, AT2. Total Tardiness Normalized to Entry Delay, and Energy Consumption Normalized to the FIFO Case

	AT1		AT2	
	Total Tardiness	Energy Consumption	Total Tardiness	Energy Consumption
FIFO	66.6%	100%	76.7%	100%
TMS	53.0%	89.1%	60.2%	90.8%

Timetables in which trains are planned to travel at less than maximum speed make possible to speed-up late trains in order to recover delays, thus increasing the probability of arriving at destination on time. At the same time, when trains are on time, considerable energy savings can be achieved by letting them travel at lower speed. As pointed out by Kraay and Harker (1995), "*planning at maximum velocity does not provide this flexibility.*"

7 Conclusions

In this paper we discussed models and algorithms capable of describing a rail network equipped both with fixed block and moving block signaling safety systems.

Performance tests were aimed at showing whether advanced optimization algorithms are useful to manage railway traffic. Results showed that the optimization algorithms turned out valuable advantages in terms of better punctuality and energy saving, when compared with simple dispatching rules, whenever appropriate slacks are present in the train timetables.

Acknowledgments: This work was partially supported by the European Commission, Grant number TR 4004, project COMBINE (enhanced COntrol center for a Moving Block sIgNalling systEm).

References

Adenso-Díaz, B., González, M. O., and González-Torre, P. (1999). On-line timetable re-scheduling in regional train services. *Transportation Research, Part B*, **33**, 387–398.

Balas, E. (1979). Disjunctive programming. *Annals of Discrete Mathematics*, **5**, 3–51.

Cai, X., Goh, C. J., and Mees, A. I. (1998). Greedy heuristics for rapid scheduling of trains on a single track. *IIE Transactions*, **30**, 481–493.

Dorfman, M. J. and Medanic, J. (2004). Scheduling trains on a railway network using a discrete event model of railway traffic. *Transportation Research, Part B*, **38**, 81–98.

Hallowell, S. F. and Harker, P. T. (1996). Predicting on-time line-haul performance in scheduled railroad operations. *Transportation Science*, **30**, 364–378.

Hallowell, S. F. and Harker, P. T. (1998). Predicting on-time performance in scheduled railroad operations: methodology and application to train scheduling. *Transportation Research, Part A*, **32**, 279–295.

Higgins, A., Kozan, E., and Ferreira, L. (1997). Modelling the number and location of sidings on a single line railway. *Computers & Operations Research*, **24**(3), 209–220.

Kraay, D. R. and Harker, P. T. (1995). Real-time scheduling of freight railroads. *Transportation Research, Part B*, **29**, 213–229.

Mascis, A. (1997). *Optimization and simulation models applied to railway traffic (in Italian)*. Ph.D. thesis, University of Rome La Sapienza.

Mascis, A. and Pacciarelli, D. (2002). Job shop scheduling with blocking and no-wait constraints. *European Journal of Operational Research*, **143**(3), 498–517.

Mascis, A., Pacciarelli, D., and Pranzo, M. (2002). Models and algorithms for traffic management of rail networks, technical report DIA-74-2002. Dipartimento di Informatica e Automazione, Università Roma Tre.

Sahin, I. (1999). Railway traffic control and train scheduling based on inter-train conflict management. *Transportation Research, Part B*, **33**, 511–534.

Team-Oriented Airline Crew Rostering for Cockpit Personnel

Markus P. Thiel

Decision Support and Operations Research Laboratory, and International Graduate School Dynamic Intelligent Systems, University of Paderborn, Warburger Str. 100, D-33100 Paderborn, Germany, thiel@dsor.de

Summary. Airline crew scheduling is a comparably well-studied field in operations research. An increasing demand for higher crew satisfaction arises; especially after most relevant cost factors have been optimized to their greatest extent, mostly with secondary or little regard on quality-of-life criteria for the involved crew members. One such criterion is *team orientation*. Independent from the chosen assignment strategy (bidline systems, personalized rostering or preferential bidding), current approaches do not consider frequently occurring changes within daily or day-by-day team compositions. By this, crew members rarely know with whom they work for the next flight(s) and/or day(s), respectively. In case of overnight stays outside their individual home base, crew members easily experience themselves having to find their ways to the booked hotels on their own. The avoidance of both aspects is highly appreciated by the crew as well as by the airlines, and will be addressed in the *Team-oriented Rostering Problem*. In this work we present a first interpretation of *Team-oriented Rostering* for cockpit crew, namely captains and first officers which can be implemented via two dedicated optimization models: *Extended Rostering Model* and *Roster Combination Model*. Due to the high combinatorial complexity, certain strategies are applied during roster generation and roster combination in order to solve mid-sized instances based on a European tourist airline setting. As a result, the implied trade-off curve between operational cost and the number of team changes will be discussed.

1 Introduction

Numerous factors influence the performance of an airline company. After fuel, the second highest expense known is personnel, especially for onboard crew. Hence crew scheduling aims to utilize crew members in such a way that their cost is minimized while ensuring the implementation of the given flight plan.

Recent approaches have focused on the pure cost perspective which is even emphasized by the strong competitiveness of the global, meanwhile also continental and domestic, air traffic markets. After all, the resulting cost-minimized crew schedules could turn out to be less satisfactory for crew members. Although all governmental restrictions, union agreements, and airline specific rules are obeyed, cost-intensive disturbances of the schedule occur frequently due to absent or sick crew members.

Based on the commonly known positive correlation between employees' satisfaction and their absence rate, we define the *Team-oriented Rostering Problem (ToRP)* as the consideration of teams within the crew rostering process. In this approach, we address a usually unconsidered factor to increase crew satisfaction, namely the avoidance of frequent team changes. This factor turns out to be notably important because of the high inherent stress level associated with it. Imagine a crew member working his/her onboard shift (or flight duty) for up to 14 hours every day, afterwards having to find the reserved hotel on his/her own in a possibly even unknown town. Or within the day, communication and companionship among crew members is hardened, if those people that just worked together get separated several times a day, always being in a hurry to arrive at the next scheduled location right in time. Additionally, the *National Transportation Safety Board* (NTSB) conducted a study on the circumstances for cockpit crew of U.S. carriers which experienced major accidents over a period of 15 years, see NTSB (1994). According to their findings, 73% of all incidents took place during the crew's first day, and 44% occurred even during the initial flight of a newly formed crew.

This paper presents techniques for two alternative optimization models treating the ToRP for cockpit crew. It is specifically tailored to the needs of European airlines with their distinct *fair-and-equal share* interpretation of workload in terms of, e.g., flight hours – as opposed to the more frequently examined U.S. systems (bidline system or preferential bidding, see Section 2.2). Both models have been formulated as a *set partitioning problem (SPP)*. Due to the high combinatorial complexity for considering roster combinations instead of "just" single rosters, a set of strategies is applied to enable appropriate solving.

The paper is structured as follows. We first give a brief survey on the airline crew scheduling problem. In Section 3 an introduction to the general ToRP follows, and, in particular, special characteristics for cockpit crew. In Section 4 we present and discuss two possible mathematical formulations for the team-oriented cockpit crew rostering. The two main tasks, roster generation and roster selection, are addressed in Section 5 by a variety of implementation methods. Some computational results based on the setting of a European tourist airline follow in Section 6. We close with a summary and outlook.

2 Airline Crew Scheduling

A general formulation for the airline *crew scheduling problem (CSP)* can be paraphrased as follows. Given the published flight schedule of an airline, the key task is to assign all necessary crew members of cockpit and cabin crew in such a way that the airline is able to operate all its flights at minimal expense for personnel. This assignment has to consider all restrictions forced by governmental regulations, union agreements, and company-specific rules. In addition, time- and location-dependent crew availabilities have to be accounted for, especially in a setting where crew is stationed at one of multiple airports (called *home bases*).

The cost of such a crew schedule is determined by two figures: crew salary and (planned) operational cost. Whereas crew salary at most European airlines is handled as a stepwise linear function (fixed salary for about 2/3 of the contracted flight hours, stepwise higher hourly rate(s) for the rest if needed), North American airlines apply a system called *pay-and-credit* which refers to the difference between the number of hours that a crew member is paid for and the actual hours of flying (see Gerhkoff (1989)). Furthermore, operational cost has to be minimized – in detail: expenses for hotel stays and for proceeding crew members from/to their current/next scheduled location (*taxiing*).

The general CSP as introduced above is known to be very hard to solve due to its combinatorial complexity (see, e.g., Barnhart *et al.* (2003), Suhl (1995)). Thus, it is usually decomposed into several sub-problems and even sub-steps: Firstly, cockpit and cabin crew types are separated, usually even to the level of their crew functions. By this, for cockpit crew, we have a dedicated CSP for the *captain (CP)* or pilot and one for the *first officer (FO)* or co-pilot. Each problem is divided into the *crew pairing problem (CPP)* and the *crew assignment problem (CAP)* which are usually solved sequentially for every examined instance, see also Fig. 1.

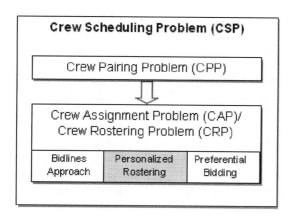

Fig. 1. Tasks of Airline Crew Scheduling

Before we describe the two scheduling steps, some basic terms used throughout the paper have to be defined as follows:

A *flight leg* is a non-stop flight from a departure airport to its destination airport. A *flight duty* is a series of flight legs that can be serviced by one crew member within a workday (24 hours). Such a flight duty is surrounded (before and after) by rest periods, whereas the off-time duration depends, e.g., on the start of the first flight leg and the number of flights serviced. If the crew members' time-dependent location does not equate to the next scheduled location, they need a pre-proceeding in case that this relocation is required in advance of servicing this flight duty, and a post-

proceeding for its succeeding occurrence. Those proceedings (or taxiing) are usually realized via public transportation (e.g., bus, taxi or train), or via passive flight legs serviced by the airline itself, called *deadheading*.

The next aggregation level is a *pairing* which starts from and returns to the crew member's home base without any further overnights at their home domicile. Therefore, *hotel stays* become necessary, if crew members have to spend their daily rest periods outside of their home base. *Pre-scheduled activities* like vacation, requested and granted off-periods, office, simulator/training, medical examination etc. represent activities that a crew member has to fulfill. Since those activities are determined in advance of the scheduling process, overlapping flight duties are not allowed. After a maximum of up to five working days that can be filled by flight duties or pre-scheduled activities, a full two-day off as the weekly rest period is required.

A *roster* (or *line-of-work*) represents a potential crew schedule for a dedicated crew member. It consists of his or her pre-scheduled activities and assigned flight duties, and it incorporates all governmental-, union- and company rules as well as the crew member's individual work history and remaining contracted flight/work hours. A *null-roster* represents a roster without any assigned flight legs.

2.1 Crew Pairing

As mentioned above, crew pairing is the first step of the solution process for the CSP. The aim of the CPP is to find a set of pairings that covers, at minimum cost, all flights of the considered, usually (semi-)monthly, planning period. Whereas those pairings themselves have to be compliant to the multitude of regulations as already described, they are still anonymously built without consideration of a crew member's individual needs or desires. Therefore, the CPP is usually solved on the level of flight legs for the entire crew, instead of considering selected crew types and/or functions (see Mellouli (2003)).

Nevertheless, the high combinatorial complexity of most solution approaches focus on the process of pairing generation on the one hand, and pairing selection of a least-cost subset on the other (see, e.g., Anbil *et al.* (1991), Graves *et al.* (1993)). The selection process then is realized via an SPP or a *set covering problem (SCP)* (see, e.g., Bixby *et al.* (1992), Hoffman and Padberg (1993)), meanwhile mostly being solved by applying the column generation approach (see, e.g., Desaulniers *et al.* (1997), Lavoie *et al.* (1988), Vance *et al.* (1997)). Alternatively, network flow models are applied (see, e.g., Guo *et al.* (2006), Mellouli (2001), Mellouli (2003), Yan and Tu (2002)), but also modern heuristics such as genetic algorithms (see, e.g., El Moudani *et al.* (2001)).

2.2 Crew Assignment / Crew Rostering

The second step of the CSP is called crew assignment or rostering. In contrast to the first step, the CAP/CRP is solved for individual crew members. The set of pairings created during the CPP is assigned in a way that considers all governmental rules, union- and company agreements as well as pre-scheduled activities, e.g., simulator or vacation, for each individual, also known as *fingerprint* (see Mellouli (2001)) or *skeleton roster* (see Barnhart *et al.* (2003)), whereas all flights are properly staffed with all onboard crew functions. This assignment is also realized with decomposed sub-instances of the CAP, e.g., by crew types (cockpit, cabin), crew functions (captain, first officer etc.), and fleet (see Ryan (1992)).

Among all airlines the individual aims of the CAP/CRP might differ, but in general it can be expected that they consist of two contrary goals: cost minimization for the airline and maximization of *quality-of-life criteria* for crew. There are three different concepts to address quality-of-life criteria, e.g., by considering crew requests or their preferences during the scheduling process. *Bidline systems* are widely applied in the US. They generate anonymous lines-of-work which are assigned after an elaborated bidding process to the crew members based on strict seniority. In Europe, *personalized rostering*, also known as *fair-and-equal share*, is more commonly used where fairness of workload among crew members replaces seniority almost completely. Therefore, the system accepts or rejects crew requests and outputs the optimal schedule considering a high degree of expressed preferences. During the last decade, a third concept called *preferential bidding* has become more popular since it bypasses the drawbacks of other methods. Preferential bidding considers crew preferences up to a certain degree, such as regularly pre-scheduled weekends or working with specific colleagues; but in case of conflicts, the seniority principle is applied. Bidline systems are treated in, e.g., Campbell *et al.* (1997), Jarrah and Diamond (1997); personalized rostering has been examined by Day and Ryan (1997), Gamache *et al.* (1999), Kohl and Karisch (2004), Nicoletti (1975), Strauss (2001); and solution methods for preferential bidding are given in Gamache *et al.* (1998), among others.

3 Team-oriented Rostering

In this section we introduce the ToRP in general, and for cockpit crew in particular. This approach is understood as an enhancement to the personalized rostering concept, see Section 2.2, where automated crew schedules are created that reveal a certain team orientation. This team orientation intends to grant higher crew satisfaction in terms of quality-of-life criteria. The basic idea is – in addition to the objectives of the airline CRP – the consideration of team orientation by avoidance of frequent changes in the composition of a servicing or operating onboard team.

Why is team orientation so important? It is known that crew satisfaction is highly dependent on the colleagues someone works with (see Strauss (2001)). In current approaches some crew members may prefer to exclusively work with the same colleague(s) over a long time period (e.g., *married couples* or *must-fly-together* restrictions (see Kohl and Karisch (2004)). The realization of such a highly restrictive approach remains theoretically simple, but it is almost impossible to implement without great financial losses because of different, non-overlapping pre-scheduled activities at most airlines. Therefore, teams should be kept as flexible as possible. On the other hand, aircraft security as well as quality-of-service for passengers are directly at risk in cases of disharmonies within and among operating cockpit and servicing cabin crew. Especially, team changes were identified to have a negative impact on the individual crew satisfaction, e.g., being left alone in a non-domicile town after work or giving up harmonizing working teams.

In order to fully explain the approach, some additional definitions become necessary:

- A *team* is to be understood as a group of different crew members with, if required, different crew functions and quantities in such a way that a single (or a series of) flight leg(s) is staffed adequately. Crew members of such a team may origin from different home bases, but they all share the minimum qualification for the fleet to be operated.
- A *team change* occurs if at least one crew member is scheduled to service the next flight activity together with a different team composition (other colleagues). Team changes may occur due to the obeyed rule set (e.g., a crew member has reached his maximum of daily working hours), or by very strict fair-and-equal share of workload; but so far, the main reason for team changes is that they are simply not considered at all. (For bidline systems it is left up to the crew member to manually choose with a colleague two corresponding rosters as far as possible. Preferential bidding allows announcing preferences also for colleagues, but team changes themselves are usually not prevented by this.)
- A *shared flight activity (SFA)* is defined to be the smallest unit that is considered in this approach. Such an activity is serviced by a team without any team change. It may be a single (or multiple) flight leg(s), flight duties, a single (or even several complete) pairing(s). SFAs can be extracted directly from the generated pairings of the CPP.

Since the ToRP approach described here aims to minimize the number of team changes we introduce so-called *team change penalties*. Such penalties are usually chosen as positive values. In contrast to this, negative team change penalties (or bonuses) can be applied for benefits of servicing as a team while, e.g., saving operational cost by sharing a taxi.

We distinguish between two kinds of team changes:

- The *type of a team change* expresses *when* and *where* the team change occurs. It can happen within the day, over night, both at the home base and outside, or after the weekly rest period at the home base. A team change within the day is the most undesired, especially in combination with an outside location. Therefore, we propose a clear hierarchy among those listed instances with decreasing penalty values for each type.
- The *degree of a team change* refers to *how* the team composition is changed. Having, e.g., three crew members that constitute a team, there are exactly two different ways to get separated: A (1-1-1)-change means that every crew member will follow his/her own way afterwards, whereas a (2-1)-change indicates that two of them will continue working together for the next SFA(s). A higher degree

of splitting is less preferable by the crew and should therefore receive a higher penalization value.

The focus of this work lies on the ToRP for cockpit teams. A cockpit team usually consists of one captain and one first officer. In the rare case of *downgrading*, a captain works in the function of a first officer. The resulting team of two captains is also valid, but two first officers are not allowed. All three types of team changes (as introduced above) can occur frequently to cockpit crew, whereas the degree of team changes is limited to (1-1)-changes.

In order to evaluate the quality of a crew schedule according to ToRP, we have to evaluate roster combinations, since all team members follow their assigned rosters when the team changes happen. In Fig. 2, some roster combinations among a single captain and several first officers are given: Whenever a shared time period is terminated, a team change takes place. (For better understanding shared flight activities are given as flight duties in this example.) On day 8 there is a team change after the weekly rest period (two consecutive OFF-days). The captain presented here experiences a total of five team changes. Team changes are only counted for one crew function as shown in the example.

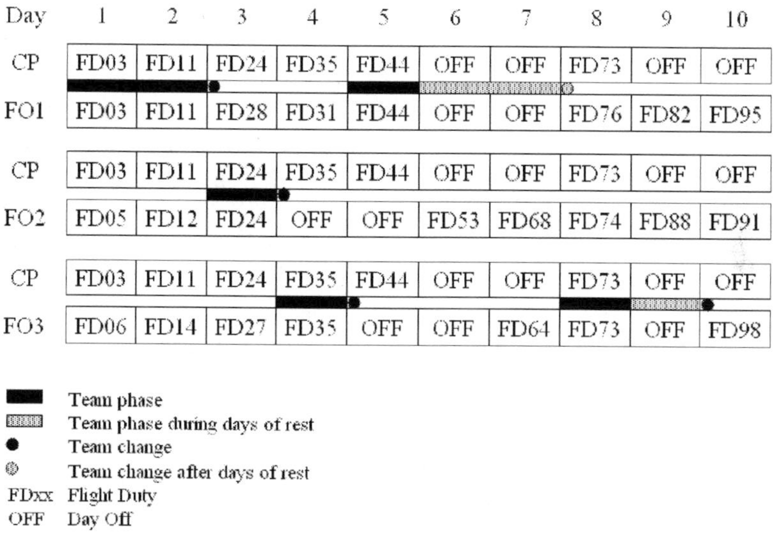

Day	1	2	3	4	5	6	7	8	9	10
CP	FD03	FD11	FD24	FD35	FD44	OFF	OFF	FD73	OFF	OFF
FO1	FD03	FD11	FD28	FD31	FD44	OFF	OFF	FD76	FD82	FD95
CP	FD03	FD11	FD24	FD35	FD44	OFF	OFF	FD73	OFF	OFF
FO2	FD05	FD12	FD24	OFF	OFF	FD53	FD68	FD74	FD88	FD91
CP	FD03	FD11	FD24	FD35	FD44	OFF	OFF	FD73	OFF	OFF
FO3	FD06	FD14	FD27	FD35	OFF	OFF	FD64	FD73	OFF	FD98

▬	Team phase
▨	Team phase during days of rest
●	Team change
◉	Team change after days of rest
FDxx	Flight Duty
OFF	Day Off

Fig. 2. Team Changes Between Roster Combinations

We finally discuss the main disadvantage of the ToRP approach. Of course, a crew schedule that focuses additionally on the minimization of team-changes is most likely more cost intensive compared to other requirements, e.g., without team orientation. In general, there is a trade-off between the minimization of operational cost and the minimization of team changes. Team change penalties may result in outweighing operationally less expensive rosters in preference to those with higher team orientation, e.g., involving fewer team changes.

Nevertheless, for certain business settings, such as for our cooperation partner, the reduction of team changes may pay out financially at a certain point. Having fixed rates for taxi proceedings within the home country of the airline, the breakeven for dedicated trips is sometimes reached even at less than four crew members. Working as a team, they are able to share their chauffeured vehicle (sometimes having a capacity of up to eight people) instead of deploying per-seat tickets for rail or air transportation.

Due to the penalization of team changes among roster combinations, the aim of the ToRP is hereby defined as the search for an appropriate set of individual rosters (one roster for each crew member) such that all given flights are covered properly at minimum cost with a socially and economically reasonable reduction of team changes (in comparison to the classical rostering process, separated by crew functions).

For a more detailed problem analysis we refer to Thiel (2005).

4 Mathematical Formulation

After introducing the idea and some basic concepts of the ToRP for cockpit crew, this section discusses two distinct mathematical formulations. First, we introduce all variables required, followed by two different approaches: the Extended Rostering Model and the Roster Combination Model. A review on both approaches discusses their pros and cons at the end of this section. Further approaches are presented in Thiel (2005).

4.1 Notations

Before presenting the two optimization models, commonly used variables and parameters are defined as follows:

F represents the number of SFAs f to be serviced.

K indicates the total number of crew members. Captains are enumerated starting from 1 to k^{CP} and first officers start from $k^{CP} + 1$ to K.

R^k expresses the total number of rosters for crew member k being considered in the model.

$R = \sum_{k=1}^{K} R^k$ gives the overall number of all rosters among all crew members, where $r^{CP} = \sum_{k=1}^{k^{CP}} R^k$ is the number of all captain rosters, first officer rosters have the indices from $r^{CP} + 1$ to R.

r^k is the index of the first roster for crew member k with $r^1 = 1$ and $r^k = \sum_{i=1}^{k-1} R^i + 1 \forall k \in \{2, \ldots, K\}$. The special case $k = K + 1$ is defined as $r^{K+1} = R + 1$.

c_{r1} represents overall cost for roster $r1$. (Those are characterized by operational cost – here, hotel and taxiing expenses – as well as deviation penalties from planned flight time or contract usage for the individual crew member to facilitate fair-and-equal share.)

$c_{r1,r2}$ indicate team change penalties of the chosen roster combination $(r1, r2)$ (see Section 3).

$a_{r,f}^{CP}$ and $a_{r,f}^{FO}$, each equals 1, if a SFA f is included in roster r as a *captain* or *first officer* activity, 0 otherwise.

x_r $\in \{0; 1\}$ equals 1, if roster r is chosen, 0 otherwise.

$x_{r1,r2} \in \{0; 1\}$ equals 1, if a specific roster combination $(r1, r2)$ is chosen by $x_{r1} = 1 \wedge x_{r2} = 1$, 0 otherwise.

$x_f^{ECP}, x_f^{EFO} \in \{0; 1\}$ equals 1, if a SFA f for a captain or a first officer is unassigned, 0 otherwise.

c_f^E points out the (virtual) cost for unassigned SFAs. (Those cases are absorbed by the usage of the identity matrix E.)

4.2 Extended Rostering Model

The key concept of the Extended Rostering Model can be depicted as a strict extension of the basic set partitioning model for the airline CRP in such a way that it handles penalties for team changes via additional rows and columns. In this model $x_{r1,r2}$ is defined as indicator variable. The resulting model can be formulated as follows:

$$\min \sum_{r=1}^{R} c_r x_r + \sum_{r1=1}^{r^{CP}} \sum_{r2=r^{CP}+1}^{R} c_{r1,r2} x_{r1,r2} + \sum_{f=1}^{F} c_f^E (x_f^{ECP} + x_f^{EFO}) \quad (1)$$

Subject to:

$$\sum_{r=r^k}^{r^k+R^k-1} x_r = 1 \ \forall k = \{1, ..., K\} \quad (2)$$

$$\sum_{r=1}^{r^{CP}} a_{r,f}^{CP} x_r + x_f^{ECP} = 1 \ \forall f \in \{1, ..., F\} \quad (3)$$

$$\sum_{r=r^{CP}+1}^{R} a_{r,f}^{FO} x_r + x_f^{EFO} = 1 \ \forall f \in \{1, ..., F\} \quad (4)$$

$$x_{r1} + x_{r2} - x_{r1,r2} \leq 1 \ \forall r1 \in \{1, ..., r^{CP}\} \forall r2 \in \{r^{CP} + 1, ..., R\} \quad (5)$$

If $c_{r1,r2} < 0$, then include

$$x_{r1,r2} \leq x_{r1} \ \forall r1 \in \{1, ..., r^{CP}\} \forall r2 \in \{r^{CP} + 1, ..., R\} \quad (6)$$

$$x_{r1,r2} \leq x_{r2} \ \forall r1 \in \{1, ..., r^{CP}\} \ \forall r2 \in \{r^{CP} + 1, ..., R\} \tag{7}$$

The objective function (1) consists of three parts: The first addend of the minimization function summarizes the required operational roster cost, whereas the second covers the corresponding team change penalties when captain rosters ($=r1$) and first officer rosters ($=r2$) are combined. The third part ensures the solvability by treating unassigned SFAs with special cost.

Restrictions (2) to (4) guarantee the regular CRP requirements, whereas the remaining focus on the consideration of team-orientated characteristics. In (2) exactly one roster is assigned to each crew member k. All captain activities are covered by crew members of this crew function or by the identity matrix in (3); respectively, all first officer activities in (4). In (5) all required team change penalties for a roster combination $(r1, r2)$ occur only in the case that both rosters are chosen. Restrictions (6) and (7) assume that negative team change penalties (or bonuses) are only selected in the solution if rosters $r1$ and $r2$ themselves are chosen, 0 otherwise.

The model structure is given in Fig. 3. The first six columns show the captain rosters (three for each), followed by (not necessarily) the same amount of rosters for each first officer (FO). For instance, the second roster of CP1 (second column of the data matrix) contains $SFA1$, $SFA2$ and $SFA5$, whereas in the third CP1 roster (third column), $SFA1$, $SFA4$ and $SFA5$ are included. Here, every first roster of a crew member is a null-roster to grant feasibility. All other columns are introduced to handle roster combinations and unassigned SFAs. The first row indicates the column's influence on the objective function (1), followed by a block of rows for restrictions (2) to (4). Since not all team change penalties in this example are positive, restrictions (6) and (7) become necessary for roster combination $(CP2\ R3, FO1\ R3)$ or $(R6, R9)$ to guarantee in addition to (5) the appropriate consideration of team change penalties where necessary. All team change penalties were set to exemplary values ahead of the model creation.

4.3 Roster Combination Model

In contrast to this, the *Roster Combination Model* follows the idea of directly considering roster combinations instead of single rosters for each individual crew member. Therefore, all columns in this model directly represent a roster combination for two crew members (CPx, FOx'), independent of whether they share any SFA or not. Such roster combinations are based on all available rosters for each individual crew member. For a better comparison of both models in Section 4.2 and Section 4.3, let $\tilde{c}_{r1} = \frac{c_{r1}}{K - k^{CP}}$ and $\tilde{c}_{r2} = \frac{c_{r2}}{k^{CP}}$ (operational cost for a captain roster is divided by the number of first officers and vice versa). Here $x_{r1,r2}$ is used as the decision variable. The resulting model can be formulated as:

$$\min \sum_{r1=1}^{r^{CP}} \sum_{r2=r^{CP}+1}^{R} (\tilde{c}_{r1} + \tilde{c}_{r2} + c_{r1,r2}) x_{r1,r2} + \sum_{f=1}^{F} c_f^E (x_f^{ECP} + x_f^{EFO}) \tag{8}$$

Subject to:

		Roster											Roster Combination										Identity Matrix											
		CP1			CP2			FO1			FO2																									
Obj		0	4	6	0	6	6	0	5	4	2	4	7	2	2	1	1	2	1	1	1	2	-1	50	50	50	50	50	50	50	50	50	50	=>	min
		1	1	1																														=	1
					1	1	1																											=	1
								1	1	1																								=	1
											1	1	1																					=	1
SFA for CP		1	1																				1											=	1
		1				1																		1										=	1
					1	1																			1									=	1
			1			1																				1								=	1
		1	1																								1							=	1
SFA for FO									1	1																		1						=	1
						1				1																			1					=	1
					1	1																								1				=	1
								1	1																						1			=	1
					1	1																										1		=	1
Roster Combination		1							1					-1																				<=	1
		1										1			-1																			<=	1
		1										1				-1																		<=	1
			1									1					-1																	<=	1
			1									1						-1																<=	1
					1				1										-1															<=	1
					1							1								-1														<=	1
					1							1									-1													<=	1
								1	1													-1												<=	1
								1		1													-1											<=	1
						-1																		1										<=	0
									-1																1									<=	0

Fig. 3. Schematic View on Extended Rostering Model

$$\sum_{r1=r^{k1}}^{r^{k1+1}-1} \sum_{r2=r^{k2}}^{r^{k2+1}-1} x_{r1,r2} = 1 \tag{9}$$

$$\forall k1 \in \left\{1,...,k^{CP}\right\} \forall k2 \in \left\{k^{CP}+1,...,K\right\}$$

$$\sum_{r1=1}^{r^{CP}} a_{r1,f}^{CP} x_{r1,r2} + (K-k^{CP})x_f^{ECP} = K - k^{CP} \tag{10}$$

$$\forall r2 \in \left\{r^{CP}+1,...,R\right\} \forall f \in \{1,...,F\}$$

$$\sum_{r2=r^{CP}+1}^{R} a_{r2,f}^{FO} x_{r1,r2} + k^{CP} x_f^{EFO} = k^{CP} \tag{11}$$

$$\forall r1 \in \left\{1,...,r^{CP}\right\} \forall f \in \{1,...,F\}$$

$$\sum_{r2=r^{k2}}^{r^{k2+1}-1} x_{r1,r2} - \sum_{r2'=r^{k2'}}^{r^{k2'+1}-1} x_{r1,r2'} = 0 \tag{12}$$

$$\forall (k1, k2): \quad k1 \in \left\{ 1, ..., k^{CP} \right\} k2, k2' \in \left\{ k^{CP} + 1, ..., K \right\} : k2 \neq k2' r1 \in$$
$$\left\{ r^{k1}, ..., r^{k1+1} - 1 \right\}$$

$$\sum_{r1=r^{k1}}^{r^{k1+1}-1} x_{r1,r2} - \sum_{r1'=r^{k1'}}^{r^{k1'+1}-1} x_{r1',r2} = 0 \tag{13}$$

$$\forall (k1, k2): \quad k1, k1' \in \left\{ 1, ..., k^{CP} \right\} : k1 \neq k1' k2 \in \left\{ k^{CP} + 1, ..., K \right\} r2 \in$$
$$\left\{ r^{k2}, ..., r^{k2+1} - 1 \right\}$$

As mentioned above, this model already considers roster combinations. Here operational roster cost and team change penalties are processed simultaneously within the objective function (8), whereas the second part summarizes the unassigned shared flight activities. A special characteristic of this modeling approach is the fact that every selected captain roster of the solution is combined with all selected first officer rosters of the solution. As a consequence, in order to remain consistent with the objective value of the Extended Rostering Model above, all cost factors for each captain roster c_{r1} are divided by the number of first officers $K - k^{CP}$, the same for first officer roster cost and the utilization of the identity matrix for unassigned SFAs (see definition of $\tilde{c}_{r1}, \tilde{c}_{r2}$ and (8)).

All restrictions satisfy the consistency of the chosen solution: Out of each (CPx, FOx')-combination exactly one corresponding roster combination $(CPxRy, FOx'Ry')$ has to be selected by (9). That is the reason why in (10) all captain SFAs have to be assigned exactly as often as there are first officers in the model. (Every SFA is still covered exactly once by a single captain CPx; but – since there are CPx times FO combinations – every SFA needs to be covered as often as first officers are available.) In (11) all SFAs for first officers are treated analogously.

In the solution a set of roster combinations is selected; each roster combination implies that a specific captain executes a selected roster $(CPx\ Ry)$, the same does the designated first officer $(FOx'\ Ry')$. Since we consider all possible roster combinations among captains and first officers, restriction (12) ensures that the chosen captain roster $(r1)$ is selected within all other chosen roster combinations among this captain $(k1)$ and all other first officers $(k2$ and $k2')$; restriction (13) does the same in a similar way for the determined roster of every first officer.

In Fig. 4 the structure of the Roster Combination Model is illustrated. Every column represents a roster combination $(CPx\ Ry, FOx'\ Ry')$ for each possible (CPx, FOx') cockpit team followed by columns that handle unassigned SFAs (like above in the Extended Rostering Model via the identity matrix). Below the first row for the objective value, restrictions (9) to (11) are realized in each row block. The synchronous arrangements in the lower half of the figure implement the set of restrictions for (12) and (13) for a consistent treatment of roster combinations. Note that the operational cost and team change penalties are taken from the example introduced by Fig. 3.

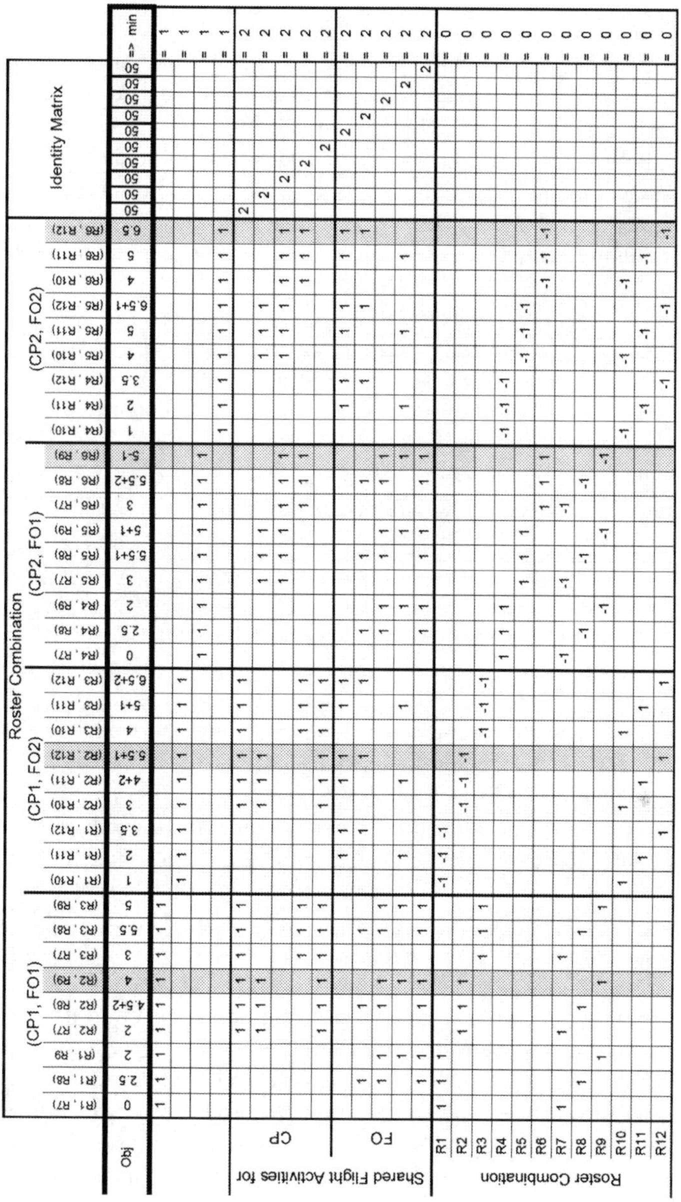

Fig. 4. Schematic View on Roster Combination Model

4.4 Model Comparison

After describing both distinctive modeling approaches from the mathematical point of view the important characteristics of both models are reviewed in this subsection.

The Extended Rostering Model formulated in Section 4.2 as a binary IP model penalizes each roster combination by additional columns and rows. The number of those possible combinations increases dramatically with regard to the number of crew members and their rosters. Considering all of them outranges rather soon the computable limitations for model generation and solution. Therefore, it is important to choose an appropriate penalization strategy which should result in relatively few penalized roster combinations with $c_{r1,r2} \neq 0$, and by this, only a small amount of additional columns in the model. As given in (5), such roster combinations require a single additional restriction to be applied properly, but in case of negative penalties, two further rows become necessary which may lead to a tremendous growth of the amount of rows for the model. For that reason the model size increases almost proportionally to the number of penalized roster combinations, which is highly influenced by the chosen penalization strategy. This leads usually to a high number of columns and rows.

On the other hand, the Roster Combination Model in Section 4.3 considers team change penalties simultaneously with operational cost. Since this binary IP-model here explicitly builds all possible roster combinations, its proposed size remains fixed independent from the chosen penalization strategy. For comparably small instances where all $c_{r1,r2} < 0$ (as the worst case for the Extended Rostering Model), this model demonstrates great advantages because the identical problem can be expressed by a much smaller model, e.g., for an instance of thirteen SFAs with five captains with a sum of 763 rosters and six first officers with totally 468 rosters, both models are almost equal regarding the number of columns (around 350,000), but the Extended Rostering Model requires more than 1 million rows whereas all restrictions of the Roster Combination Model only demand around 5,700 rows. Nevertheless, the sheer model size does not justify a selection among both alternatives. For the Roster Combination Model the selection of the optimal solution is much harder (due to the doubled amount of SFAs closely considered throughout the roster combinations). In contrast to this, the Extended Rostering Model can be characterized by handling two almost separate sets of SFAs which are more loosely linked by the team change penalty restrictions.

A further practical requirement is downgrading, where for cockpit crew a captain operates one or multiple SFAs in the function of a first officer. For the Extended Rostering Model those cases are relatively easy to implement by inserting additional columns, where a valid roster is modified in such a way that a subset of the included SFAs is shifted to the position of first officer SFAs. Solvability is not endangered by this action, but in order to consider also team changes of two captains (CPx, CPx'), several modifications become necessary for the range of the sums in the objective function and the affected restrictions of the model. For the more compressed Roster Combination Model it is very hard to realize downgrading without restructuring the complete formulation. An overall comparison of both modeling approaches is given in Table 1.

Again, the key characteristic of the ToRP is the consideration of roster combinations instead of single rosters. The *quadratic assignment problem (QAP)* handles this special aspect already. In the QAP, quadratic formulations, e.g., $x_{r1}x_{r2}$, are allowed.

Table 1. Comparison of Extended Rostering and Roster Combination Model

	Extended Rostering Model	Roster Combination Model
Basic idea	• Columns represent single rosters or penalized roster combinations. • Operational cost and team change penalties are treated separately.	• Columns represent roster combinations. • Operational cost and team change penalties are considered simultaneously.
Max. model size	Columns (all $c_{r1,r2} \neq 0$): $R + r^{CP}(R - r^{CP}) + 2F$ Rows (all $c_{r1,r2} < 0$): $K + 2F + 3r^{CP}(R - r^{CP})$	Columns: $r^{CP}(R - r^{CP}) + 2F$ Rows: $k^{CP}(K - k^{CP}) + 2F + r^{CP}(K - k^{CP} - 1) + (R - r^{CP})(k^{CP} - 1)$
Model growth	Strongly depends on penalty strategy chosen	Independent of penalty strategy
Downgrading	Yes (model modifications for indices required)	No

Therefore, the objective function of the Extended Rostering model in (1) can be expressed as follows:

$$\min \sum_{r=1}^{R} c_r x_r + \sum_{r1=1}^{r^{CP}} \sum_{r2=r^{CP}+1}^{R} c_{r1,r2} x_{r1} x_{r2} + \sum_{f=1}^{F} c_f^E (x_f^{ECP} + x_f^{EFO}) \quad (14)$$

Both variables in the product of the binary decision variables have to equal one in order to enforce the team change penalty for the selected roster combination. All restrictions (5) – (7) become obsolete. Nevertheless, instances with more than 10,000 binary variables (here: rosters) are still almost impossible to solve today (see Anstreicher *et al.* (2002), Caprara (2004)). This makes the application for most real-life instances of the ToRP impossible and results in the deployment of the above models as appropriate alternatives.

Typical applications for the QAP are efficient wiring problems (e.g., *Steinberg Wiring problem*) or layout problems for hospitals and production lines (see, e.g., Commander (2003) for further examples). To the knowledge of the author, there is no application reported for personnel scheduling so far.

5 Implementation

In this section we describe some of the concepts applied to solve several test instances in a team-oriented way. With regard to the real-life requirements for a successful application of the described ToRP approach, it has to be acknowledged that model size, as the most decisive model criterion, indicates that the Extended Rostering Model is the preferred basis for the upcoming implementation and computational experiments.

Independent from the mathematical formulation chosen, the ToRP implies two major problems: roster generation and roster combination. As already pointed out in the state-of-art in Section 2.2, the airline CRP is known as a source of huge SPP models. Hence we introduce briefly our approach to address those problems adequately.

5.1 Generating Rosters

The implementation of the roster generation is realized in analogy to the recursive approach presented in Kohl and Karisch (2004). The first run of the algorithm starts with the null-roster, which is filled step-by-step with the remaining SFAs until it becomes illegal due to incompliance with governmental rules, union agreements and/or airline specific rules. If so, the last element is replaced by the next one on the SFA list. By this, even a small number of SFAs may produce a high number of legal rosters for each crew member. In case of no pre-scheduled activities, every single included SFA needs to be considered, by which we quite soon reach several million rosters due to combinatorial possibilities.

To address this fact we propose to reduce the set of SFAs called the *roster combination basis* in an appropriate way. As given above, especially individual crew members having a high availability (usually with few or no pre-scheduled activities during the examined time period) are very flexible and produce the highest amount of legal rosters. Although they are capable to service theoretically on every single day, they also need their weekly rest periods. Therefore, we propose to review the supply and demand for every day and home base of the data set. On some days, we may observe an oversupply where those crew members are most likely not necessarily required, and their SFAs for this day can be removed from their roster combination basis without notable impact on the solution.

For settings with multiple home bases another aspect should be reviewed. All SFAs are initially assigned to home bases by the CPP due to cost minimization. It is an advantage to keep those pairings primarily at the originally chosen home base, and in combination with a local gap of personnel, all additional SFAs from other home bases can be neglected.

Furthermore, it makes sense sometimes to reduce the number of rosters according to given quality criteria, such as limits on overall operational cost, the number of hotel stays and/or proceedings etc. Those additional restrictions or their combinations can be applied within defined rule sets, which unfortunately have to be re-evaluated for every single instance. Furthermore, individually calculated target flight times can be applied to the roster selection to assure higher fair-and-equal share assignments.

We are aware that whatever roster pre-selection takes place to filter "good" ones out of valid rosters may greatly influence the quality of the solution. Nevertheless, the application of some of the proposed strategies remains necessary; however, they have to be chosen very carefully.

5.2 Combining Rosters

As discussed in Section 4.4, the Extended Rostering Model has to cope with a high amount of potentially penalized roster combinations. Because of this, the strategy for penalizing team changes becomes quite critical, since it very much determines the overall size of the model. Therefore, we propose a strategy that on the one hand considers as many roster combinations as possible, but on the other hand penalizes only quite few of them. We achieved a very low rate for penalties by linking them only to roster combinations if there is a team change within a working week. So team changes over the weekly rest period are neglected. Further on, we recommend choosing a strategy with few or even no negative penalization values to prevent a high increase in the amount of rows for the model.

So far, we allow all pre-selected rosters of each individual crew member to be considered in the model. The amount of resulting roster combinations that require a penalization is still too high to be handled appropriately (usually several million). One alternative is a strict pre-selection of the rosters for one crew function, namely first officers. (This is done because captains can be downgraded to first officers in case that their time and location dependent capacities are not sufficient to cover all their SFAs.) From those pre-selected rosters, build the starting point for all further roster combinations with the opposite crew function(s). They are determined by solving the general CRP for first officers as often as the rostering loop counter is set up; thereby the best roster solution found so far is explicitly excluded from the solution space.

6 Computational Experiences

In order to evaluate the effectiveness and efficiency of the Team-oriented Rostering approach proposed here, a series of computational experiments was conducted. First key results are presented in this section.

All experiments were realized on a PC with an Intel Pentium IV, 2.26 GHz CPU with 2.0 GB RAM, operating on Microsoft Windows XP Professional. The prototype is implemented in Visual C++ 6.0 and considers only valid rosters for each individual crew member. All models were solved using CPLEX, version 9.0 (see CPLEX Optimization Inc. (2003)). Time measurements are given in CPU seconds. The considered data set originates from a European tourist airline. The instances examined below are based on two typical holiday periods which represent high demand periods of the year 2002.

Each instance below is described by the time period chosen, the number of home bases (HB), the amount of captains (CP), first officers (FO), and considered SFA. Further parameters that have been considered are the maximum number of elements in the roster combination basis, the maximum number of disposable working days within the period, the chosen rule set, an indicator whether other airports are serviced, the penalty value for a single team change (TP), the number of rostering loops (RL), and an indicator whether downgrading is considered in this model or not. The resulting integer programming models have been implemented following the Extended Rostering Model description in (4.2). The model characteristics include the number of rosters generated for captains (RGCP) and first officers (RGFO), the time for their generation (RGT), the number of rosters included in the model (RCP and RFO), the model size in rows and columns, the number on non-zeros (NZ), the duration for solving (ST), the operational cost (OC), and the number of team changes (TC). (All parameters above which are not mentioned in the tables are unchanged for all examined instances.)

6.1 Team Change Penalties

The consideration and appropriate setting of team change penalties are the major aspect within the approach presented in this paper. Therefore, a set of test runs was conducted with different penalty values for team changes (100, 200, 300, 500 and 1,000) on the same instance in comparison to the conventional approach without any penalization as documented in Table 2.

For all instances the conventional approach ($TP = 0$) offers the cheapest solution in terms of operational cost (OC), but with very frequent team changes. In contrast to that, we observe a tendency for a monotonously slightly increasing operational cost for all listed ToRP variants ($TP \neq 0$) which can be explained by the amplified trade-off between operational cost and increasing team change penalties. As the instances proved, simply applying the ToRP approach manages to dramatically reduce the number of team changes for the new crew schedules at the expense of slightly higher operational cost. All instances were solved with the same amount of unassigned SFAs. In Table 2, the significant difference regarding model size between the conventional and ToRP variants becomes quite obvious. It is caused by the additional columns and rows for team change penalization as discussed earlier.

In Fig. 5, the decreasing amount of team changes is visualized for the different team change penalties, where two pre-selected rosters for each first officer are considered (RL = 2, see Section 6.2) and downgrading is enabled.

6.2 Rostering Loops

The second set of test runs to be presented is the performance of the so-called *rostering loops* introduced for the ToRP. Following a sequential procedure which solves the original CSP only for one crew function several times, we get a set of only few (first officer) rosters to be pre-selected instead of including all of them in the model. This time the ToRP variants examined differ by the number of resulting rosters based on those rostering loops as they are set to values from 1 to 5. Due to the model size, only the two small instances (July 1-15 with one home base and December 16-31 with two home bases) have been realized without pre-selected rosters for reference.

As shown in Table 3, the application of such rostering loops for their the roster pre-selection turned out to be quite valuable for both, solution quality and model size. Although only few (instead of all) rosters have been chosen, the gap in terms

Table 2. Results for Different Team Change Penalties (TP)

Period	HB	CP	FO	SFA						
Jul 1-15, 2002	2	8	10	31						

TP	RGCP	RGFO	RGT	RCP	RFO	Rows	Cols	NZ	ST	OC	TC
0	11103	12803	11:06	11103	20	80	11185	69232	00:01	4685	10
100	11103	12803	11:06	11103	20	34851	45956	208316	00:27	4925	4
200	11103	12803	11:06	11103	20	34851	45956	208316	01:05	5110	3
300	11103	12803	11:06	11103	20	34851	45956	208316	01:19	5386	2
500	11103	12803	11:06	11103	20	34851	45956	208316	03:10	5386	2
1000	11103	12803	11:06	11103	20	34851	45956	208316	01:49	5386	2

Period	HB	CP	FO	SFA						
Jul 1-15, 2002	4	24	22	78						

TP	RGCP	RGFO	RGT	RCP	RFO	Rows	Cols	NZ	ST	OC	TC
0	126504	169680	236:17	126576	42	202	126774	802220	01:04	13649	30
100	126504	169680	236:17	126576	42	543205	669777	2974232	240:25*	14084	11
200	126504	169680	236:17	126576	42	543205	669777	2974232	240:19*	13875	15
300	126504	169680	236:17	126576	42	543205	669777	2974232	240:23*	14090	13
500	126504	169680	236:17	126576	42	543205	669777	2974232	240:19*	14421	14
1000	126504	169680	236:17	126576	42	543205	669777	2974232	480:24**	14200	11

Period	HB	CP	FO	SFA						
Jul 1-15, 2002	6	29	27	99						

TP	RGCP	RGFO	RGT	RCP	RFO	Rows	Cols	NZ	ST	OC	TC
0	102281	121177	223:32	103375	52	254	103625	629153	00:34	19196	51
100	102281	121177	223:32	103375	52	433512	536883	2362185	240:15*	19529	34
200	102281	121177	223:32	103375	52	433512	536883	2362185	240:11*	19735	32
300	102281	121177	223:32	103375	52	433512	536883	2362185	240:11*	19735	32
500	102281	121177	223:32	103375	52	433512	536883	2362185	240:11*	19787	33
1000	102281	121177	223:32	103375	52	433512	536883	2362185	240:12*	19603	32

Period	HB	CP	FO	SFA						
Dec 16-31, 2002	6	33	26	44						

TP	RGCP	RGFO	RGT	RCP	RFO	Rows	Cols	NZ	ST	OC	TC
0	14163	10904	20:40	14163	48	147	14299	72973	00:01	10532	9
100	14163	10904	20:40	14163	48	36604	50756	218801	02:01	10828	0
200	14163	10904	20:40	14163	48	36604	50756	218801	01:05	10828	0
300	14163	10904	20:40	14163	48	36604	50756	218801	01:10	10828	0
500	14163	10904	20:40	14163	48	36604	50756	218801	00:55	10828	0
1000	14163	10904	20:40	14163	48	36604	50756	218801	01:25	10828	0

* Abortion of Optimization after 240 minutes. Usage of the best IP-Solution found.
** Abortion of Optimization after 480 minutes. Usage of the best IP-Solution found.

of operational cost between their complete consideration in the model and an obviously appropriate pre-selection appears to be quite low. It has to be noticed that
the number of team changes tend to decrease with a higher number of pre-selected
rosters. In addition, a significant reduction of model size is accomplished, indicated
by the comparison rate (MR) giving the proportions of model sizes with and without those pre-selected rosters. (All instances were computed with $TP = 300$ and
enabled downgrading.)

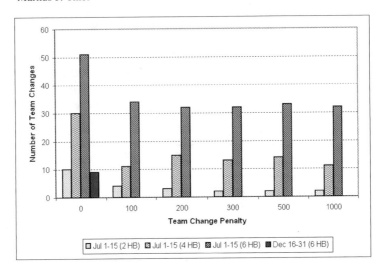

Fig. 5. Development of Team Change Count for Different Penalty Values

6.3 Further Results

Further test runs were exhaustively conducted on all available parameters. They greatly confirm the following two additional statements:

1. Restrictions on the combinatorial basis for each crew member have to be chosen very carefully (see Section 5.2). If the number of SFAs is too small, multiple SFAs remain unassigned, but considering too many of them makes the model itself impossible to handle.
2. Restrictions regarding roster acceptance within the roster generation part (see Section 5.1) show that a significant reduction of rosters via stricter rule sets (e.g., limits for cost, hotel stays) trades-off with the quality of the solution.

Table 3. Results for Pre-selected Roster Variants

Period		HB	CP	FO	SFA							
Jul 1-15, 2002		1	4	3	16							
BC	RGCP	RGFO	RGT	RCP	RFO	MR	Rows	Cols	NZ	ST	OC	TC
1	682	2440	01:06	599	3	99,89%	891	1486	6752	00:00	3128	4
2	682	2440	01:06	770	6	99,72%	1851	2620	11384	00:00	2965	0
3	682	2440	01:06	770	9	99,58%	2830	3602	15321	00:00	2965	0
4	682	2440	01:06	883	12	99,36%	3901	4789	20263	00:01	2965	0
5	682	2440	01:06	883	15	99,20%	4844	5735	24055	00:00	2965	0
-	682	2440	01:06	682	2440	0,00%	513829	516944	2073733	01:17	2965	0

Period		HB	CP	FO	SFA							
Jul 1-15, 2002		2	8	10	31							
BC	RGCP	RGFO	RGT	RCP	RFO	MR	Rows	Cols	NZ	ST	OC	TC
1	11103	12803	11:06	10905	10	99,92%	18168	29065	140807	00:01	4909	5
2	11103	12803	11:06	11103	20	99,84%	34851	45956	208316	03:14	5386	2
3	11103	12803	11:06	11103	30	99,77%	53643	64758	283535	24:27	5386	2
4	11103	12803	11:06	11103	40	99,69%	72720	83845	359894	28:09	5386	2
5	11103	12803	11:06	11103	50	99,61%	91627	102762	435573	24:20	5386	2

Period		HB	CP	FO	SFA							
Dec 16-31, 2002		2	10	9	12							
BC	RGCP	RGFO	RGT	RCP	RFO	MR	Rows	Cols	NZ	ST	OC	TC
1	149	141	00:07	99	9	95,76%	96	185	696	00:00	5504	3
2	149	141	00:07	149	16	88,65%	184	330	1235	00:00	5611	1
3	149	141	00:07	164	23	82,05%	240	408	1543	00:01	5611	1
4	149	141	00:07	164	30	76,58%	307	482	1836	00:00	5611	1
5	149	141	00:07	175	38	68,35%	361	555	2121	00:00	5592	1
-	149	141	00:07	149	141	0,00%	1249	1520	5938	00:01	5535	0

Period		HB	CP	FO	SFA							
Dec 16-31, 2002		6	33	26	44							
BC	RGCP	RGFO	RGT	RCP	RFO	MR	Rows	Cols	NZ	ST	OC	TC
1	14163	10904	20:41	13438	26	99,77%	17618	31023	140380	00:03	11684	0
2	14163	10904	20:41	14163	48	99,56%	36604	50756	218801	01:10	10828	0
3	14163	10904	20:41	14163	69	99,37%	55333	69506	293803	01:34	10828	0
4	14163	10904	20:41	14163	92	99,16%	73501	87697	366565	02:03	10828	0
5	14163	10904	20:41	14163	115	98,95%	89571	103790	430935	08:49	10828	0

7 Summary and Outlook

In this work we defined and presented the new Team-oriented Rostering Problem in
the context of airlines – an approach within the crew assignment phase for onboard
crew scheduling. The ToRP focuses on the minimization of team changes within the
cockpit crew. Based on a setting with time and location dependent crew availabilities
several strategies addressing the high combinatorial complexity were discussed and
implemented, accounting for roster combinations instead of single rosters.

Two distinct mathematical formulations were given to realize the ToRP approach,
whereas for real-life instances the Extended Rostering Model was proven to be more
applicable than the Roster Combination Model. Although some problem characteris-

tics are literally shared by the widely examined quadratic assignment problems, the proposed IP models are comparably easier to solve in terms of size and time.

Several implementation techniques were tested on various instances, each with different parameters. One of the key objectives of this study was to show the effects of ToRP that result in a trade-off between operational cost on the one hand, and the number of team changes on the other. Especially the pre-selection of good rosters for one crew function – here, first officers due to downgrading – turned out to result in high model size reduction rates without a notable lack of solution quality.

It is acknowledged that only relatively small instances (with less than 1,000 flight legs) are solved within an acceptable time frame so far. Therefore, further research will especially concentrate on this drawback of our approach. Firstly, we suggest appropriate penalization strategies, since their setting is tightly linked to the mostly critical size of the model. By this, even the application of a Branch-and-Cut approach may turn out to be suitable. Secondly, a great benefit will arise when defining SFAs properly already during the pairing generation phase. Although this requirement implies a modification of the models and techniques applied currently for the CPP, the generation of thousands of potential rosters (consisting of short SFAs) is prevented and, as a result, larger instances can be solved.

Another option appeared after analyzing our results. We noticed that (1) several crew members share exactly identical sets of rosters and (2) the majority of staff (>70%) never experiences any team change in the final crew schedule. Therefore, a great model reduction can be achieved by grouping crew members with identical rosters, and, if possible, by building "pre-defined" groups already for (potential) teams, where there will be no team change at all. By this we get a hybrid IP model; see also Thiel (2005), where the residual problem can be solved, e.g., by the Extended Rostering Model as described previously.

References

Anbil, R., Gelman, E., Patty, B., and Tanga, R. (1991). Recent advances in crew-pairing optimization at American Airlines. *Interfaces*, **21**(1), 62–74.

Anstreicher, K., Brixius, N., Goux, J. P., and Linderoth, J. (2002). Solving large quadratic assignment problems on computational grids. *Mathematical Programming, Series B91*, pages 563–588.

Barnhart, C., Cohn, A. M., Johnson, E. L., Klabjan, D., Nemhauser, G. L., and Vance, P. H. (2003). Airline crew scheduling. In R. Hall, editor, *Handbook of Transportation Science*, pages 517–560. Kluwer, Boston, 2nd edition.

Bixby, R., Gregory, J., Lustig, I., Marsten, R., and Shano, D. (1992). Very large-scale linear programming: A case study in combining interior point and simplex methods. *Operations Research*, **40**, 885–897.

Campbell, K. W., Durfee, R. B., and Hines, G. S. (1997). FedEx generates bid lines using simulated annealing. *Interfaces*, **27**(2), 1–16.

Caprara, A. (2004). The basic approach to 0-1 quadratic programs. In A. Agnetis and G. Di Pillo, editors, *Modelli e Algoritmi per l'Ottimizzazione di Sistemi Complessi*, pages 71–89. Pitagora Editrice.

Commander, C. W. (2003). A survey on the quadratic assignment problem, with applications, undergraduate honors thesis. Technical report, Department of Mathematics, University of Florida, Gainsville.

CPLEX Optimization Inc. (2003). Using the CPLEX callable library. Technical report, CPLEX Optimization, Incline Village.

Day, P. R. and Ryan, D. M. (1997). Flight attendant rostering for short haul airline operations. *Operations Research*, **45**(5), 649–661.

Desaulniers, G., Desrosiers, J., Dumas, Y., Marc, S., Rioux, B., Solomon, M. M., and Soumis, F. (1997). Crew pairing at Air France. *European Journal of Operational Research*, **97**, 245–259.

El Moudani, W., Cosenza, C. S. N., and de Coligny, M. (2001). A bi-criterion approach for the airline crew rostering problem. In *Lecture Notes in Computer Science*, volume 1993, pages 486–500. Springer, Heidelberg.

Gamache, M., Soumis, F., Villeneuve, D., Desrosiers, J., and Gélinas, É. (1998). The preferential bidding system at Air Canada. *Transportation Science*, **32**(3), 246–255.

Gamache, M., Soumis, F., Marquis, G., and Desrosiers, J. (1999). A column generation approach for large-scale aircrew rostering problems. *Operations Research*, **47**(2), 247–263.

Gerhkoff, I. (1989). Optimizing flight crew schedules. *Interfaces*, **19**(4), 29–43.

Graves, G., McBride, R., Gershkoff, I., Anderson, D., and Mahidhara, D. (1993). Flight crew scheduling. *Management Science*, **39**, 736–745.

Guo, Y., Mellouli, T., Suhl, L., and Thiel, M. P. (2006). A partially integrated airline crew scheduling approach with time-dependent crew capacities and multiple home bases. *European Journal of Operational Research*, **171**, 1169–1181.

Hoffman, K. L. and Padberg, M. (1993). Solving airline crew scheduling problems by branch and cut. *Management Science*, **39**, 657–682.

Jarrah, A. I. Z. and Diamond, J. T. (1997). The problem of generating crew bidlines. *Interfaces*, **27**(4), 49–64.

Kohl, N. and Karisch, S. E. (2004). Airline crew rostering: Problem types, modeling, and optimization. *Annals of Operations Research*, **127**, 223–257.

Lavoie, S., Minoux, M., and Odier, E. (1988). A new approach for crew pairing problems by column generation with an application to air transportation. *European Journal of Operational Research*, **35**, 45–58.

Mellouli, T. (2001). A network flow approach to crew scheduling based on an analogy to a train/aircraft maintenance routing problem. In S. Voß and J. Daduna, editors, *Computer-Aided Scheduling of Public Transport, LNEMS 505*, pages 91–120. Springer, Berlin.

Mellouli, T. (2003). Scheduling and routing processes in public transport systems. Habilitation thesis. University of Paderborn.

Nicoletti, B. (1975). Automatic crew rostering. *Transportation Science*, **9**, 33–42.

NTSB (1994). A review of flightcrew-involved major accidents of U.S. air carriers, 1978 through 1990. Technical report, National Transportation Safety Board, Washington, D.C.

Ryan, D. M. (1992). The solution of massive generalized set partitioning problems in aircrew rostering. *Journal of the Operational Research Society*, **43**(5), 459–567.

Strauss, C. (2001). *Quantitative Personaleinsatzplanung im Airline Business*. Peter Lang Publishing Group, Frankfurt am Main.

Suhl, L. (1995). *Computer-Aided Scheduling: An Airline Perspective*. Deutscher Universitätsverlag, Wiesbaden.

Thiel, M. P. (2005). *Team-oriented Airline Crew Scheduling and Rostering: Problem Description, Solution Approaches, and Decision Support*. Ph.D. thesis, University of Paderborn. http://ubdata.upb.de/ediss/05/2005/thiel.

Vance, P. H., Atamtürk, A., Barnhart, C., Gelman, E., Johnson, E. L., Krishna, A., Mahidhara, D., Nemhauser, G. L., and Rebello, R. (1997). A heuristic branch-and-price approach for the airline crew pairing problem. Working paper. Auburn University.

Yan, S. and Tu, Y. (2002). A network model for airline cabin crew scheduling. *European Journal of Operational Research*, **140**, 531–540.

Part II

Routing and Timetabling

The Modeling Power of the Periodic Event Scheduling Problem: Railway Timetables – and Beyond

Christian Liebchen and Rolf H. Möhring

TU Berlin, Institut für Mathematik, Straße des 17. Juni 136, D-10623 Berlin, Germany
{liebchen,moehring}@math.tu-berlin.de

Summary. In the planning process of railway companies, we propose to integrate important decisions of network planning, line planning, and vehicle scheduling into the task of periodic timetabling. From such an integration, we expect to achieve an additional potential for optimization.

Models for periodic timetabling are commonly based on the Periodic Event Scheduling Problem (PESP). We show that, for our purpose of this integration, the PESP has to be extended by only two features, namely a linear objective function and a symmetry requirement. These extensions of the PESP do not really impose new types of constraints. Indeed, practitioners have already required them even when only planning timetables autonomously without interaction with other planning steps. Even more important, we only suggest extensions that can be formulated by mixed integer linear programs.

Moreover, in a self-contained presentation we summarize the traditional PESP modeling capabilities for railway timetabling. For the first time, also special practical requirements are considered that we prove not being expressible in terms of the PESP.

1 Introduction

Traditionally, the planning process of railway companies is subdivided into several tasks. From the strategic level down to the operational level, the most prominent subtasks are network planning, line planning, timetable generation, vehicle scheduling, crew scheduling, and crew rostering, see Fig. 1.

For a detailed description of these planning steps, as well as for an overview of solution approaches, we refer to Bussieck *et al.* (1997). Notice that network planning and line planning are of course part of the strategic planning process of public transportation companies. In contrast, vehicle scheduling and crew scheduling are of operational nature. In between, timetabling forms the linkage between service and operation. An important reason for the division into at least five subtasks is the high complexity of the overall planning process (Bussieck *et al.* (1997), Grötschel *et al.* (1997)).

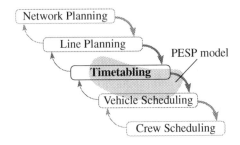

Fig. 1. Planning Phases Covered by the PESP Beforehand

During the last years, a trend towards the integration of several planning steps has emerged. For example, vehicle and crew scheduling were successfully combined by Borndörfer *et al.* (2002) and by Haase *et al.* (2001). Similarly, a combination of line planning and network planning is the objective of Borndörfer *et al.* (2007).

Periodic timetabling has also served as a starting point for such attempts. Nachtigall (1998) computes timetables that require only few rolling stock for a specific vehicle schedule. Engelhardt-Funke and Kolonko (2004) consider investments into infrastructure by using multi-criteria optimization. Lindner (2000) integrates the choice of rolling stock types in a non-linear model. Liebchen and Peeters (2002) provide a linear model that serves as a good approximation for minimizing rolling stock while optimizing periodic timetables.

In this paper, we demonstrate how periodic timetable construction can be combined with other planning steps. Further, we incorporate other practical conditions on timetables such as timetable symmetry, line planning, and even infrastructure decisions. We show that this can in fact be achieved with only slight variations of the commonly used model for periodic timetable construction, the PESP model introduced by Serafini and Ukovich (1989). The variations keep much of the properties of the PESP model and are again mixed integer programs over a feasibility domain with essentially the same structure as the original PESP. In particular, all of the valid inequalities for the PESP stay valid, and some of the new formulations even speed up the solution time with standard MIP solvers. But there have also been proposed other solution techniques for PESP instances: constraint programming (Schrijver and Steenbeek (1993)) and genetic algorithms (Nachtigall and Voget (1996)). Hence, in this paper we will restrain ourselves to the pure modeling capabilities of the general PESP model – with only two small exceptions. But these exceptions have already been asked explicitly by practitioners for their own sake.

In the discussion of these modeling features, we will also lay out large parts of the map of the borderline between what still fits into the traditional PESP model, and what requires new features, and at what cost. To this end, we also review the traditional PESP modeling issues, thus altogether providing a self-contained presentation of the PESP modeling capabilities and its extensions to symmetry, line planning, and network planning. Any of our suggestions for integrating these features can be formulated as a MIP, in particular not involving any quadratic terms.

The paper is organized as follows. Section 2 introduces the PESP. It presents its main formulations as a graph theoretic potential problem and as a mixed integer program, and reports on its complexity and a useful characterization of periodic timetables.

Section 3 discusses requirements for cyclic timetables that can be met by the PESP. These include simple requirements such as collision-free traffic on single tracks and headway between successive trains, but also more sophisticated ones such as bundling of lines, train coupling and sharing, fixed events in connection with hierarchical planning, and also disjunctive constraints and soft constraints.

Section 4 is devoted to timetable requirements that are beyond the scope of the traditional PESP, such as balanced reduction of service and symmetry of timetables. We show that the PESP or its MIP model only needs to be extended slightly in order to accommodate symmetry requirements.

Finally, in Section 5, we consider the integration of aspects of other planning steps into periodic timetable construction, in particular vehicle scheduling (minimization of rolling stock), line planning (simultaneous construction of line plan and timetable), and network planning (making infrastructure decisions). This integration makes essential use of the flexibility of the PESP, in particular disjunctive constraints, uses symmetry, and – as a new technique – integrates aspects of graph techniques into the PESP in order to handle line planning.

All model features are illustrated by examples from our practical experience with timetable construction at Deutsche Bahn AG, S-Bahn Berlin GmbH, and BVG (Berlin Underground).

2 The Periodic Event Scheduling Problem

Serafini and Ukovich (1989) introduced the PESP, by which periodic timetabling instances may be formulated in a very compact way. Since then, this model has been widely used (Schrijver and Steenbeek (1993), Nachtigall (1994), Odijk (1996), Lindner (2000), Peeters (2003)). In the PESP, we are given a period time T and a set V of events, where an event models either the arrival or the departure of a directed traffic line at a certain station. Furthermore, we are given a set of constraints A. Every constraint $a = (i, j)$ relates a pair of events i, j by a lower bound ℓ_a and an upper bound u_a.

A solution of a PESP instance is a node assignment $\pi : V \mapsto [0, T)$ that satisfies

$$(\pi_j - \pi_i - \ell_a) \bmod T \le u_a - \ell_a, \ \forall a = (i, j) \in A, \tag{1}$$

or $\pi_j - \pi_i \in [\ell_a, u_a]_T$ for short. We call a feasible node potential π a feasible *timetable*. Notice that we can scale an instance such that $0 \le \ell_a < T$, and for the span $d_a := u_a - \ell_a$ of a *feasible interval* $[\ell_a, u_a]_T$ we may assume w.l.o.g. $d_a < T$. Furthermore, for every fixed event i_0, every fixed point of time $t_0 \in [0, T)$, and every feasible timetable π there exists an equivalent timetable π' with $\pi'_{i_0} = t_0$. This is achieved by performing the simple shift $\pi'_i := (\pi_i - (\pi_{i_0} - t_0)) \bmod T$. Let us denote by $D = (V, A, \ell, u)$ the *constraint graph* modeling a PESP instance.

There are several practical aspects of periodic timetabling which profit from the presence of a linear objective function of the form

$$\sum_{a=(i,j)\in A} w_a \cdot (\pi_j - \pi_i - \ell_a) \bmod T,$$

with weights w_a. In our opinion, the most striking one is the integration of central aspects of vehicle scheduling, cf. Section 5.1.

Another perspective on periodic scheduling can be obtained by considering tensions instead of potentials. In a straightforward way, define for a given node potential π its *tension*

$$\hat{x}_a := \pi_j - \pi_i, \ \forall a = (i,j) \in A.$$

We call a set of edges $C \subseteq A$ an *oriented cycle* if re-orienting a subset of its edges yields a directed circuit. The *incidence vector* γ_C of an oriented cycle C is a vector in $\{-1,0,1\}^A$, where the entry minus one indicates a backward arc of the oriented cycle. The cycle space $\mathcal{C}$ of a directed graph D is defined as

$$\mathcal{C} := \mathrm{span}\{\gamma_C \mid C \text{ oriented cycle in } D\}.$$

Recall that a vector $\hat{x}$ is a tension (or potential difference), if and only if for some cycle basis B of $\mathcal{C}$, and each of its oriented cycles $C \in B$ with incidence vectors γ_C it holds that $\gamma_C \hat{x} = 0$ (e.g., Bollobás (2002)). This yields the following MIP formulation

$$
\left.
\begin{array}{ll}
\min \ c^t(\hat{x} + pT) & \min \ c^t x \\
\text{s.t.} \ \ \Gamma \hat{x} = 0 \qquad \text{or} & \text{s.t.} \ \ \Gamma(x - pT) = 0 \\
\qquad \ell \le \hat{x} + pT \le u & \qquad \ell \le x \le u \\
\qquad p \in \mathbb{Z}^A, & \qquad p \in \mathbb{Z}^A,
\end{array}
\right]
\tag{2}
$$

where $\Gamma \in \{-1,0,1\}^{(|A|-|V|+1)\times|A|}$ denotes the cycle-arc incidence matrix (*cycle matrix*) of some cycle basis of the directed graph D. The x variables are in fact a *periodic tension*, which we formally define for a given node potential π to be

$$x_{ij} := (\pi_j - \pi_i - \ell_{ij}) \bmod T + \ell_{ij}.$$

Sometimes, it is useful to define *slack variables* $\tilde{x}_a := x_a - \ell_a$.

Recall that cycle matrices are totally unimodular (Schrijver (1998)). This is the main observation to prove the following lemma.

Lemma 1 (Odijk (1994)). *Let $\mathcal{I}$ denote an instance of PESP with integral vectors ℓ and u and an integer period time T. If $\mathcal{I}$ admits some feasible timetable $\pi \in [0,T)^V$, then it also admits an integral feasible timetable $\pi' \in \{0,\dots,T-1\}^V$.*

Already Serafini and Ukovich made the following simple but useful observation.

Lemma 2 (Serafini and Ukovich (1989)). *If we relax the requirement $\pi \in [0,T)^V$ to $\pi \in \mathbb{Q}^V$, then for every spanning tree H and every feasible timetable π there exists an equivalent feasible timetable π' which induces $p_a = 0$ for $a \in H$.*

Notice that we may interpret the remaining non-zero integer variables as the representants of the elements of a (strictly) fundamental cycle basis. A generalization to integral cycle bases yields many variants of Formulation (2), some of which are easier to solve for MIP solvers (Liebchen (2003)).

Periodic tensions can be characterized similarly to classic aperiodic tensions.

Lemma 3 (Cycle Periodicity Property). *A vector $x \in \mathbb{Q}^A$ is a periodic tension, if and only if for every cycle C with incidence vector $\gamma_C \in \{-1, 0, 1\}^A$, there exists some $z_C \in \mathbb{Z}$, such that*

$$\gamma_C x = z_C T. \tag{3}$$

The PESP is $\mathcal{NP}$-complete, since it generalizes Vertex Coloring (Odijk (1994)). To see this, orient the edges of a Coloring instance arbitrarily and assign feasible periodic intervals $[1, T - 1]_T$ to each of them. Solution methods for the PESP include Constraint Programming (Schrijver and Steenbeek (1993)), Genetic Algorithms (Nachtigall and Voget (1996)), and of course integer programming techniques. For a computational study in which these substantially different approaches are compared to each other, we refer to Liebchen *et al.* (2007). For the MIP approach, a very important ingredient is

Theorem 1 (Odijk (1996)). *An integer vector p allows a feasible solution for the MIP (2), if and only if for every oriented cycle C of the constraint graph, the following cycle inequalities hold*

$$\underline{p}_C := \left\lceil \frac{1}{T} \left(\sum_{a \in C^+} \ell_a - \sum_{a \in C^-} u_a \right) \right\rceil \leq \sum_{a \in C^+} p_a - \sum_{a \in C^-} p_a \leq \left\lfloor \frac{1}{T} \left(\sum_{a \in C^+} u_a - \sum_{a \in C^-} \ell_a \right) \right\rfloor =: \overline{p}_C, \tag{4}$$

where C^+ and C^- denote the forward and the backward arcs of the cycle C.

We close this section by listing other totally different practical applications which can be modeled via the PESP (Serafini and Ukovich (1989)). The most prominent ones are the scheduling of systems of traffic lights and periodic job shop scheduling.

3 Timetabling Requirements Covered by the PESP

This section gives a broad overview of the timetable modeling capabilities of the PESP. Contrary to the following sections, practical requirements to be modeled are limited to those arising in periodic timetabling. Nevertheless, there are many facts we have to discuss in order to give a self-contained overview.

However, let us start by naming two facts which are definitely beyond the scope of the PESP: routing of trains through stations or even alternative tracks, and routing of the passenger flow. Hence, throughout this paper we assume fixed routes for both trains and passengers. A short motivation for these assumptions will be given at the beginning of Section 4.

For the vast majority of practical requirements to be modeled, we provide examples which are close to practice. However, in particular time and track information might not always reflect practice exactly. Depending on the fact to be modeled, we provide a track map, a line plan, a visualization (In German: "Bildfahrplan") of the timetable of a given track by means of a time-space diagram, and last but not least the resulting PESP subgraph. For readers not familiar with the first three types of charts, we refer to any textbook on railway engineering.

Most of our real-world examples are taken from the surroundings of the station Köln-Deutz (Cologne), which is part of the German ICE/IC-network. Fig. 2 displays the general track map of Köln-Deutz. Unless stated otherwise, we assume a period time of $T = 60$ minutes.

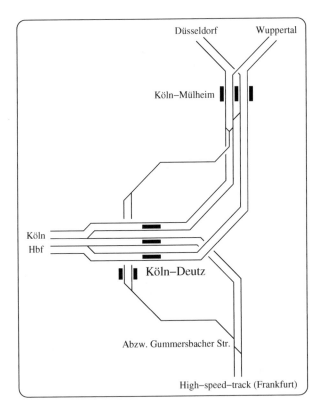

Fig. 2. Track Map of Köln-Deutz (Cologne) – Based on Leuschel (2002)

3.1 Elementary Requirements

Both for the sake of completeness and in order to introduce the notation used in the following figures, we start by modeling the three most elementary actions within public transportation networks: trips, stops, and changeovers.

In Fig. 3 (**a**), we highlight the tracks used by two lines which cross at Köln-Deutz. The lines themselves are given in Fig. 3 (**b**). Finally, we provide the constraint graph which models running, stopping, and changeover activities of these lines at Köln-Deutz in Fig. 3 (**c**) as PESP constraints. For instance, the trip arc with the constraint $[4, 4]_{60}$ ensures a trip time of precisely four minutes from Köln-Deutz to Köln Hbf. Within Köln Hbf, the minimum stopping time is set to three minutes such that passengers can board and alight the train. Finally, the increase of travel time for passengers that stay within the train is bounded by additional five minutes, providing an upper bound of $3 + 5 = 8$.

Notice that we ensure changeover quality by linearly penalizing changeover times which exceed a certain minimal changeover time required for changing platforms. In our example, a minimal changeover time of six minutes is assumed when connecting from Dortmund to Frankfurt. Using this approach, changeover arcs typically have a wide span.

An alternative way of modeling changeovers is to require some important ones not to exceed a maximal amount of effective waiting time. Then, we end up with rather small spans for changeover arcs. Schrijver and Steenbeek (1993) follow this approach, which seems to be very suitable for constraint programming solvers.

Stopping arcs typically have very small span. In rather unimportant stations, in general it is a good choice to fix the span to zero, in particular if there is neither a junction of tracks, nor a single track, nor any changeovers.

Just as trip arcs, stopping arcs with span zero constitute redundancies which can be eliminated very efficiently in a preprocessing step. For example, one can contract any *fixed arc*, i.e. having zero span, together with its target node. Doing so, the arcs which were incident with the contracted target node only have to be redirected to the source node of the contracted arc, after having shifted their feasible intervals appropriately. Moreover, an arc being (anti-) parallel to another one can be eliminated, if its feasible interval is a superset of the other arc. In addition to nodes with degree at most two, Lindner (2000) gives further situations in which the graph can be simplified.

If there are several lines using the same track into the same direction, sometimes a balanced service might be required. For n lines, this can easily be achieved by introducing arcs with feasible interval $[\frac{T}{n}, T - \frac{T}{n}]_T$ between any unordered pair of events that represent the departure at the first station of the common track. Certainly, strict balance may be relaxed by increasing the feasible interval.

Safety Requirements. If, in contrast to the previous discussion, there is no need for a balanced service, then at least a minimal headway h between any two trains has to be ensured. In the easiest case, the lines are operated with the same type of trains, and their running time is fixed. Then, we can sufficiently separate any two lines by introducing constraints similar to the above ones, having feasible interval $[h, T-h]_T$. These can be inserted either at the beginning or at the end of their common track. The more sophisticated constellation of trains involving different speeds will be discussed in Section 3.2.

But two trains may also use the same track in opposite directions. This is mainly the case for single tracks, see Fig. 4 (**a**). Obviously, a train may not enter the single track until the train of the opposite direction has left it. In Fig. 4 (**b**), we give a

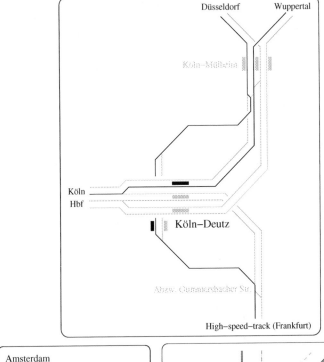

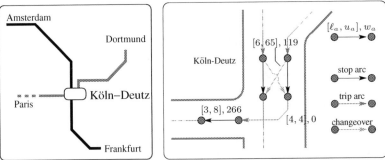

Fig. 3. Modeling Elementary Requirements: (**a**) Two Disjoint Routes of Lines Serving Köln-Deutz (**b**) The Corresponding Line Plan (**c**) PESP Constraints Modeling Running Activities, Stopping Activities, and Changeover Activities

timetable visualization that is extremely useful in particular for single tracks. We assume a fixed local signaling, and the grey boxes visualize the time a train blocks a certain part of the track. Surprisingly, there is only one single constraint needed to prevent two trains of opposite directions from colliding within the single track, as can be seen in Fig. 4 (**c**). To that end, consider the western entry point to the single track. A train may only enter the single track after a train of the opposite direction

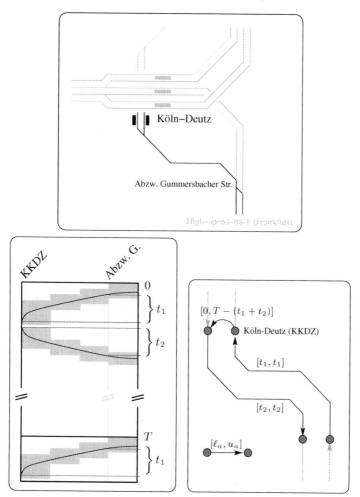

Fig. 4. Modeling Single Tracks: (**a**) A Single Track South of Köln-Deutz (**b**) Visualization of a Feasible Timetable for that Track (**c**) PESP Constraints Ensuring Safety Distance for a Single Track

has left ($\ell_a = 0$). But it also must have left the single track before the next train of the opposite direction may enter the single track ($u_a = T - (t_1 + t_2)$).

Note that so far we did not care about any buffer times and blocking times when setting the feasible interval to $[0, T - (t_1 + t_2)]_T$. Assuming a *minimal crossing time b* at both endpoints of the single track, i.e., the time that has to pass from a train leaving the single track until a train in opposite direction may enter, we obtain the following feasible interval

$$[b, T - (t_1 + t_2 + b)]_T.$$

Again, if there are several lines that have to be scheduled on a single track, one constraint for every unordered pair of opposite directions is needed.

Some authors (Krista (1997)) consider situations at crossings, where trains are shortly using the track of the opposite direction (cf. Fig. 5), as another modeling feature. But this is just a special case of single tracks, if the network is modeled at an

Fig. 5. Crossing of Track of the Opposite Direction South of Köln-Deutz

appropriate granularity. Abzw. Gummersbacher Straße has to be split into a northern station and a southern station which are linked by an eastern and a western track, where the western track can be traversed in both directions.

3.2 More Sophisticated Requirements

Whereas the practical requirements discussed in the previous section might arise in almost every railway network, the following aspects are of a more specialized nature.

Fixed Events. When planning a timetable hierarchically, e.g. from international trains down to local trains, one has to consider the fixed settings of previous hierarchies without replanning their times. Hence, the capability to fix an event to a certain point of time is another important modeling feature.

Fortunately, due to the periodic nature of the PESP, we may shift every feasible timetable such that a fixed event i_0 is fixed to a desired point in time $t_0 \in [0, T)$, i.e. $\pi_{i_0} = t_0$, and the objective value remains unchanged. By defining one of the events to be fixed as a kind of "anchor" event, we can easily relate the other events i_j to be fixed to certain points of time t_j by introducing arcs $a_j = (i_0, i_j)$ with $\ell_{a_j} = u_{a_j} = t_j - t_0$.

Bundling of Lines. Hierarchical planning gives rise to a further challenging aspect of timetabling. Notice that if a track is used by trains of different speeds, the capacity of that track significantly depends on the ordering of the trains. The first two parts of Fig. 6 visualize this effect. In the first scenario, slow and fast trains alternate, which implies that only two hourly lines of each of the two train types can be scheduled. However, if lines are bundled with respect to their speeds, three lines of the same two types of trains can be scheduled without having to invest into infrastructure, cf. Fig. 6 (**b**).

On the one hand, when only planning the high-speed lines in the first step of a hierarchical approach, it may happen that decisions on a higher level result in infeasibility on a lower level. On the other hand, hierarchical decomposition might have been chosen because an overall plan was considered to be too complex.

In order to keep the advantage of decomposition but limit the risk of infeasibility on lower levels, we propose to only bundle the lines of the current level of hierarchy. Fig. 6 (**c**) gives the complete set of lines which should be operated on the track in question. In Fig. 6 (**d**), we provide the PESP graph for the ICE/IC network. To bundle the three active lines, we introduce an artificial event and require each of the departure events to be sufficiently close to that artificial event. Hereby, the departure events will be close to each other as well.

In particular, we must not choose one of the existing events as "anchor", because this would predict the corresponding line to be the head of the sequence of bundled lines. This must definitely be avoided, because – contrary to assumptions made by Krista (1997) – the ordering of lines is indeed a major result of timetabling. Finally, based on profound estimates on passengers' behavior the management has to decide whether it is more important to operate as many trains as possible – and hereby bundle the trains of the same type – or whether a balanced service within the different types of trains should be preferred.

Train Coupling/Train Sharing. During the last decade, in railway passenger traffic a trend emerged towards train units which can easily be coupled and shared. Doing so, more direct connections can be offered without increasing the capacity of some bottleneck tracks.

In Fig. 7 (**a**), we display a line which is operated by two coupled train units between Berlin and Hamm. They split in Hamm to serve the two major routes of the Ruhr area, hereby offering direct connections from Berlin to the most important cities of that region. Still, this line occupies, e.g., the high-speed track between Berlin and Hannover only once per hour.

In Fig. 7 (**b**), we provide PESP constraints which ensure the time for splitting the two train units in Hamm to be at least five minutes. Furthermore, for the two departing trains, a safety distance of four minutes is guaranteed. Notice that we do not need to specify which train should leave Hamm first. This decision will be made implicitly, and in an optimized way, by the PESP solver.

Variable Trip Times. As long as trip times are fixed, a usual safety constraint prevents two identical trains from overtaking each other. With h being the minimal headway for the track, we put an arc with feasible interval $[h, T - h]_T$ between the two events of entering the common track. If the line at the tail of the constraints is by f time units faster than the line at the tail of the constraints, overtaking can be

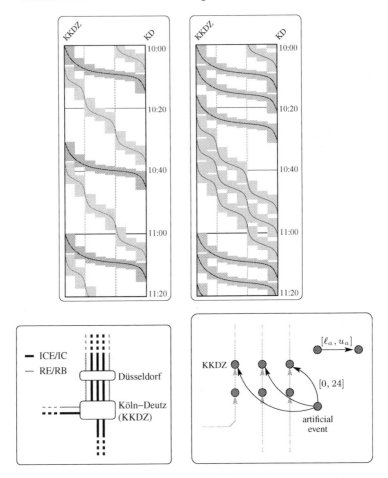

Fig. 6. Bundling of Lines: (**a**) Poor Capacity if Slow and Fast Trains are Alternating (**b**) Capacity Increase by Bundling Trains of the Same Type (**c**) Complete Line Plan for All the Types of Lines (**d**) PESP Constraints Ensuring Enough Capacity for RE/RB Lines Already when Planning Only ICE/IC Lines Within the First Step of a Hierarchical Planning

prevented by modifying the constraint to $[h + f, T - h]_T$. This can be understood easily by having again a look at the corresponding situation in Fig. 6 (**a**).

But this is no longer guaranteed if the model includes variable trip times. Even ensuring the minimal headway at the end of the track, too, does no longer prevent overtaking (even of trains of the same type) if the span in the trip times is at least twice the safety distance h, i.e. $u_a - \ell_a \geq 2h$. Schrijver and Steenbeek (1993), Lindner (2000), and Kroon and Peeters (2003) tackle this phenomenon by adding extra constraints on the integer variables of the MIP formulations. Hereby, they leave the PESP model. In addition, Kroon and Peeters (2003) provide some sufficient condi-

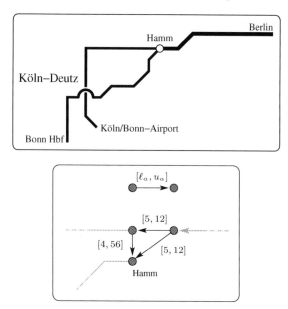

Fig. 7. Modeling Train Sharing: **(a)** Line Plan for the Line Berlin-Hamm-{Bonn Hbf | Köln/Bonn-Airport} **(b)** PESP Constraints Ensuring Safety Distance and Time to Split Train Units, but not Specifying the Ordering of Departures

tions on trip times, safety distance, and on the degree of flexibility of the trip times that prevent trains from overtaking.

In order to stay within the PESP model, we propose to subdivide[1] an initial trip arc into new smaller ones such that $u_a - \ell_a < 2h$ for every new trip arc. For an example, we refer to Fig. 8, where **bold** arcs represent arcs of the spanning tree for which we set $p_a = 0$, cf. Lemma 2, and $3r$ is the minimum running time for the track.

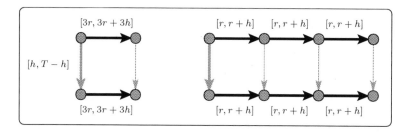

Fig. 8. Overtaking and Variable Trip Times: **(a)** Standard Granularity does not Prevent Overtaking **(b)** Finer Granularity Prevents Overtaking

---

[1] This approach has also been discussed by Peeters (2000, 2003) several years ago.

Although this might seem to expand the model, the approach behaves rather well. More precisely, in every feasible timetable, the integer variables which we have to introduce for our additional arcs are in fact fixed to zero. This can simply be seen by applying the cycle inequalities (4) to any of the three squares in Fig. 8 (**b**),

$$\underline{p} = \left\lceil \frac{1}{T}(r + h - (T - h) - (r + h)) \right\rceil = \left\lceil \frac{h - T}{T} \right\rceil = 0,$$

$$\overline{p} = \left\lfloor \frac{1}{T}((r + h) + (T - h) - h - r) \right\rfloor = \left\lfloor \frac{T - h}{T} \right\rfloor = 0.$$

Notice that the corresponding bounds for the initial formulation are only -1 and 1. But this is very natural, because there are three different types of timetables possible, of which we have to cut off two. The value one, e.g., models the fact that the second (lower) train is overtaking the first (upper) train.

Although we showed that the inconveniences caused by flexible running times can be overcome, we will assume fixed running times throughout the remainder of this paper.

3.3 General Modeling Capabilities

There are also important non-timetabling features which can be modeled by the PESP in a very elegant way. The types of such constraints are disjunctive constraints and soft constraints. Although they were originally introduced for their own sake, they turn out to be very useful for even more specialized requirements, which practitioners require to be modeled.

Disjunctive Constraints. The feasible region of MIPs are commonly given as the intersection of finitely many half-spaces, plus some integrality conditions. If disjunctive constraints have to be modeled, usually artificial integer variables are introduced. However, the PESP offers a much more elegant way.

When introducing the PESP, Serafini and Ukovich (1989) already made the important observation that the intersection of two PESP constraints is not always again a single PESP constraint. Rather, the feasible interval for a tension variable can become the *union* of two PESP constraints, e.g.,

$$\pi_j - \pi_i \in [\ell_1, u_1]_T \cap [\ell_2, u_2]_T \Leftrightarrow \pi_j - \pi_i \in [\ell_1, u_2]_T \cup [\ell_2, u_1]_T.$$

We illustrate their observation in Fig. 9. Nachtigall (1998) observed that any union of k PESP constraints can be formulated as the intersection of at most k PESP constraints.

As an immediate practical application of disjunctive constraints, we consider optional operational stops. Long single tracks with no stop may cause the timetable of a line to be fixed within only small tolerances. In such a situation, Deutsche Bahn AG considers the option of letting the ICE/IC trains of one direction stop somewhere, although there is no ICE/IC station. In the current timetable, this takes places on the line between Stuttgart and Zurich, at Epfendorf.

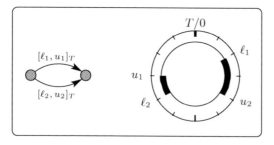

Fig. 9. Disjunctive Constraints

If we want periodic timetable optimization to be competitive, we should enable the PESP to introduce an additional stop as well. We do so by introducing a pair of disjunctive constraints. The first constraint is a usual stop arc a_1. We set the lower bound ℓ_{a_1} to zero, which models the option of not introducing an additional stop. The upper bound u_{a_1} is set to the sum of the minimal increase b of travel time occurring from braking and accelerating, plus the maximal amount of stopping time s at the station. For the effected increase $\tilde{x}_a$ of travel time, this translates to

$$\tilde{x}_a \in \{0\}_T \cup [b, b + s]_T,$$

which is a disjunctive constraint. Notice that additional waiting time should be penalized in this situation similarly to an extension of a regular service stop. Moreover, if there are other lines operating on the same track, we have to take precautions that were discussed in the paragraph on variable trip times. However, optional operational stops make most sense within long single tracks. In many cases there are not several lines using that large bottleneck.

Obviously, the introduction of an additional stop can also be due to the construction of a new station. Since such decisions are a part of network planning, we postpone this discussion until Section 5.3.

Soft Constraints. Nachtigall (1996) investigated the combination of two antiparallel arcs $a_1 = (i, j)$ and $a_2 = (j, i)$. If they have an identical coefficient in the objective function and if neither of them can become infeasible for any vector π, or x respectively, then they model a *soft constraint*.

Classically, if a certain tension value x_a does not satisfy a given PESP constraint $[\ell_a, u_a]_T$, one would declare the complete timetable as infeasible. But sometimes, it can be an alternative only to produce a significant penalty in the objective function, if a constraint is not satisfied.

To that end, we relax the upper bound of the original constraint to $\ell + T - 1$ – we may assume the instance being scaled such that the precondition of Lemma 1 is satisfied. Further, we introduce a new antiparallel arc with feasible interval according to Fig. 10. Then, these two constraints yield a piecewise constant behavior of the objective function, which serves as an indicator for the violation of the original constraint, but without guaranteeing feasibility. For an initial constraint $x_a \in [\ell_a, u_a]$ consider the corresponding pair of artificial constraints a_1 and a_2 – each of these having cost coefficient M. They contribute to the objective function

$$M \cdot (x_{a_1} + x_{a_2}) = \begin{cases} M \cdot (u - \ell) & \text{if } x_{a_1} \in [\ell_a, u_a]_T \text{, and} \\ M \cdot (u - \ell + T) & \text{otherwise,} \end{cases}$$

hereby indicating whether the original constraint a is satisfied for the tension vector x.

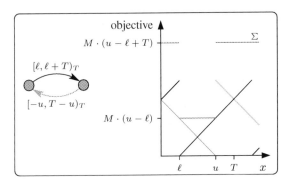

Fig. 10. Soft Constraints

In our cooperation with Berlin Underground, we were asked to construct a timetable that, among the top 50 most important connections, maximizes the number of connections having a waiting time of at most five minutes. In fact, soft constraints are well-suited for letting MIP solvers produce a timetable being optimal subject to this kind of objective function.

4 Timetabling Requirements *Not* Covered by the PESP

Although the most important practical requirements for a periodic timetable can be modeled within the PESP, we are still aware of some special features for which the PESP fails. To the best of our knowledge this is the first time that practical requirements of timetabling are proven to be beyond the scope of the PESP.

First, one may think of situations in which it is not fixed which trains are operated on which track, e.g., within stations. Consider a station having two tracks in the same direction and three lines serving that direction. Then we cannot decide *a priori* which pair of lines shall be within the station at the same time, hence omitting the sequencing constraint between these two lines. This observation is the motivation for the DONS system to be subdivided into CADANS, covering the timetabling step, and STATIONS, covering the routing aspect (van den Berg and Odijk (1994)).

Apart from the rather important routing requirement, which unfortunately is simply out of scope for the PESP, we will analyze a very special situation in more detail, namely the balanced reduction of service. Finally, we will introduce the important notion of *symmetry*. On the one hand, symmetry slightly exceeds the original PESP, but on the other hand, when added explicitly, gives rise to a mechanism to include important aspects of line planning into the very same planning step as periodic timetabling and vehicle scheduling.

4.1 Balanced Reduction of Service

The Berlin fast train company (S-Bahn Berlin GmbH) aims at operating only one timetable for one whole day. The late evening service differs from the rush hour only in that some trains are omitted. Hence, the timetable must respect the available capacity during the rush hour, and it has to offer a balanced service in the late evening as well.

From a pure operations point of view, it could seem strange to sidestep an intraday change of the timetable structure. It is for sure that the information technology available in the 21st century could cope with this. But it is still the policy of the company. It is given as a motivation that customers really expect to have only one single timetable to be kept in mind for their station.

Consider the approximately 10 km long track from Zoo station to Berlin East station. On it, a minimal headway of 2.5 minutes has to be respected. The period time is 20 minutes and eight[2] lines (having identical train types) per period and direction have to be scheduled. In the late evening service, there are four trains every 20 minutes, two of them being fixed to a 10 minute time lag. We call these two lines *core-lines*.

Of course it would be ideal to have a five minute time lag between two consecutive trains in the evening. But this is impossible because one of the evening trains is required to serve Potsdam every 10 minutes together with a rush hour train. Hence, one should ensure that the maximal time lag between two consecutive trains does not exceed 7.5 minutes.

But this simple requirement cannot be covered by the PESP. Consider the two types of timetables given in Table 1. Timetables of type 1 satisfy our requirement by bounding the maximum distance between two consecutive trains to 7.5 minutes, but type 2 does not because there we have a gap of 10 minutes.

Proposition 1. *For every set of PESP constraints either timetables of both types are feasible, or timetables of both types are infeasible.*

Proof. There are two types of constraints to be analyzed:

  i. one constraint between the two non-core lines,
  ii. four constraints between one of the two core lines and one of the two non-core lines.

---

[2] One of them only serves as a free slot for occasional non-passenger trips.

Table 1. Possible Timetables for the Late Evening Service from Zoo Station to Berlin East Station (This table only shows the core-lines that are actually running in the evenings. Each of the – entries is a wild card for a rush-hour train.)

Timetable	Departure times ($T = 20$ minutes)							
Type 1	0.0	–	–	7.5	10.0	12.5	–	– (20.0)
Type 2	0.0	2.5	–	7.5	10.0	–	–	– (20.0)

Since we must not specify the sequence of the lines in advance, only symmetric constraints $[\ell, T - \ell]_T$ make sense. Moreover, all constraints of type (ii) have to be identical for the same reason.

To guarantee feasibility of type 1 timetables, we deduce $\ell \leq 5$ for the constraint of type (i) and $\ell \leq 2.5$ for the constraints of type (ii). But then, timetables of type 2 stay feasible as well. Hence, in order to cut off timetables of type 2, we have to increment one of the given bounds. But since they are tight, this would immediately cut off timetables of type 1 as well.   □

4.2 Symmetry of a Periodic Timetable

Throughout our discussion of symmetry, we assume that for every directed line there exists another directed line serving the same stations just in opposite order. Moreover, the concept of symmetry only makes sense, if, for every traffic line, the running and stopping times of its two opposite directions are the same. Also for the minimum headways and other operational constraints we require them to be identical in both directions. Furthermore, the passenger flow is assumed to be symmetric.

First, observe that in every periodic timetable with period time T, every train meets some train of the opposite direction of its line twice within the period time – assuming the lines to have travel times of at least once the period time. In general, every line can have different times for these meetings.

A periodic railway timetable is called *symmetric with* (global) *axis s*, if at time s every train in the network meets a train of the opposite direction of its line. From the above considerations we deduce that we may assume w.l.o.g. $s \in [0, \frac{T}{2})$.

For the arrival or departure event of a directed line at a certain station, we denote by its *complementary event* the departure or arrival, respectively, of the opposite line at the same station. In the sequel, we provide two characterizations of symmetric timetables.

Lemma 4. *A timetable is symmetric with axis s if and only if for every pair i and $\bar{i}$ of complementary events there holds*

$$\frac{(\pi_i + \pi_{\bar{i}}) \bmod T}{2} = s. \tag{5}$$

Proof. Let i and $\bar{i}$ be any two complementary events. By definition, they are part of the two opposite directions of the same line. Moreover, they are located in the same station S.

In a symmetric timetable, the trains of the two opposite directions meet at times s and $s + \frac{T}{2}$. Consider two virtual events j and $\bar{j}$ of passing the meeting point M. As the trains meet there, we have $\pi_j = \pi_{\bar{j}} \in \{s, s + \frac{T}{2}\}$.

We assumed the travel times of two opposite trains to be identical and denote the travel time between S and M by t. Hence, w.l.o.g.

$$(\pi_i + \pi_{\bar{i}}) \bmod T = ((\pi_j + t) + (\pi_{\bar{j}} - t)) \bmod T = (2 \cdot \pi_j) \bmod T.$$

$\square$

To define a counterpart of condition (5) for the tension formulations (2), we define two arcs $a = (i, j)$ and $\bar{a} = (\bar{j}, \bar{i})$ to be *complementary*, if $\{i, \bar{i}\}$ and $\{j, \bar{j}\}$ are complementary, and we have $\ell_a = \ell_{\bar{a}}$ and $u_a = u_{\bar{a}}$. With these definitions at hand, we are able to define a symmetric instance of PESP: A constraint graph is called *symmetric*, if every arc connects either two complementary events, or if for every arc $a \in A$ there exists some complementary arc $\bar{a} \in A \setminus \{a\}$.

Lemma 5. *Consider an instance of PESP that is modeled by a connected symmetric constraint graph. Let π be a feasible timetable with corresponding periodic tension x. There exists some $s \in [0, \frac{T}{2})$ such that Condition (5) holds for every pair of symmetric events, if and only if every pair of complementary arcs a and $\bar{a}$ fulfills*

$$\tilde{x}_a = \tilde{x}_{\bar{a}}. \tag{6}$$

Proof. "$\Rightarrow$": Let $a = (i, j)$ and $\bar{a} = (\bar{j}, \bar{i})$ denote two complementary arcs of the constraint graph. Then, we have

$$\tilde{x}_a = x_a - \ell_a \overset{(2)}{=} (\pi_j - \pi_i - \ell_a) \bmod T$$
$$\overset{(5)}{=} (2s - \pi_{\bar{j}} - (2s - \pi_{\bar{i}}) - \ell_a) \bmod T$$
$$= (\pi_{\bar{i}} - \pi_{\bar{j}} - \ell_{\bar{a}}) \bmod T = x_{\bar{a}} - \ell_{\bar{a}} = \tilde{x}_{\bar{a}}.$$

"$\Leftarrow$": Let x be the periodic tension of some feasible timetable π. We show that there exists one global symmetry axis s such that Condition (5) is satisfied for π.

We compute s from an arbitrary fixed event, say i,

$$s := \frac{(\pi_i + \pi_{\bar{i}}) \bmod T}{2}.$$

Now, we consider an arbitrary pair of complementary events j and $\bar{j}$. Since D is connected and symmetric, there exists a path P from i to j or $\bar{j}$ that only contains arcs a such that $\bar{a} \in A \setminus \{a\}$. We assume w.l.o.g. that P starts at i and ends at j. By setting

$$x_P := \sum_{a \in P^+} x_a - \sum_{a \in P^-} x_a,$$

we obtain $\pi_j = (\pi_i + x_P) \bmod T$. As for every $a \in P$ there exists its complementary arc $\bar{a} \in A \setminus \{a\}$, the complementary path $\bar{P}$ of P from $\bar{j}$ to $\bar{i}$ is well-defined. Equation (6) ensures $x_{\bar{P}} = x_P$.

In total, we obtain

$$\frac{(\pi_j + \pi_{\bar{j}}) \bmod T}{2} = \frac{(\pi_i + x_P + \pi_{\bar{i}} - x_{\bar{P}}) \bmod T}{2} = \frac{(\pi_i + \pi_{\bar{i}}) \bmod T}{2} = s.$$

$\square$

Remark 1. If the line plan of a traffic network is connected and the constraint graph is symmetric, we are able to give an even more compact characterization of symmetry. Then, a feasible tension encodes a symmetric timetable, if and only if Condition (6) is satisfied for changeover arcs and stopping arcs. In fact, in the proof of Lemma 5 we can then find a path that only uses such arcs, plus trip arcs, which we assume to have zero span.

Surely, one can introduce a certain tolerance Δ on the symmetry requirement. But notice that in this case, condition (6) has to be expanded by a new integer variable.

Example 1 (Deutsche Bahn AG). Fig. 11 shows two real-world timetable queries for opposite directions. These are representative for large parts of central European countries, such as Germany and Switzerland, which are operated with symmetry axis zero within only minor tolerances. Hence, if not stated otherwise we assume $s = 0$ throughout this paper for ease of notation.

We check the three characterizations of symmetry. Most striking, the changeover waiting time is almost the same in both directions, cf. Remark 1 and Equation (6). To check Condition (5), we consider the arrival of ICE 952 in Köln Hbf and the complementary departure of ICE 953. The two events sum up to $(14+47) \bmod 60 \approx 0$, and the same can be observed for the Brussels trains. Finally, notice that the Berlin line has one of its meeting points between Köln-Deutz and Wuppertal Hbf, at minute zero, of course. To that end, we have to know that the trains from Berlin arrive at Köln-Deutz at minute 09, which is two minutes before its departure at minute 11.

Some practitioners consider the changeover condition in Remark 1 to be an important advantage of symmetric timetables. Even though this might depend on personal preferences, we do *not* consider this really to be a striking argument for symmetry. Actually, there are examples which prove that symmetric timetables are only suboptimal, even if the input data is symmetric (Liebchen (2004)).

Apparently there are not yet many discussions of symmetric timetables available. But among further motivations for symmetry, as they can be found in Liebchen (2004), the most convincing one seems to be that symmetry halves the complexity of an instance. This can in particular be useful if there are complex interfaces to international trains or to regional traffic, and when planning is performed manually. However, this argument should become less important in the future, as we think that PESP solvers achieve some more progress in performance, and hence find their way into practice.

To summarize, besides a linear objective function, symmetry is the second important requirement arising in the practice of periodic railway timetabling, by which

Station/Stop	Date	Time	Platform	Products	Comments
Berlin Zoologischer Garten	05.06.03	dep 09:54	4	ICE 952	InterCityExpress
Wolfsburg		dep 10:54			BordRestaurant
Hannover Hbf		dep 11:31			
Bielefeld Hbf		dep 12:24			
Hamm(Westf)		dep 12:54			
Hagen Hbf		dep 13:25			
Wuppertal Hbf		dep 13:42			
Köln-Deutz		dep 14:11			
Köln Hbf	05.06.03	arr 14:14	6		
Köln Hbf	05.06.03	dep 15:13	8	ICE 14	InterCityExpress
Aachen Hbf		dep 15:52			Onboard meeting place
Aachen Süd(Gr)					
Liege-Guillemins					
Bruxelles-Midi	05.06.03	arr 17:46			

Duration: 7:52; runs daily

All information is issued without liability. Software/Data: HAFAS 5.00.DB.4.5 - 20.05.03 [5.00.DB.4.5/v4.05.p0.13_data:59e79704]

Station/Stop	Date	Time	Platform	Products	Comments
Bruxelles-Midi	05.06.03	dep 12:16		ICE 15	InterCityExpress
Liege-Guillemins		dep 13:28			Onboard meeting place
Aachen Süd(Gr)					
Aachen Hbf		dep 14:10			
Köln Hbf	05.06.03	arr 14:46	3		
Köln Hbf	05.06.03	dep 15:47	2	ICE 953	InterCityExpress
Köln-Deutz		dep 15:51			BordRestaurant
Wuppertal Hbf		dep 16:17			
Hagen Hbf		dep 16:35			
Hamm(Westf)		dep 17:10			
Bielefeld Hbf		dep 17:37			
Hannover Hbf		dep 18:31			
Wolfsburg		dep 19:05			
Berlin Zoologischer Garten	05.06.03	arr 20:02	1		

Duration: 7:46; runs Mo - Fr, not 29. May, 9. Jun, 21. Jul, 15. Aug, 11. Nov
Hint: Prolonged stop

All information is issued without liability. Software/Data: HAFAS 5.00.DB.4.5 - 20.05.03 [5.00.DB.4.5/v4.05.p0.13_data:59e79704]

Fig. 11. Symmetric Timetables in Practice

the initial PESP model should be extended. Fortunately, in computations on real-world data sets it has been observed that MIP solvers may profit from the addition of symmetry constraints, in particular in formulation (6) (Liebchen (2004)). Such a generalized MIP model even inherits large parts of the structure of a pure PESP model. Most important, the cycle inequalities (4) remain valid.

5 Further Planning Steps Covered by the PESP

In the following, we will demonstrate that the modeling capabilities of the PESP are not limited only to periodic timetabling. Rather, central aspects of both preceding and succeeding planning steps in the sense of Fig. 1 can be integrated.

We start this discussion with the well-established technique of minimizing the number of vehicles required to operate a periodic timetable by penalizing waiting times of vehicles. Hereafter, we provide first ideas for the integration of important decisions of line planning. We close this section by proposing a way to model some specialized decisions arising in network planning.

5.1 Aspects of Vehicle Scheduling

Almost all companies in public transportation have in common that they want to minimize the amount of rolling stock required to serve their networks. Notice that the quality of the vehicle schedule for a fully periodic timetable, i.e. with no peak trips included, is largely determined by the timetable.

Consider, e.g., the hourly line displayed in Fig. 12 (**a**). Assume the minimal travel times between the two endpoints to be 235 minutes for each direction. Given strict minimal turnover times of 45 and 60 minutes, respectively, the minimal number of vehicles required to operate this line is precisely

$$N := \left\lceil \frac{1}{60}(235 + 235 + 45 + 60) \right\rceil = 10.$$

A timetable which lets the trains leave at the full hour from Frankfurt and Amsterdam can indeed be operated with only 10 trains, at least if the stopping times are extended only moderately. On the contrary, a timetable in which only the trains starting at Frankfurt depart at minute 00, but the trains from Amsterdam leave at minute 30 requires at least 11 vehicles. Hence, the amount of vehicles depends on the timetable.

We will analyze in which special cases pure PESP constraints are able to control the number of trains required. After that, we show that a linear objective function covers many more of the practical cases.

Proposition 2 (Nachtigall (1998)). *Consider a fixed traffic line with period time T. If we assume trains always serve only this line, and if we do not allow inserting additional stopping time, then there exist upper bounds u for the turnover activities, such that the only feasible timetables are those which can be operated with the minimal amount of trains.*

Proof. We present a proof of this simple fact, both in order to provide the notation used in the following paragraphs, and because it avoids modulo-notation.

Denote the endpoints of the line by A and B. Let ℓ_{AB} denote the minimal travel time from A to B, i.e. the sum of the minimal stopping and running times of the activities of this directed traffic line. Moreover, denote by ℓ_B the minimal amount of time a train has to stay in endpoint B between two consecutive trips.

The minimal number N of trains required to operate this line is precisely

$$N = \left\lceil \frac{\ell_{AB} + \ell_B + \ell_{BA} + \ell_A}{T} \right\rceil.$$

From the cycle periodicity property (3) we know that every feasible timetable x fulfills

$$x_{AB} + x_B + x_{BA} + x_A = zT, \tag{7}$$

for some $z \in \mathbb{Z}$. Hence, we must ensure $z = N$. To that end, consider the slack

$$\sigma := NT - (\ell_{AB} + \ell_B + \ell_{BA} + \ell_A) \tag{8}$$

of this traffic line, implying $(x_A - \ell_A) + (x_B - \ell_B) = \sigma$. But since $\sigma < T$, by setting

$$u_A := \ell_A + \sigma \tag{9}$$

we even ensure $x_{AB} + x_B + x_{BA} + x_A < (N+1)T$. $\qquad\square$

Let us now analyze the case in which additional stopping times may be inserted, i.e., $u_{AB} > \ell_{AB}$. We will show that together with the constraints (9), some timetables which require an additional train may become feasible.

On the one hand, consider a timetable for which we have $x \equiv \ell$ for all activities, except for the turnover time in one endpoint. This timetable can still be operated with the minimal number of trains, showing that decreasing the value (9) for u_A would cut off timetables we seek.

On the other hand, assume $x_{AB} = u_{AB}$ and $x_{BA} = u_{BA}$. If

$$(u_{AB} - \ell_{AB}) + (u_{BA} - \ell_{BA}) + \sigma \geq T, \tag{10}$$

then we can extend x to a timetable that still respects (9), but which requires at least one additional train. For instance, if inequality (10) is tight, then for $x \equiv u$ we have

$$
\begin{aligned}
x_{AB} + x_B + x_{BA} + x_A &= u_{AB} + u_B + u_{BA} + u_A \\
&\overset{(9)}{=} (u_{AB} - \ell_{AB}) + (\ell_B + \sigma) + (u_{BA} - \ell_{BA}) + \\
&\quad + (\ell_A + \sigma) + \ell_{AB} + \ell_{BA} \\
&\overset{(10)}{=} T + \sigma + \ell_{AB} + \ell_B + \ell_{BA} + \ell_A \\
&\overset{(8)}{=} (N+1)T.
\end{aligned}
$$

The above dilemma is our main motivation for the need of a linear objective function. Such a function takes advantage of equation (7): By assigning a value M to the arcs modeling a traffic line, every additional train adds $M \cdot T$ to the objective function value. Of course, it suffices to consider arcs with positive span, cf. Fig. 12 (**b**). If the value for M is chosen to be relatively large compared to the passenger weights, the

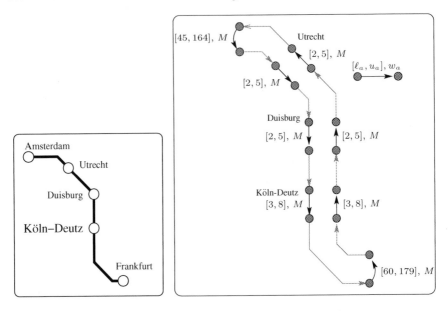

Fig. 12. Modeling Aspects of Vehicle Scheduling: (**a**) Line Plan (**b**) PESP Constraints Measuring the Number of Trains Required to Operate the Line

objective function essentially models the piecewise constant behavior of the cost of the rolling stock for operating the railway network.

From a more local perspective, we just penalize idle time of trains. But this can even be done without knowing *a priori* the circulation plan of the trains. Although a straight-forward exact model involves a quadratic objective function, Liebchen and Peeters (2002) report that a simple linear relaxation in terms of the PESP yields results of high quality.

5.2 Aspects of Line Planning

Our main idea for letting PESP solvers even take decisions of line planning is to combine – or match – pre-defined line-segments. To that end, we will make intensive use of disjunctive constraints. Unfortunately, we will only be able to ensure symmetric line plans if we require symmetry also within the stations where lines are matched.

We are aware of only one other approach for integrating the planning phases of line planning, timetabling and vehicle scheduling (Völker (2003)). Whereas that approach is based on the assumption that the line plan contains no cycles, our ideas do not require any restrictive assumptions on the topology of the network. Rather, we are able to keep even very important technical restrictions such as single tracks.

Notice that bad decisions at the level of line planning may cause very bad results also for vehicle scheduling. Consider the four line segments displayed in Fig. 13. We assume a period time of $T = 60$ minutes and a minimal turnover time of 30 minutes at each of the four terminus stations. The time for a one-way trip from the matching station to one of the endpoints is indicated at the corresponding edge.

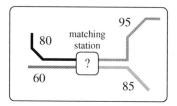

Fig. 13. Line Segments Where Only One Matching Provides Good Vehicle Schedules

In fact, the vehicle schedule is fixed due to the distinct endpoints. Combining the south-west segment with the north-east segment causes this line to require at least

$$\left\lceil \frac{1}{60}(60 + 95 + 30 + 95 + 60 + 30) \right\rceil = \left\lceil \frac{370}{60} \right\rceil = 7 \text{ trains.}$$

The other line of the same matching requires seven trains, too.

In contrast, the other matching implies seven trains only for the northern line consisting of the two top line segments. But the other line can be operated with only six trains. Hence, already the line plan has a major impact on the cost of operation. Claessens *et al.* (1998) consider this phenomenon in their approach for constructing cost-optimal line plans. However, they omit the important intermediate linking step of computing a timetable. Therefore, their approach must also consider possible constellations in which there is no feasible timetable using only six trains for the southern line. This would be the case if there was a single track with travel time 25 minutes for every direction just at the end of the south-east segment. The same holds if it is required that the two lines together form an exact half-hourly service along the backbone of the network.

We consider a track that has to be served in the same direction by n directed lines which are operated by trains of identical type. We denote the matching station by S which resides between the two endpoints of the common track. We consider n line segments $L_1^a, \ldots, L_n^a$ which have station S as their common endpoint, and n line segments $L_1^d, \ldots, L_n^d$ having station S as their common starting point. Any (bipartite) perfect matching between the arriving and the departing line segments induces a line plan.

But from the perspective of timetabling, there are only n arrival events $a_1, \ldots, a_n$ as well as n departure events $d_1, \ldots, d_n$ visible. Hence, we must deduce only from their arrival times π_{a_i} and their departure times π_{d_j} which arriving line segment L_i^a should be matched with which departing line segment L_j^d. This can be done in a canonical way, if we choose the matching station S such that it has only one track in the direction of the line segments we consider. If necessary, we add an artificial station in the middle of some track. Then, at most one train can be in S at the same time. Timetables respecting this constraint can be characterized very easily as follows.

Definition 1 (Alternating timetable). *For a fixed station S and a fixed direction, a periodic timetable π with n pairwise different arrival times $0 \leq \pi_{a_1} < \cdots < \pi_{a_n} < T$ and n pairwise different departure times $0 \leq \pi_{d_1} < \cdots < \pi_{d_n} < T$ is called* alternating *at S, if either $\pi_{a_i} \leq \pi_{d_i} < \pi_{a_{i+1}}$ for every $i = 1, \ldots, n$, or $\pi_{d_i} < \pi_{a_i} \leq \pi_{d_{i+1}}$ for every $i = 1, \ldots, n$, where we define $\pi_{\cdot_{n+1}} := \pi_{\cdot_1} + T$.*

Lemma 6. *A timetable π ensures that there is always at most one train at station S if and only if it is alternating at S.*

Hence, for an alternating periodic timetable, we combine the arriving line segment L_i^a with the departing line segment L_j^d, if and only if the latter marks the unique first possible departure. In the sequel, we will give PESP constraints ensuring every feasible timetable to be alternating at S. Thus, every feasible timetable will encode some unique matching and the associated line plan.

The first two sets of constraints ensure the minimal headway d in front of and behind the matching station S:

$$\forall\, i, j \in \{1, \ldots, n\} : \ \pi_{a_j} - \pi_{a_i} \in [d, T - d]_T, \tag{11}$$

$$\forall\, i, j \in \{1, \ldots, n\} : \ \pi_{d_j} - \pi_{d_i} \in [d, T - d]_T. \tag{12}$$

Notice that (11) and (12) can only be fulfilled if $0 \leq d \leq \frac{T}{n}$. Moreover, we relate arrival events to departure events by the following disjunctive constraints

$$\forall\, i, j \in \{1, \ldots, n\} : \ \pi_{d_j} - \pi_{a_i} \in [0, T - d + h]_T, \tag{13}$$

$$\forall\, i, j \in \{1, \ldots, n\} : \ \pi_{d_j} - \pi_{a_i} \in [d, T + h]_T, \tag{14}$$

where we denote by h the maximal stopping time for a train at station S. Together, these constraints (13) and (14) yield

$$(\pi_{d_j} - \pi_{a_i}) \bmod T \in [0, h] \,\dot\cup\, [d, T - d + h]. \tag{15}$$

Trivially, $0 \leq h < d$ is necessary for every feasible timetable π to be alternating at S.

Theorem 2. *Let π be a timetable respecting constraints (11) to (14). Then for every departure event d_j, there exists a* unique *arrival event a_i satisfying*

$$\pi_{d_j} - \pi_{a_i} \in [0, h]_T, \tag{16}$$

if and only if $h < (n + 1)d - T$.

Since $0 \leq h$, from $h < (n + 1)d - T$ we conclude $\frac{T}{n+1} < d$.

Proof. "$\Rightarrow$": We assume $h \geq (n + 1)d - T$. Since $d = \frac{T}{n}$ would imply $h \geq d$, we must only investigate the case that $d < \frac{T}{n}$. We will construct a timetable which respects the constraints (11) to (14), but which contradicts (16).

Define $\pi_{a_i} := (i - 1)d$, for all $i = 1, \ldots, n$, and $\pi_{d_j} := j \cdot d$, for all $j = 1, \ldots, n$. By construction, all the constraints are satisfied. However, since $\pi_{a_n} + h < n \cdot d = \pi_{d_n}$, for departure π_{d_n} none of the arrival events fulfills (16), q.e.d.

"⇐": We assume there exists a timetable π having one departure event d_0 such that

$$\forall i = 1, \ldots, n : (\pi_{d_0} - \pi_{a_i}) \bmod T > h,$$

but which respects the constraints (11) to (14). We may assume w.l.o.g. that for the cyclic predecessor arrival a_1 of d_0 we have $\pi_{a_1} = 0$. As π is feasible, it satisfies (15). From our assumption, we conclude $d \leq \pi_{d_0}$ and $\pi_{d_0} + (d - h) \leq \pi_{a_2}$, and hence $\pi_{a_2} - \pi_{a_1} \geq 2d - h$. Event a_1 also takes place at time T. For notational convenience, we define $\pi_{a_{n+1}} := T$. With this notation, we have $\pi_{a_{i+1}} - \pi_{a_i} \geq d$, for all $i = 2, \ldots, n$. By the definition of $\pi_{a_{n+1}}$, we know that

$$\sum_{i=1}^{n} (\pi_{a_{i+1}} - \pi_{a_i}) = \pi_{a_{n+1}} - \pi_{a_1} = T.$$

Summing up the lower bounds yields $T \geq (n + 1)d - h$, which contradicts the hypothesis of Theorem 2.   □

Corollary 1. *If $h < (n + 1)d - T$, then every timetable which respects constraints (11) to (14) is an alternating timetable.*

In Fig. 14, we provide an example for the easiest case, namely matching two lines. As usual, we assume the period time to be 60 minutes.

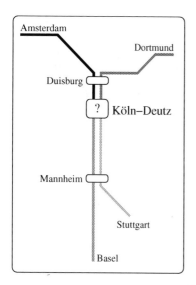
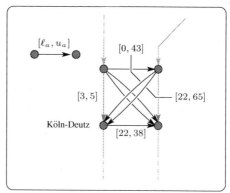

Fig. 14. Modeling Aspects of Line Planning: (**a**) Line Segments (**b**) PESP Constraints Ensuring the Segments to be Matched

Remark 2. There are of course alternating periodic timetables in the case $d \leq \frac{T}{n+1}$. PESP solvers are able to detect even those, if we were able to pre-define sufficiently many empty slots. By an "empty slot" we understand an artificial line which we have to schedule in the same way as the original lines, hereby separating the lines before and after the empty slot.

In more detail, let us assume that $\frac{T}{n^*+1} < d \leq \frac{T}{n^*}$ for some $n^* > n$, and that h satisfies the assumptions of Theorem 2 for n^*. We then introduce $n^* - n$ artificial dummy arrival and departure events a_i and d_i, $i = n+1, \ldots, n^*$. To prevent the original line segments from being matched with an artificial event, we require $\pi_{d_i} - \pi_{a_i} \in [0, h]$ for all $i = n + 1, \ldots, n^*$.

By construction, only feasible timetables let the original arrivals and departures alternate. However, perfectly balanced timetables, i.e. $\pi_{a_i} := (i-1)\frac{T}{n}$, are infeasible under these settings if $n^* < 2n$, since they do not provide $n^* - n$ empty slots.

Recall that so far we have considered only one direction. Hence, there is no mechanism yet to bind the matching of one direction to that of the opposite direction. But the matchings of opposite directions must fulfill the symmetry assumption that we gave at the beginning of Section 4.2. Otherwise, the trains from direction A could pass the matching station S in order to continue towards B, but the trains from B pass S before continuing in direction C. Thus, it would not be possible to communicate the line plan in the way customers are used to, because it may no more be visualized by an undirected graph. However, limited asymmetries in operation are accepted in practice.

Example 2 (S-Bahn Berlin GmbH). We consider the line S2 serving the route Blankenfelde-Lichtenrade-Buch-Bernau. Between Lichtenrade and Buch, a ten minute headway must be offered, for the remaining parts a 20 minute headway suffices.

In the current timetable (S-Bahn Berlin GmbH (2003)), this line is served in an asymmetric way. In order to cope with the single tracks (which are present at both endpoints) to limit the total amount of stopping time, and to ensure an efficient employment of the rolling stock, an asymmetric service is offered, and we present it in Table 2.

In order to ensure symmetric line plans, we have to guarantee the following condition. If we combine the arrival event a_i with the departure event d_j in one direction, then in the opposite direction the complementary arrival event a'_j must be combined with the departure event d'_i. More precisely, when considering the corresponding tension variables $x_{a_i d_j}$ and $x_{a'_j d'_i}$, they must fulfill

Table 2. Asymmetric Service of Line S2 (Berlin)

Blankenfelde	dep		10:09		arr	o	11:14	
Lichtenrade	dep	↓	10:15	10:25	arr	o	11:05	11:15
Buch	arr	o	11:06	11:16	dep	↑	10:14	10:24
Bernau	arr	o	11:21		dep				10:10

$$x_{a_i d_j} \in [0, h] \Leftrightarrow x_{a'_j d'_i} \in [0, h]. \tag{17}$$

In fact, this condition is quite similar to the symmetry constraints (6). What makes things more complicated is the fact that we must not predict in advance for which pairs (i, j) requirement (17) has to hold, and for which pairs it may be violated. Hence, we propose to guarantee property (17) for the matched pairs by imposing symmetry requirements on *every* pair of complementary junctions. But it is clear that this approach cuts off feasible timetables for symmetric line plans just because such timetables need not to be symmetric; see, e.g., Example 3.

Example 3 (S-Bahn Berlin GmbH). Consider the current timetable (S-Bahn Berlin GmbH (2003)) of the ring subnetwork of S-Bahn Berlin GmbH, of which we provide an excerpt in Table 3. Obviously, the line plan is symmetric. But the timetable is not

Table 3. Symmetric Line Plan but Asymmetric Timetable

Direction A						
Line	S45	S46	S8	S9	S47	S8
Origin	BFHS	BKW	BGA	BFHS	BSPF	BZN
Schöneweide dep ↓	xx:01	xx:06	xx:10	xx:13	xx:15	xx:18
Baumschulenweg arr o	xx:03	xx:09	xx:13	xx:16	xx:17	xx:21
Destination	BHMS	BGS	BPKR	BZOO	BWES	BPKR

Direction B						
Line	S8	S46	S9	S47	S8	S45
Origin	BPKR	BGS	BZOO	BWES	BPKR	BHMS
Baumschulenweg dep ↓	xx:02	xx:06	xx:08	xx:13	xx:14	xx:19
Schöneweide arr o	xx:05	xx:08	xx:10	xx:15	xx:17	xx:21
Destination	BGA	BKW	BFHS	BSPF	BZN	BFHS

symmetric. This can be seen by calculating the symmetry axes of lines S47 and S9 at station Schöneweide. Departure and arrival of line S47 sum up to 30, hence the trains of this line meet at times 5 and 15. For line S9 the sum yields 23, providing a symmetry axis of 1.5. An easier argument for asymmetry is that the sequence of the trains in Direction B is not the inverse of the one in Direction A.

There are two main objectives for the matching approach. First, we want to offer direct trips for as many passengers as possible. Second, the timetable should require only few trains for operation.

For the second criterion, in the case $h = 0$, no additional weight on arcs within the matching node is required in order to minimize the amount of rolling stock required to operate the timetable. In the case $h > 0$, one could put the vehicle weight on the arcs with feasible interval $[0, T - d + h]$. But this would no longer yield the desired exact piecewise-constant behavior of the objective, because some double counting can appear.

For maximizing the number of direct travelers, we consider the number of passengers w_{ij} starting their trip before the common track on a train covering line segment L_i^a, and finishing their trip after the common endpoint on a train covering line segment L_j^d. The value w_{ij} is added to the weight of the arc $a = (a_i, d_j)$ with $\ell_a = 0$ and $u_a = [0, T - d + h]$. The resulting cost coefficients in the objective function make sense even for pairs of line segments which are not matched, because long changeover times of many passengers are penalized.

Notice that the values w_{ij} are only well-defined if the two line segments do not serve a second matching station. This shows that the decisions to be taken within a matching station are of a rather local nature.

Summarizing, there are important scenarios in which the PESP can integrate relevant aspects of line planning into a model suited for timetabling and key issues of vehicle scheduling. This is in particular the case if symmetric timetables and balanced sequences along the common tracks, i.e. $d > \frac{T}{n+1}$, are requested for their own sake. Moreover, we observed that the larger the distance between two matching stations, the more reliable the passenger weight that we propose.

We think that fast train networks of European agglomerations, such as Frankfurt, Munich, or Paris (RER), are well-suited candidates for this approach. There, many passengers might have their origin or destination somewhere on the backbone route, and balanced sequences must be ensured due to the large number of lines per period.

5.3 Aspects of Network Planning

We propose to also model two questions which arise in network planning within the PESP: the extension of existing tracks, and thus lines, beyond their current endpoints, and the construction of faster tracks as substitutes for existing ones. Taking into account that, in these questions, we have to select one option out of a small number of disjoint options, it is evident that we will make intensive use of disjunctive constraints, cf. Section 3.3. Recall that there, we already discussed the introduction of optional additional stops. With appropriate weights that reflect amortization – see below – these may also cover the construction of new stations along an existing track.

We only discuss the construction of faster tracks in detail. But the reader will have no difficulty to adapt our suggestions to the very similar task of the extension of tracks.

In Fig. 15, we provide a constraint graph which offers the option of a new track between Aachen and Köln (Engl.: Cologne), being then part of the European high-speed line PBK (Paris-Brussels-Köln). We provide the status quo, with one intermediate stop, only for illustration purposes. In the future, we have the option to either use the current tracks, thus keeping a trip time of 38 minutes, or to establish the new high-speed track, hereby reducing the trip time down to 26 minutes.

To define appropriate weights for the arcs, we have to take into account three different types of objectives: The number of customers c who profit from a new track by shorter travel times, the trip times of the trains which may allow to reduce the number of trains required (M, cf. Section 5.1), and the cost M' of the investment.

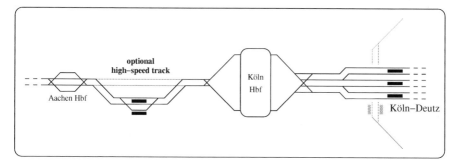

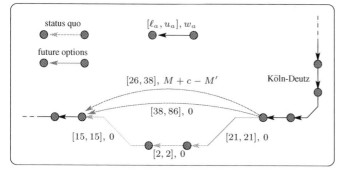

Fig. 15. Modeling Aspects of Network Planning: (**a**) Infrastructure Including Optional High-speed Track (**b**) PESP Constraints Taking into Account the Two Infrastructural Alternatives

One can imagine that it is a non-trivial management decision to derive an hourly weight M' from the total cost of the investment.

Similarly to line planning, investments into infrastructure will only make sense if they are effected for both directions at the same time. Again, we ensure symmetric investments by requiring the timetable to be symmetric.

Let us now analyze the situation in which several lines have the option of using the same new, faster track. Of course, we want to ensure that infrastructure is only paid once in terms of the objective function. Hence, we have to partition the total cost onto all of the concerned lines. But what if in a solution of a PESP instance only one line is routed over the new track?

But a reasonable allocation of the total costs is only possible if we know in advance how many lines will have to use the new track. Unfortunately, we are only able to ensure this with constraints of the types already introduced, if *all* the lines must use the same track. This would, e.g., be the case when analyzing two mutually exclusive variants of constructing a new track.

We can guarantee that all the lines use the same track simply by enforcing the same running time for each line. This is achieved by introducing constraints of type (6). However, notice that we cheat a bit in this case, because those constraints no longer relate only pairs of complementary arcs to each other. Nevertheless, the

MIP formulation of this even slightly more extended model incorporates many of the computational aspects of the pure PESP model.

6 Conclusion

Our discussion of the PESP model shows that it has a great modeling power and extendibility. We have demonstrated that many non-standard requirements for periodic timetables and also important aspects of other – traditionally separate – planning phases can be integrated into the PESP. Fig. 16 displays the gain by this modeling power over the traditional use of the PESP displayed in Fig. 1.

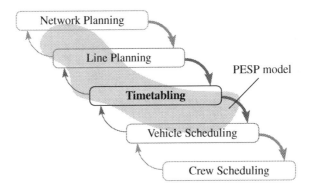

Fig. 16. Planning Phases Covered by the PESP with Our Contribution

Interestingly, this integration into the PESP has been possible without seemingly complicating it too much. In all cases, we obtained mixed integer programs that still have the characteristics of a PESP. Hence we believe that these extended models stay computationally tractable also for networks of relevant sizes. So far, our belief is confirmed by a confidential study for S-Bahn Berlin GmbH for two of its three major subnetworks.

We therefore hope that these models, through their integrative approach to vehicle scheduling, timetabling, line planning, and infrastructure planning, will eventually lead to better decision making in practice.

Acknowledgments: We want to thank the staff of Deutsche Bahn AG, S-Bahn Berlin GmbH, and Berliner Verkehrsbetriebe (BVG) for providing us with both real-world data and very detailed requirements of their specific periodic timetabling problems. Moreover, we thank the referees for their *very* detailed suggestions. This work has been supported by the DFG Research Center "Mathematics for key technologies" in Berlin.

References

Bollobás, B. (2002). *Modern Graph Theory*, volume 184 of *Graduate Texts in Mathematics*. Springer. 2nd printing.

Borndörfer, R., Löbel, A., and Weider, S. (2002). Integrierte Umlauf- und Dienstplanung im öffentlichen Nahverkehr. In *HEUREKA '02: Optimierung in Transport und Verkehr, Tagungsbericht*, number 002/72, pages 77–98. FGSV Verlag.

Borndörfer, R., Grötschel, M., and Pfetsch, M. E. (2007). Models for line planning in public transport. This volume.

Bussieck, M. R., Winter, T., and Zimmermann, U. (1997). Discrete optimization in public rail transport. *Mathematical Programming B*, **79**, 415–444.

Claessens, M., van Dijk, N., and Zwaneveld, P. J. (1998). Cost optimal allocation of rail passenger lines. *European Journal of Operational Research*, **110**(3), 474–489.

Engelhardt-Funke, O. and Kolonko, M. (2004). Analysing stability and investments in railway networks using advanced evolutionary algorithms. *International Transactions in Operational Research*, **11**, 381–394.

Grötschel, M., Löbel, A., and Völker, M. (1997). Optimierung des Fahrzeugumlaufs im öffentlichen Nahverkehr. In K. Hoffmann, W. Jäger, T. Lohmann, and H. Schunck, editors, *Mathematik - Schlüsseltechnologie für die Zukunft*, pages 609–624, Berlin. Springer.

Haase, K., Desaulniers, G., and Desrosiers, J. (2001). Simultaneous vehicle and crew scheduling in urban mass transit systems. *Transportation Science*, **35**(3), 286–303.

Krista, M. (1997). *Verfahren zur Fahrplanoptimierung am Beispiel der Synchronzeiten*. Ph.D. thesis, Technische Universität Braunschweig. In German.

Kroon, L. G. and Peeters, L. W. (2003). A variable trip time model for cyclic railway timetabling. *Transportation Science*, **37**, 198–212.

Leuschel, I. (2002). Der Fernverkehrsfahrplan 2003 der Deutschen Bahn AG. *Eisenbahntechnische Rundschau*, **51**(7–8), 452–464. In German.

Liebchen, C. (2003). Finding short integral cycle bases for cyclic timetabling. In G. D. Battista and U. Zwick, editors, *ESA*, volume 2832 of *Lecture Notes in Computer Science*, pages 715–726. Springer.

Liebchen, C. (2004). Symmetry for periodic railway timetables. *Electronic Notes in Theoretical Computer Science*, **92**, 34–51.

Liebchen, C. and Peeters, L. (2002). Some practical aspects of periodic timetabling. In P. Chamoni, R. Leisten, A. Martin, J. Minnemann, and H. Stadtler, editors, *Operations Research Proceedings 2001*, pages 25–32. Springer, Berlin.

Liebchen, C., Proksch, M., and Wagner, F. H. (2007). Performance of algorithms for periodic timetable optimization. This volume.

Lindner, T. (2000). *Train Schedule Optimization in Public Rail Transport*. Ph.D. thesis, Technische Universität Braunschweig.

Nachtigall, K. (1994). A branch and cut approach for periodic network programming. Hildesheimer Informatik-Berichte 29, Universität Hildesheim.

Nachtigall, K. (1996). Cutting planes for a polyhedron associated with a periodic network. Institutsbericht IB 112-96/17, Deutsche Forschungsanstalt für Luft- und Raumfahrt e.V.

Nachtigall, K. (1998). *Periodic Network Optimization and Fixed Interval Timetables.* Habilitation thesis, Universität Hildesheim.

Nachtigall, K. and Voget, S. (1996). A genetic algorithm approach to periodic railway synchronization. *Computers & Operations Research,* **23**(5), 453–463.

Odijk, M. A. (1994). Construction of periodic timetables, Part 1: A cutting plane algorithm. Technical Report 94-61, TU Delft.

Odijk, M. A. (1996). A constraint generation algorithm for the construction of periodic railway timetables. *Transportation Research B,* **30**(6), 455–464.

Peeters, L. W. (2000). Personal Communication.

Peeters, L. W. (2003). *Cyclic Railway Timetable Optimization.* Ph.D. thesis, Erasmus Universiteit Rotterdam.

S-Bahn Berlin GmbH (2003). S-Bahn-Fahrplan (gültig ab 16. Juni 2003).

Schrijver, A. (1998). *Theory of Linear and Integer Programming.* Wiley, 2nd edition.

Schrijver, A. and Steenbeek, A. G. (1993). Dienstregelingontwikkeling voor Nederlandse Spoorwegen N.S. Rapport Fase 1, Centrum voor Wiskunde en Informatica.

Serafini, P. and Ukovich, W. (1989). A mathematical model for periodic scheduling problems. *SIAM Journal on Discrete Mathematics,* **2**(4), 550–581.

van den Berg, J. and Odijk, M. A. (1994). DONS: Computer aided design of regular service timetables. In T. Murthy, B. Mellitt, C. Brebbia, G. Sciutto, and S. Sone, editors, *Computers in Railways IV (COMPRAIL) – Vol. 2: Railway Operations.* WIT Press.

Völker, M. (2003). Ein multikriterieller Algorithmus zur automatisierten Busliniennetzplanung. Lecture on the OR Workshop Optimierung im öffentlichen Nahverkehr.

Performance of Algorithms for Periodic Timetable Optimization

Christian Liebchen[1], Mark Proksch[2], and Frank H. Wagner[3]

[1] TU Berlin, Institut für Mathematik, Straße des 17. Juni 136, D-10623 Berlin, Germany
liebchen@math.tu-berlin.de
[2] intranetz GmbH, Bergstraße 22, D-10115 Berlin, Germany
mark@mark-proksch.de
[3] Deutsche Bahn AG, Konzernentwicklung, Potsdamer Platz 2, D-10785 Berlin, Germany
Frank.H.Wagner@bahn.de, from June 2005 on Frank.Geraets@bahn.de

Summary. During the last 15 years, many solution methods for the important task of constructing periodic timetables for public transportation companies have been proposed. We first point out the importance of an objective function, where we observe that in particular a linear objective function turns out to be a good compromise between essential practical requirements and computational tractability. Then, we enter into a detailed empirical analysis of various Mixed Integer Programming (MIP) procedures – those using node variables and those using arc variables – genetic algorithms, simulated annealing and constraint programming. To our knowledge, this is the first comparison of five conceptually different solution approaches for periodic timetable optimization.

On rather small instances, an arc-based MIP formulation behaves best, when refined by additional valid inequalities. On bigger instances, the solutions obtained by a genetic algorithm are competitive to the solutions CPLEX was investigating until it reached a time or memory limit. For Deutsche Bahn AG, the genetic algorithm was most convincing on their various data sets, and it will become the first automated timetable optimization software in use.

1 Introduction

The central task in the planning process of a large public transport company is timetabling. So far this is done mostly manually, using computers as clever editors – if at all. At Deutsche Bahn AG, being the major supplier of railway transport in Germany, the amount of people and time spent on this task is enormous, e.g., some hundreds of people are working on it in the year. Roughly speaking the timetabling task discussed here consists of finding periodic completely regular timetables (no exceptions on weekends, in the night, on the borders, etc.) given the infrastructure, a line system, and the amount of changing travelers between the lines (Bussieck *et al.* (1997)). The optimization goals are minimizing the travel times and the amount of rolling stock needed, i.e., satisfying the needs of the customers and the company.

There have been various approaches proposed for solving this very hard problem (cf. MIPLIB, Liebchen and Möhring (2003)). These include mixed-integer programming and constraint propagation, but also genetic algorithms and simulated annealing. Nevertheless, there are no computational studies available that compare at least two of these techniques on the very same data set. As Deutsche Bahn AG aims at automating at least parts of the timetabling process in the near future – i.e. within the next few years – we perform an extensive computational study to examine the above-mentioned algorithms in detail.

In Section 2 we present the Periodic Event Scheduling Problem (PESP) which is the model of our choice for periodic railway timetabling. For a detailed description of its very rich modeling capabilities, we refer to Liebchen and Möhring (2007). In Section 3 we derive several equivalent MIP formulations for the PESP. This step is very important as there are immense differences in the performance of the various MIP formulations – e.g. the most intuitive one does not behave best.

After a short sketch of some refinements of the general methods (Section 4), we start our computational study in Section 5 by giving detailed information of the three data sets to which we apply the algorithms. Our program makes use of CPLEX as a MIP solver, ILOG Solver for constraint programming (CP), and the prosim Express optimization workbench for local optimization algorithms. The latter has been developed beforehand in order to deal with other optimization tasks within the Deutsche Bahn. It is a toolbox of general purpose optimization algorithms. Combining these with a problem specific interface makes it easy to tackle a problem with different algorithms.

There will be a certain focus on MIP techniques. This is because these offer the most variety of parameters in conjunction with three different problem formulations which can be sharpened by making use of five kinds of valid inequalities which are defined for every elementary cycle of the constraint graph. The impacts of these numerous adjusting crews becomes most visible on our medium size instance, cf. Section 5.2. Here, on the one hand, the best parameter settings provide solution times which are not too short for identifying significant differences. On the other hand, solution times are not too long to try out a large number of different parameter settings.

On small and medium sized problems, we will observe that CPLEX is able to terminate with a provably optimum solution within the time and memory limits that we define. Only on the smallest instance, the other algorithms are able to construct (almost) optimum solutions. This might not be considered very astonishing. Instead, on bigger instances, where CPLEX fails to terminate, we were surprised that in particular the quality of the solutions obtained by the genetic algorithm is still competitive. If we run CPLEX at default parameter settings, even when refining the most promising problem formulation with additional valid inequalities CPLEX gets outperformed by our genetic algorithm. Only with some variations to the parameter settings of CPLEX, the picture changes slightly. This shows that our earlier parameter testing was worthwhile.

2 Modeling Periodic Railway Timetables

Serafini and Ukovich (1989) introduced the periodic event scheduling problem (PESP), by which instances of periodic timetabling may be formulated in a very compact way. Since then, this model has been widely used (Schrijver and Steenbeek (1993), Nachtigall (1994), Lindner (2000)). In the PESP, we are given a period time T and a set V of events, where an event models either the arrival or the departure of a directed traffic line at a certain station. Furthermore, we are given a set of constraints A. Every constraint $a = (i, j)$ relates a pair of events i, j by a lower bound ℓ_a and an upper bound u_a.

A solution of a PESP instance is a node assignment $\pi : V \mapsto [0, T)$ that satisfies

$$(\pi_j - \pi_i - \ell_a) \bmod T \leq u_a - \ell_a, \ \forall a = (i, j) \in A, \tag{1}$$

or $\pi_j - \pi_i \in [\ell_a, u_a]_T$ for short. Notice that we may assume w.l.o.g. that $0 \leq \ell_a < T$ and $u_a - \ell_a < T$. The PESP is $\mathcal{NP}$-complete, since it generalizes Vertex Coloring (Odijk (1997)): Orient the edges of a Coloring instance arbitrarily and assign feasible periodic intervals $[1, T - 1]_T$ to each of them.

At the end of this section, we will give several motivations why we consider an objective function to be important. On the one hand, a linear objective function is rich enough to model the most important features. On the other hand, a linear objective function permits to include powerful MIP solvers, in particular CPLEX, into our study. Hence, we add a linear objective function of the form

$$\sum_{a=(i,j)\in A} c_a \cdot (\pi_j - \pi_i - \ell_a) \bmod T$$

with costs c_a.

The PESP yields the capability to model manifold practical requirements arising in periodic railway timetabling. To name just a few, we will give only three examples. We model a trip of t time units of a directed line from station D to station A by requiring $\pi_a - \pi_d \in [t, t]_T$. To separate two lines sharing a common track by a safety distance of d time units, we require $\pi_{d_j} - \pi_{d_i} \in [d, T - d]_T$. Finally, we are going to model the quality of changeovers. Notice that a timetable is still feasible from an operational point of view, even though it may offer very long waiting times for changeovers. Hence, we only introduce "loose constraints," i.e. we set $u_a := \ell_a + (T - 1)$, where ℓ_a models the minimal amount of time required for changing trains. By setting the cost coefficient of such a loose constraint to the number of passengers on that specific connection, we are able to guarantee good timetables by minimizing the total changeover waiting time. For further practical requirements, we refer to Liebchen and Möhring (2007).

In our dialogue with practitioners of both national railway companies and urban transportation companies, the following three features turned out to be important:

- simultaneous minimization of the amount of rolling stock required to operate the timetable (Nachtigall (1998) and Liebchen and Peeters (2002b))

- minimization of passenger waiting time with no risk of overdetermining the system by the definition of maximal changeover times which are too tight
- maximization of the number of connections not exceeding a certain waiting time by making use of so-called soft constraints, cf. Liebchen and Möhring (2007), Nachtigall (1996).

Fortunately, all these can easily be expressed by means of a linear objective function.

Whereas the way of modeling changeover activities can be seen to depend only on the flavor of each individual company, almost all companies have in common that they want to minimize the amount of rolling stock. In fact, this requirement has to be seen as an input for timetabling, because the quality of the vehicle schedule, being the next planning step in the classical hierarchical approach, is largely determined by the timetable. For example, during the off-peak traffic time, in which still a 10 minute headway is offered, the Berlin Underground strictly rejects timetables which require 75 trains or more, because only 68 are technically necessary and the salaries form a considerable portion of the operational costs. In order to get an acceptable situation for changing passengers, about 70 trains suffice.

Consider the very special case where the vehicle schedule is fixed *a priori and* the stopping times are fixed, too. Here, Nachtigall (1998) identified PESP constraints that ensure that only periodic timetables remain feasible, that can be operated with the minimum number of trains. However, in the more general case, Liebchen and Möhring (2007) show these constraints to no longer work. More generally, either we had to cut off timetables that we initially seek for, or timetables that require additional trains become feasible.

This dilemma is our main motivation for the need of an objective function, at least for a linear one. Such a function takes advantage of Equation (7) on p. 139 in Liebchen and Möhring (2007): By assigning a value M to the arcs modeling a traffic line, every additional train pays $M \cdot T$ to the objective function value. If the value for M is chosen to be relatively large compared to the passenger weights, the objective function essentially models the piecewise constant behaviour of the cost of the rolling stock for operating the train network.

From a more local perspective, we just penalize idle time of trains. But this can even be done without knowing *a priori* the circulation plan of the trains. Although an exact model involves a quadratic objective function, Liebchen and Peeters (2002b) report that a linear relaxation yields results of high quality.

But there is even another problem with forcing lines to be operated with the minimal number of trains. In Berlin, e.g., the two underground lines U6 and U7 are required to meet at Mehringdamm, because there they share a common platform. But due to the existing running times, turnover times, and minimal changeover times, this simple requirement yields an inconsistent constraint system, as long as we require *both* lines to be operated with the minimal number of trains. However, we do not want to take the decision in advance, on which line to add the extra train. Hence, every feasible constraint system must contain timetables which require an additional train for both lines. Whereas the pure PESP has to fail, already by the means of a linear objective function we are able to prefer timetables which require only one extra train in total.

3 Mixed Integer Programming Formulations

Recall the initial definition (1) of the PESP in the previous section. We can interpret the variables π as a *node potential*, which periodically satisfies the given constraints. Notice that if we omit the modulo operator in (1), we obtain the more restrictive Feasible Differential Problem (FDP), which can be solved easily by network flow techniques.

The initial formulation (1) will immediately serve as input for the Constraint Programming formulation, as well as for the local search procedures we are going to examine. But in order to get to an MIP formulation, we must resolve the modulo operator by integer variables. The original constraint (1) translates to

$$\ell_a \leq \pi_j - \pi_i + p_a T \leq u_a,$$

where p_a is required to be integer. Here, the integer variables permit to shift potential differences into the target interval $[\ell_a, u_a]$, where the pure aperiodic difference fails. We obtain the first MIP formulation:

$$\left. \begin{array}{rl} \min & \sum\limits_{a=(i,j)\in A} c_a \cdot (\pi_j - \pi_i + p_a T) \\ \text{s.t.} & \ell \leq B^t \pi + pT \leq u \\ & p \in \mathbb{Z}^A \\ & \pi \in [0, T)^V, \end{array} \right] \qquad (2)$$

where B denotes the node-arc incidence matrix of the directed (multi-) graph $D = (V, A)$. Notice that for every feasible solution, we are able to guarantee $p_a \in [0, \overline{p_a}] \cap \mathbb{Z}$, with

$$\overline{p_a} = \begin{cases} 1, & \text{if } u_a < T, \\ 2, & \text{otherwise.} \end{cases} \qquad (3)$$

Obviously, for a fixed vector p, the feasible region of (2) is precisely the FDP, showing that indeed the integer variables form the core of the model. Notice that for a fixed spanning tree H, we may fix $p_a = 0$ for every $a \in H$, if we relax $\pi \in \mathbb{Q}^V$ (Serafini and Ukovich (1989)), which yields a formulation that we call (2a).

Another perspective of periodic scheduling can be obtained by considering tensions instead of potentials. In a straightforward way, define for a given node potential π its tension

$$\hat{x}_a := \pi_j - \pi_i, \ \forall a = (i, j) \in A.$$

Recall that a vector $\hat{x}$ is a tension, if and only if for an arbitrary cycle basis $\mathcal{C}$, $\gamma_C \hat{x} = 0$ for every cycle $C \in \mathcal{C}$ with incidence vector $\gamma_C \in \{-1, 0, 1\}^A$. This yields the second MIP formulation:

$$\left.\begin{array}{ll} \min c^t(\hat{x} + pT) & \min c^t x \\ \text{s.t. } \Gamma \hat{x} = 0 & \text{s.t. } \Gamma(x - pT) = 0 \\ \ell \leq \hat{x} + pT \leq u & \ell \leq x \leq u \\ p \in \mathbb{Z}^A, & p \in \mathbb{Z}^A, \end{array}\right] \quad (4)$$

where $\Gamma \in \{-1, 0, 1\}^{(|A|-|V|+1) \times |A|}$ denotes the cycle-arc incidence matrix (*cycle matrix*) of some cycle basis $\mathcal{C}$ of the graph D. Of course, the box constraints (3) apply to formulation (4) as well.

We are able to reduce the number of integer variables from $|A|$ down to $|A| - |V| + 1$, by introducing periodic tensions. For a given node potential π, we define the corresponding *periodic tension* x as

$$x_{ij} := (\pi_j - \pi_i - \ell_{ij}) \bmod T + \ell_{ij}.$$

Periodic tensions can be characterized similarly to classic aperiodic tensions.

Lemma 1 (Cycle Periodicity Property). *A vector $x \in \mathbb{Q}^A$ is a periodic tension if and only if for every cycle C with incidence vector $\gamma_C \in \{-1, 0, 1\}^A$, there exists some $z_C \in \mathbb{Z}$, such that*

$$\gamma_C x = z_C T. \quad (5)$$

By extending an approach of Nachtigall (1994), Liebchen and Peeters (2002a) proved that it suffices to ensure equation (5) only for the elements of an integral cycle basis of the directed graph, which leads to the third MIP formulation

$$\left.\begin{array}{l} \min c^t x \\ \text{s.t. } \Gamma x = zT \\ \ell \leq x \leq u \\ z \in \mathbb{Z}^{|A|-|V|+1}. \end{array}\right] \quad (6)$$

Here, Γ denotes the cycle matrix of an integral cycle basis. By defining *slack variables* $\tilde{x}_a := x_a - \ell_a$, we obtain formulation (6a), which turns out to be slightly easier to solve for CPLEX.

But there is even a problem with formulation (6a): its LP-relaxation has minimal value 0, because a fractional vector z is always able to compensate any vector $\tilde{x}_a$, thus in particular $\tilde{x} = \mathbf{0}$. Hence, additional valid inequalities are essential for obtaining good lower bounds.

Theorem 1 (Odijk (1997)). *An integer vector p allows a feasible solution for the MIP (4), if and only if for every oriented cycle C of the constraint graph, the following cycle inequalities hold*

$$\left\lceil \frac{1}{T} \left(\sum_{a \in C^+} \ell_a - \sum_{a \in C^-} u_a \right) \right\rceil \leq \sum_{a \in C^+} p_a - \sum_{a \in C^-} p_a \leq \left\lfloor \frac{1}{T} \left(\sum_{a \in C^+} u_a - \sum_{a \in C^-} \ell_a \right) \right\rfloor, \quad (7)$$

where C^+ and C^- denote the forward and the backward arcs of the cycle C.

Of course, there is a reformulation of the valid inequalities (7), such that they apply to formulations (6) and (6a) as well. In these formulations, they immediately yield box constraints $\underline{z}_C \leq z_C \leq \overline{z}_C$ for every integer variable z_C, when applied to the corresponding cycle C of the cycle matrix in the problem formulation. Defining $\tilde{z}_C := z_C - \underline{z}_C$ provides formulation (6b), in which we may declare certain variables to be binary, which is preferred by the MIP solvers as well.

Furthermore, for a fixed cycle C, the span between lower and upper bound of a pair of cycle inequalities (7) behaves similarly to the value $\sum_{a \in C}(u_a - \ell_a)$. In order to have only a few choices for the integer variables, we are looking for an integral cycle basis $\mathcal{C}$, which minimizes

$$\sum_{C \in \mathcal{C}} \sum_{a \in C} d_a, \tag{8}$$

where we define $d_a := u_a - \ell_a$ to be the *span* of arc a. More precisely, Liebchen (2003) reports a correlation of about 0.5 between the width

$$\prod_{C \in \mathcal{C}} (\overline{z}_C - \underline{z}_C + 1) \tag{9}$$

and the solution time of CPLEX on formulation (6b).

Minimizing (8) for arbitrary cycle bases is just the minimal cycle basis problem (MCB), for which Horton (1987) designed a polynomial time algorithm. However, the complexity of minimizing (8) only for integral cycle bases is unknown to the authors. Finding minimal strictly fundamental cycle bases – which are a very special subclass of integral cycle bases – has been proven to be $\mathcal{NP}$-hard; see Deo et al. (1982). Nevertheless, there are powerful heuristics available for constructing both short strictly fundamental cycle bases and short integral cycle bases; see Deo et al. (1982), Deo et al. (1995), Liebchen (2003).

We propose to use a variant of the cycle inequalities (7) as well. From formulation (6), one can see that the integer variables can be expressed by sums of tension variables. After only a few elementary transformations, an original cycle inequality (7) in terms of the integer variables z becomes a valid inequality (7a) in terms of the tension variables. Nachtigall (1996) introduced further inequalities in terms of the tension variables.

Theorem 2 (Nachtigall (1996)). *For every elementary cycle C, define* $b := (\sum_{a \in C^-} \ell_a - \sum_{a \in C^+} \ell_a) \bmod T$. *If $b > 0$, then*

$$(T - b)(\sum_{a \in C^+} \tilde{x}_a) + b(\sum_{a \in C^-} \tilde{x}_a) \geq b(T - b) \tag{10}$$

is a facet defining inequality for the polyhedra defined by the mixed integer linear programs (6a) and (6b), in terms of slack variables.

4 Exhausting the Problem Formulations

In any of the MIP formulations, we have to decide for which cycles to add their cycle inequalities (7), occasionally in their tension variant (7a). In addition, we may add change cycle inequalities (10) to formulations (4) and (6b). Of course, problem formulation (6b) is most challenging, because there we may even choose an integral cycle basis. However, this choice makes it very difficult to compare formulation (6b) for different cycle bases, in particular if we add cycle inequalities (10), as their formulation essentially depends on the integer variables being available in the specific formulations.

After occasionally having added some of these valid inequalities by iterated calls to separation heuristics, we transfer the instance to the MIP solver of CPLEX (*Cut and Branch*).

Since there are no polynomial separation algorithms available for the valid inequalities that we consider, and since both kinds of valid inequalities are defined for oriented cycles of the directed graph, we heuristically generate cycles. Apart from the fundamental cycles of minimal spanning trees (MST) subject to random edge weights, we use the following four heuristics:

- fundamental cycles of minimal spanning trees subject to the values x^* in an optimal solution of the current LP relaxation,
- fundamental cycles of minimal spanning trees subject to the integral gap $|p_a^* - \text{round}(p_a^*)|$ in an optimal solution of the current LP relaxation[4],
- the up to $|A| \cdot |V|$ candidate cycles of Horton's polynomial MCB algorithm (Horton (1987)) subject to the integral gap in an optimal solution of the current LP relaxation, and
- the up to $|A| \cdot |V|$ candidate cycles of Horton's polynomial MCB algorithm subject to the arc spans d.

The cycle bases that we consider in formulation (6b) are

1. MST span: the fundamental cycles of an MST subject to edge weights d_a,
2. MST nspan: the fundamental cycles of an MST subject to edge weights $T - d_a$,
3. NT: the fundamental cycles obtained by the NT heuristic (non-tree edges) of Deo *et al.* (1995),
4. UV one: the fundamental cycles obtained by the UV heuristic (unexplored vertices) of Deo *et al.* (1995),
5. UV span: the fundamental cycles obtained by the UV heuristic, in which we introduced the values d_a as edge weights,
6. UV nspan: the fundamental cycles obtained by the UV heuristic, in which we introduced the values $T - d_a$ as edge weights, and
7. Horton: the minimal cycle basis obtained by Horton's algorithm, given that it produces an integral cycle basis.

---

[4] In formulation (6b), it makes only sense to identify the components of p^* with the (non-tree) arcs of the digraph, if we use strictly fundamental cycle bases.

To any of the heuristics (1) to (6), we apply fundamental improvements (see Liebchen (2003)), as they have been proposed by Berger (2002).

For the genetic algorithm approach we are going to follow Nachtigall and Voget (1996) who proposed to encode a timetable by storing, for each event i, at which point of time $\pi_i \in \{0, \ldots, T - 1\}$ it should take place. Moreover, they proposed to apply a local improvement heuristic to every new individual, which is obtained by a mutation or a crossover operation. In this local improvement step, they subsequently consider every event i, and compute for every point of time $t \in \{0, \ldots, T - 1\}$ the (local) objective value along the arcs in the cutset induced by node i, and set π_i such that the minimum is attained. Notice that this procedure depends heavily on the time precision that is chosen for the computation.

We propose two modifications which make this approach more efficient. First, in our practical data sets, there are several arcs $a = (i, j)$ with $u_a - \ell_a \ll T$, in particular stopping activities. Since in such a situation, only few pairs $(\pi_i, \pi_j) \in \{0, \ldots, T - 1\} \times \{0, \ldots, T - 1\}$ satisfy constraint a, we propose to encode for event j only its *offset relative to* π_i. Second, we profit from the fact that we only consider linear objective functions. Hence, for every feasible timetable π, there exists a timetable π' having objective value not bigger than π, but in that for every node i, there exists an arc $a = (i, j) \in \delta^{\text{out}}(i)$ or an arc $b = (k, i) \in \delta^{\text{in}}(i)$, such that $\pi'_i \in \{\pi'_k + \ell_b, \pi'_k + u_b, \pi'_j - \ell_a, \pi_j - u_a\} \bmod T$. Using this property, we propose to consider only these *tightening values* during the local improvement step. Doing so, the running time of the local improvement step becomes independent from the time precision, i.e., it is not a big difference anymore, whether one time unit represents 60 seconds ($T = 120$), or only 6 seconds ($T = 1200$), where only the latter is the standard for tactical internal documents of Deutsche Bahn.

In contrast to solving LPs, we do not use well known standard software for local search. Therefore, we should spend some more words on this topic. For the tests of the genetic algorithm we use a very simple version of the algorithm with only a few parameters ($p, g \in \mathbb{N}^+, m \in \mathbb{R}^+$):

1. Create an initial population of p random individuals.
2. Repeat g times:
   a) Pair the p individuals randomly to $\lfloor p/2 \rfloor$ pairs. Create 2 children from every couple by recombination.
   b) Create $\lceil m \cdot p \rceil$ mutants of the p individuals by the mutation operator. This is done by first creating $\lfloor m \rfloor$ mutants from every individual. Afterwards $\lceil m \cdot p \rceil - \lfloor m \rfloor \cdot p$ individuals are randomly selected to create another mutant each.
   c) Remove duplicate individuals.
   d) Compute the cost function for all individuals (given generation, children and mutants). Select the p best individuals to form the new generation.
3. Select the best individual of the last generation as the result of the algorithm.

This and some more elaborated versions of the genetic algorithm are discussed in Mühlenbein (1997). Notice that the best individual of every generation is better or equal to the one of the previous generation. Therefore this version of the genetic algorithm implements an improvement only strategy.

Surely, we are aware that constraint programming algorithms originally were not designed to solve optimization problems. Nevertheless, the discussion in Section 2 explains why we have to insist on an objective function. As other researchers reported to us that they successfully applied constraint programming to the feasibility variant of periodic timetabling, we are giving it a try.

In order to help the constraint programming approach in the optimization context, we strengthen some constraints with large span and big objective value. In more detail, for the 15 arcs a with biggest objective value and $d_a > \frac{T}{2}$, we set $u'_a :=$ $\ell_a + \frac{T}{2}$. But we also try to prevent the problem from getting over-determined. Hence, we effect this strengthening only if for every cycle of the constraint graph the sum of the spans of its arcs remains at least as large as the period time T (Laube (2004)).

5 Computational Results

We perform our computations on three data sets. This small number is motivated by two facts. Firstly, there are no collections of timetabling instances publicly available, mostly because companies consider these data very sensitive. Secondly, already the combination of these three data sets with different families of algorithms – each with a considerable number of major parameters to be set – leads to a substantial amount of data, of which we hope to give the reader an accurate overview. We first give a short description of the real-world problems on which we perform the computations. Then, we will report the behaviour of the algorithms, where we start each time with the various MIP formulations. There, besides problem specific parameters, out of the huge number of CPLEX parameters we follow suggestions of Bixby (2003) and vary on the following MIP strategies:

- variable selection strategy: default or strong branching (ILOG SA (2004))
- MIP emphasis (ILOG SA (2004)): default, integer feasibility, or optimality
- MIP cuts: default or aggressive cut generation
- user cuts: add valid inequalities as full constraints or only as user cuts (ILOG SA (2004))

All computations which involve CPLEX are carried out on Intel Pentium 4 machines with 2.8 GHz and 1024MB RAM.

For the genetic algorithm, the algorithmic behaviour does not change over the generations. Hence, the total number of generations g is not an interesting parameter. The result of any test with a large number of generations can be used to analyze a smaller one, just by cutting off the appropriate number of generations.

The two remaining parameters – population size p and mutation intensity m – are the subject of our tests. Since both parameters affect the number of produced individuals per generation and thus the run time, we coordinated them to get almost the same number of individuals in every test run.

Emden-Weinert and Proksch (1999) and Proksch (1997) successfully used MIR (*Multiple Independent Runs*) on Simulated Annealing for the airline crew scheduling problem. Here we try MIR on U Berlin and ICE small. In addition we test two different versions of it on ICE small, which will be described there. We further test the simulated annealing algorithm with the geometric cooling schedule. Since the results are rather poor, we do not present parameter studies, but only some numbers. All computations for genetic algorithms and simulated annealing are carried out on the same machine as those for CPLEX (Intel Pentium 4, 2.8 GHz and 1024MB RAM).

The constraint programming parameters we are going to adjust are the variable selection strategy and the domain reduction policy. Other experimental studies (Laube (2004)) showed that for timetable optimization instances, the forward checking (FC) policy (Barták (1999)) and the so-called "look ahead" (LA) policy (Barták (1999)) perform best. Moreover, it seems to be worth trying to proceed with the variable having minimal current domain. Unfortunately, an ILOG Solver license is available to us only on a SUN UltraSPARC-IIi at 333 MHz.

In contrast to the other two approaches, local search procedures – like the genetic algorithm and simulated annealing – are randomized algorithms which cannot be judged by a single run. Thus, we always start a number of runs with identical parameter settings and average their results. Such a group of single runs is named "test run" in the subsequent text.

The deviation of the results within one test run turn out to be very high, especially on ICE small. When dealing with large deviations on randomized algorithms, a promising idea is to start a couple of those algorithms and take the best result as the output of the whole process. In the special case of genetic algorithms, the selection of the best result can be done by collecting the individuals of all runs to a common population, on which a final collecting run is started. In doing so the genetic algorithm has the chance to combine different good solutions to a possibly better one. We try this approach on U Berlin and ICE small. In addition we test two different versions of this approach on ICE small, which will be described there.

5.1 Solving U Berlin

The first data set models the Berlin Underground. In the evening hours and on weekends, the period length is $T = 10$ minutes. During this *off-peak traffic time*, with only one small exception, each of the nine lines is operated on its own track. The only safety conditions to be obeyed are crossings of tracks in front of terminal stations, in case that no depot is located behind the station.

There are several objectives to pursue. First, if different lines share a platform, then a good cross-wise correspondence has to be ensured. Second, the number of trains required to operate the network has to be minimized. Third, out of the about 170 changeover relations[5], the 48 TOP connections must not offer effective waiting time of more than five minutes. Fourth, out of the next 36 relations, for a

---

[5] These relations include ten important connections to the fast train network, which we assume to be fixed.

maximal number of connections the five minute criterion should hold as well. Finally, the minimal average changeover waiting time has to be minimized. To that end, we allow to insert additional stopping times at the eight most important correspondence stations, which involve 34 stopping activities in total.

After redundancies are eliminated, the contracted digraph has 40 nodes and 240 arcs. There are 157 arcs with $d_a = T - 1$, and 40 arcs with $d_a \leq 0.2 \cdot T$. The average span is 73.25%.

MIP Formulations: Among the three types of MIP formulations, we start with the integral cycle basis formulation (6b). Since this formulation will allow very short solution times for most integral cycle bases and CPLEX parameter settings, we only give a very compact summary in Table 1.

First, for every integral cycle basis, we give its width (9) and the optimal value of the LP relaxation of system (6b) (relative to the optimal value) with cycle inequalities (7) added as box constraints on the integer variables. We add up to 250 further valid inequalities or none, and varied the two CPLEX parameters variable selection and MIP emphasis.

Table 1. Solution Times on U Berlin for Various Cycle Bases

Tree	MST		MST		UV		NT		UV		UV		Horton
Weight	nspan		span		one		one		span		nspan		span
Fund. improve	no	yes	no	yes	no	yes	no	yes	no	yes	no	yes	—
Width	10^{108}	10^{49}	$\mathbf{10^{65}}$	10^{46}	10^{74}	10^{48}	10^{74}	10^{48}	10^{78}	10^{49}	10^{71}	$\mathbf{10^{45}}$	10^{40}
LP relax (%)	8.0	25.1	**18.9**	24.5	7.9	25.5	8.1	26.1	6.7	24.9	17.7	**35.7**	24.7
Min time (s)	25	1	1	1	1	1	1	1	1	1	1	1	1
Min param	(1)	div.	div.	div.	div.	div.	div.	div.	div.	div.	div.	div.	div.
Max time (s)	tilim	11	9	2	2	2	2	3	4	7	2	3	1
Max param	div.	(2)	(2)	(2)	(2)	div.	div.	div.	(2)	(2)	(2)	(2)	div.
(1): strong branching, emphasize optimization													
(2): no additional inequalities, emphasize integer feasibility													

Table 1 shows that on our smallest instance, we may use almost every integral cycle basis in formulation (6b). Only if we put the arcs with largest spans into a spanning tree, we really get a significantly worse problem formulation. However, there are parameter settings for which even this formulation can be solved. In particular, strong branching and an emphasis on optimization are a good choice, after we add additional valid inequalities ((7) and (10)) by iterated calls to separation heuristics. These push the LP relaxation up to 67.8% of the optimal value. For any of the other formulations, the longest solution times are attained when we do not add additional valid inequalities, do not activate strong branching, but put an emphasis on integer feasibility.

But switching to the node-oriented formulation (2) or to the arc-oriented formulation (4), the picture changes completely. Table 2 shows very impressively that neither formulation (2) nor formulation (4) are able to attain a solution behaviour

Table 2. Solution Times on U Berlin for Formulations (2) and (4)

Formulation	(2)	(2)	(2a)	(4)	(4)	(4)
Valid inequalities	none	(7)	none	none	(7)	all
LP relaxation (%)	0		0	0	82.6	87.7
Min time (s)	295	**146**	7190	(22%)	(90%)	1538
default time (s)	181	2155	3329	(28%)	(86%)	340

which would be competitive to reasonable formulations in terms of integral cycle bases (6b), as they can be found in Table 1. Although after at most 90 minutes an optimal solution is found with formulation (4) even when no cuts are added, the lower bound is less than 30% when the memory limit of 512 MB has been reached. We may only summarize that among these formulations, the node-oriented variant (2) behaves *least bad*, and it profits from the addition of cycle inequalities in their pure form (7).

As in some spot tests on instance ICE small we observed a similar behaviour, we do not follow these alternative formulations in our further considerations.

Local Search Procedures: For evaluating the genetic algorithm on U Berlin, we start a number of test runs with different parameter settings for the population size and the mutation intensity.

Consider Fig. 1. Every function plot represents the cost function by the runtime, averaged over 30 single runs of the genetic algorithm on a certain parameter set. For every single run the cost function of the best individual and the run time is taken after every generation. The run times after every generation are averaged among the 30 runs to get the x-value. The cost function as the y-value is not the average, but the median of the corresponding values. See the discussion below.

For every used parameter setting, the cost function reaches 12% above the optimum within the first 40 seconds, and on settings with small mutation intensity and bigger population size even faster. On the two settings (pop 100, mut 0) and (pop 50, mut 1) a de facto optimum (about 0.02% above the optimum) is reached after 64 and 84 seconds, respectively. The other settings perform worse. While (pop 2, mut 49) stays at about 9.2% above the optimum within the given runtime, the remaining two settings reach about 1.2% above the optimum.

We conclude that a small mutation intensity in connection with a large population size performs best on this data set. De facto optimal solutions can be obtained on those settings with high probability in a short runtime.

Since the feasibility of a solution is relaxed as a part of a cost function, finding a feasible solution can not be guaranteed. While among all of the above runs only one does not satisfy all technical constraints, a small number of solutions with violated service constraints can be found in almost every test run. Those infeasibilities are penalized with a high cost value (577% of the optimum for every infeasibility) and hence have a strong influence on an averaged cost value. Since one infeasible solution more or less in each test run changes the average cost dramatically, those averaged values have no significance. Using the median instead solves that problem.

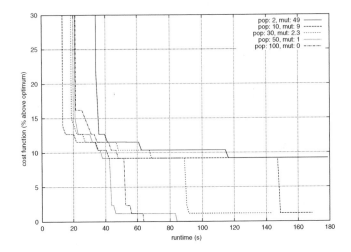

Fig. 1. Runtime Behaviour of the Genetic Algorithm on U Berlin (Every Plot is Averaged over 30 Single Runs)

Constraint Programming: On U Berlin, also the third class of algorithms is able to construct a (de facto) minimal solution. For the strengthened instance, the ILOG Solver does not exceed the time limit[6] of one hour and thus provided an optimality proof.

Table 3 shows that our heuristic of tightening ten heavy constraints supports the work of the solver considerably. However, notice that after the strengthening operation has been applied, the optimal solution value increases slightly, from 1732571 to 1732708. The results on this particular instance suggest that the combination of the "look ahead" (LA) propagation strategy with the selection of the variable with minimum current domain is a good choice.

Table 3. Solution Times on U Berlin for Constraint Programming

Strengthening	no			yes		
Propagation	LA	LA	FC	LA	LA	FC
Variable Selection	default	MinDom	MinDom	default	MinDom	MinDom
First solution (s)	< 1	< 1	–	< 1	< 1	224
First solution (%)	116.1%	125.5%	–	101.2%	100.1%	100.1%
Best solution (s)	1745	889	–	21	< 1	230
Best solution (%)	100.0%	110.5%	–	100.0%	100.0%	100.0%
Total time	tilim	tilim	tilim	603	**172**	1603

---

[6] An entry "tilim" in our tables indicates that the corresponding algorithm has been interrupted after the time limit had been reached.

If we do not expect the constraint programming algorithm to terminate with an optimality proof, then on our smallest instance there exist parameter settings such that it is really competitive to the other algorithms – even though optimization is conceptually out of scope for constraint programming.

Summary: On U Berlin, any of the algorithms is able to construct an optimal solution. With respect to both computation time and the ability to provide a proof of optimality, it is by far the best choice to solve a MIP in the cycle formulation (6), where almost every cycle basis can be used.

5.2 Solving ICE small

The data sets ICE small and ICE big share the same basic network. In particular, ICE small is a subset of ICE big, resulting from the deletion of certain traffic lines. In turn, the lines contained in ICE big are a subset of a strategic planning scenario of Deutsche Bahn AG. Beyond the 31 pairs of directed two-hourly traffic lines which are contained in ICE big, it consists of seven additional pairs of two-hourly lines, as well as several four-hourly variants. Hence, ICE small and ICE big share large parts of their structure. Thus, we give the classification numbers for both data sets together at this point. However, since the underlying infrastructure has the same capacity for the two scenarios, it shall be easier to construct a feasible timetable for ICE small than for ICE big. ICE small is designed such that most parameter settings for CPLEX yield a provably optimal solution within a reasonable time limit. In contrast, ICE big is designed such that even with the best parameter combinations that we investigate, CPLEX will not be able to prove optimality of a solution. However, it should be noted that even this data set is not yet a complete practical scenario.

The real-world instances are described in Table 4. Notice that two lines, which shall be synchronized to a frequency of $\frac{T}{2}$ are synchronized explicitly at *every* station, where an extension of minimal stopping time is allowed. Thus, there are still some lines in ICE small which are not synchronized with any other line.

We obtain our data by some train network planning and analysis software. Naturally, there are many redundancies in the resulting digraph associated with the PESP instance. These can be eliminated in a preprocessing phase that "contracts" the graph. For example, nodes with degree at most one as well as arcs with span equal to zero can be contracted. Table 5 describes the effect of this contraction step for the digraphs. Let us mention that the size of the initial digraphs essentially depends on how safety arcs are generated. They are needed to ensure a safety distance between two consecutive trains. If two trains share five consecutive tracks, this could be translated into five safety arcs. However, our preprocessing method only creates one single safety arc in this case.

Compared to the timetab-instances (Liebchen and Möhring (2003)) of the latest MIPLIB, it might seem that already ICE small has a complexity comparable to the bigger instance timetab2. However, it appears that CPLEX has even less difficulties in solving ICE small than in solving the smaller MIPLIB instance timetab1.

Table 4. Classification Numbers of the Real-world Problems

Quantity	ICE small	ICE big
Pairs of traffic lines	11	31
Change activities	30	101
Stopping activities with extension of minimal stopping time allowed	80	164
Number of pairs of directed lines synchronized to a frequency of $\frac{T}{2}$	40	56
Number of sets of four lines synchronized to a frequency of $\frac{T}{4}$	8	8
Number of pairs of lines coupled on some track	2	8
Turnover activities	22	62

Table 5. Classification Numbers of the Digraphs

	Quantity	ICE small	ICE big
Original Digraph	Nodes	6592	14516
	Arcs	7571	17836
	Run/stop arcs	6570	14454
	safety arcs	488	1660
Contracted Digraph	Nodes	69	173
	Arcs	347	1234
	– with $d_{ij} = T - 1$	43	132
	– with $d_{ij} \geq 0.9 \cdot T$	256	1016
	– with $d_{ij} \leq 0.1 \cdot T$	59	137
	average span	76.7%	84.2%

We suppose that this is due to the fact that in ICE small there are much fewer change activities and turnover activities than in timetab1. Since these are typically the only arcs with non-negative objective value – apart from stopping activities – this might be a significant simplification for CPLEX. Nevertheless, the instance ICE big is apparently at least as difficult to solve for CPLEX as timetab2, for which so far no solution has been proven to be optimal.

MIP Formulations: The instance ICE small poses more difficulties even to the cycle basis formulation (6b). Hence, we have to analyze the influence of the three main ingredients for CPLEX:

- Which cycle basis shall we use?
- Which and how many valid inequalities shall we add to the problem formulation?
- Which parameter settings shall we select for CPLEX?

Obviously, it is not reasonable to consider combinations of *each* possible choice for the above settings. Hence, we decided to proceed as follows.

First, we compute the width (9) of the 13 integral cycle bases we consider throughout this paper, as well as the objective values of their LP relaxations. In order to get a more precise feeling for the different cycle bases, we add to any of the formulations fixed sets of change cycle inequalities (10) in their original formulation. For every cycle basis, we solve the original formulation as well as the refined ones. Next, we focus on the types of valid inequalities to add. To the three most promising cycle bases, we add up to 1000 valid inequalities in any combination of the available types, in order to obtain the largest lower bounds. Then, we investigate how many valid inequalities are necessary, again to get very good lower bounds. We perform these tests with three different parameter sets for the cutting plane pool and for the 13 integral cycle bases. Finally, we ran CPLEX with different values for its MIP emphasis, its variable selection strategy, and its strategies for cuts, both user cuts and CPLEX MIP cuts (ILOG SA (2004)). These experiments are performed for the cycle bases with smallest search space, shortest solution times in the previous cycle basis test, and for the cycle bases with biggest lower bound after the previous phase.

Which cycle basis? We start by computing the integral cycle bases for any of the heuristics that we mentioned in Section 4. Furthermore, we ran our cutting plane algorithm, in order to detect good sets of valid change cycle inequalities (10), i.e. sets which induce big lower bounds. This is performed nine times each for different sizes of the cutting plane pool.

The overall best set of change cycle inequalities has cardinality 243. Besides this, we considered the best sets of change cycle inequalities having 100 and 200 cuts, respectively. Notice that we construct these sets such that every valid inequality is tight for the LP relaxation.

We add these three fixed sets of valid inequalities – as well as the empty set – to formulation (6b), for each of the 13 integral cycle bases. These formulations are solved by CPLEX with strong branching as a variable selection strategy and with a time limit of 2.5 hours. Notice that we add the three non-empty fixed sets of valid inequalities as pure constraints, as well as user cuts. Hence, for each of the 13 cycle bases, we perform seven runs of the MIP solver.

Table 6 shows that only for the cycle bases induced by a minimal spanning tree subject to the arcs' spans, and for a minimal cycle basis, CPLEX is able to solve ICE small to optimality for *any* of the seven settings for valid inequalities. Apart from these cycle bases, CPLEX is only able to solve the UV formulation to optimality, if we turned off the fundamental improvements to spanning trees. Notice that this cycle basis has smallest width among the strictly fundamental cycle bases, but implies only a very poor LP relaxation.

After having applied the fundamental improvement heuristic, for every such cycle basis there is a parameter setting such that CPLEX is able to solve that formulation to optimality. In most cases, the quickest solution times are attained by adding our best set of valid inequalities as pure constraints to the original formulation.

Notice that the pure MIP formulation, i.e., without any valid inequality added, is only solved for those cycle bases which are solved for *any* set of additional inequalities. Moreover, in all of these three cases, the solution time for the pure formulation

Table 6. Solution Times on ICE small for Cycle Bases and Valid Inequalities

Tree	MST		MST		UV		NT		UV		UV		Hort
Weight	nspan		span		one		one		span		nspan		span
Fund. improve	no	yes	no	yes	no	yes	no	yes	no	yes	no	yes	—
Width	10^{178}	10^{71}	10^{122}	10^{73}	$\mathbf{10^{109}}$	$\mathbf{10^{70}}$	10^{117}	10^{72}	10^{112}	10^{72}	10^{132}	10^{71}	10^{67}
LP relax (%)	5.3	23.3	2.1	26.3	1.5	**43.5**	5.2	32.1	1.4	10.0	4.2	19.2	37.9
# opt	0	2	**7**	**7**	1	2	0	1	0	2	0	2	**7**
Min time (s)	tilim	880	258	178	4748	697	tilim	5831	tilim	1355	tilim	365	**161**
Min cuts	–	(0)	(0)	(0)	(0)	(1)	(2)	(0)	(2)	(1)	(1)	(0)	(0)
(0): best 243 change cycle inequalities as additional rows													
(1): best 243 change cycle inequalities as user cuts													
(2): 100 change cycle inequalities as additional rows													

is longer than those for the formulations with 200 or our best set of 243 valid inequalities added.

Which cuts? To analyze which types of cuts contribute a sufficient benefit to formulation (6b), we run the cutting plane algorithm for three very promising integral cycle bases: MST span with and without fundamental improvements, because in the previous step, each of the seven runs was successful; and, UV one with fundamental improvements, because this yields the best LP relaxation.

For these three cycle bases and any combination of classes of valid inequalities (7), (7a), and (10), we launched the cutting plane algorithm nine times. In any of these runs, we held up to 1000 valid inequalities in the pool. In every iteration up to 100 inequalities could be added, and after every iteration, weak cuts are deleted, if the cutting plane pool is full. Table 7 presents average and extremal values for the lower bounds of the refined LP relaxations, when only one type of cut is added. Notice that we only add cycle inequalities in their original formulation (7) to strictly fundamental cycle bases.

Table 7. Lower Bounds on ICE small for Classes of Valid Inequalities

Tree	MST span			MST span		UV one	
Fund. improve	no			yes		yes	
Cuts	(7)	(7a)	(10)	(7a)	(10)	(7a)	(10)
Minimum (%)	44.4	50.8	43.8	55.6	72.6	55.6	80.7
Average (%)	49.4	57.9	55.1	59.3	73.0	61.1	**81.5**
Maximum (%)	60.0	66.0	57.6	66.1	73.2	68.6	82.1

Whereas for different cycle bases the lower bounds do not differ much for cycle inequalities (7a), the lower bounds attained by change cycle inequalities(10) essentially depend on the cycle bases: The better the LP relaxation of the cycle basis, the better the LP relaxation after adding cuts (10).

Our explanation for these phenomena is the following. On the one hand, the initial LP relaxation is only different from zero because we add box constraints of type (7) on the integer variables. As the box constraints are only a fixed number of constraints, it is very important to select a very good set of cycles to contribute their cycle inequalities, i.e., to select a very short cycle basis, in order to obtain a big objective value for the initial LP relaxation. But if we are free to add to the problem formulation any other cycle inequality that we are able to separate, there is no more need to have chosen the *best* cycles already for the cycle basis. Hence, it is plausible that although the initial LP relaxations of the three cycle bases differed much, after adding further cycle inequalities (7a), similar lower bounds are attained.

On the other hand, adding change cycle inequalities (10) provides completely new information to the problem, since these inequalities can be considered to be complementary to cycle inequalities (Liebchen and Peeters (2002a)). Roughly speaking, the headstart of short cycle bases is kept when adding change cycle inequalities.

In Table 8, we consider combinations of types of valid inequalities to be added. One can observe that the best lower bounds are achieved, when at most the cycle inequalities in their original formulation (7) are excluded. Moreover, the levels of the final LP relaxations approach each other.

Table 8. Lower Bounds on ICE small for Combinations of Classes of Inequalities

Tree Fund. improve Cuts	MST span no not (7)	not (7a)	not (10)	all	MST span yes not (7)	UV one yes not (7)
Minimum (%)	84.0	80.4	43.9	79.4	83.6	85.4
Average (%)	84.8	83.1	57.3	84.3	84.2	**86.6**
Maximum (%)	89.3	84.6	66.4	88.2	85.0	89.1

Finally, it is interesting that if we omit change cycle inequalities (10), then it makes no big difference, whether we add only the tension formulation (7a) of the cycle inequalities, or their original counterpart as well. However, it is somehow surprising to us that formulation (7a) is slightly – but still significantly – superior to formulation (7).

How many cuts? Now, we want to investigate how many valid inequalities we should separate both in total and in every iteration of the cutting plane algorithm, in order to obtain the best lower bounds. To that end, we ran the cutting plane algorithm for each of the 13 cycle bases we consider, nine times for each of the following parameter settings: We consider pool sizes of 200, 350, 500, 650, and 800 cuts. Moreover, we separated 10 or 100 cuts per iteration. Finally, when adding only 10 cuts per iteration, we (dis-) allow old cuts to be removed from the current LP, if they are no longer tight.

The best results by far are attained by adding up to 100 valid inequalities per iteration, and hence, by removing weak cuts. The best lower bounds are attained with pool sizes of 500 or more, which is approximately twice the number of rows of the initial MIP formulation.

For each of the different cycle bases, their three best runs yield similar lower bounds: The NT cycle basis with fundamental improvements applied achieves the three worst lower bounds (85.0%–85.1%). The UV one cycle basis with fundamental improvements applied as well, leads to three out of the four best lower bounds (89.2%–89.5%). For strictly fundamental cycle bases, the best lower bound is 89.4%, which is attained with the UV nspan basis. But it is somehow interesting that there are ten cycle bases whose best lower bounds are superior to the best lower bound computed with a minimal cycle basis.

Which CPLEX parameters? Based on observations of the previous tests, we are now ready to examine under which parameter settings CPLEX behaves best for periodic timetabling instances. We will perform runs on a minimal cycle basis and on the two bases stemming from a minimal spanning tree subject to the arcs' spans, because CPLEX behaves well on those bases, see Table 6. Furthermore, we consider the improved UV one basis, because it yields the best initial lower bound. Finally, the UV nspan tree is considered, because after adding valid inequalities it allows the best lower bound for a strictly fundamental cycle basis. We add the specific sets of cuts, which lead to the largest lower bounds in our previous experiments.

With a time limit of six hours, we solve ICE small for any of the five cycle bases and any of the 24 combinations for the parameters we analyze, cf. Section 5. There is only one combination, where the memory limit of 512MB applied after 1.5h (basis UV span, user cuts, emphasis on optimization, and aggressive cut generation), and the best solution still has objective value 109% of the minimal solution.

In Table 9, we report the average and the extremal running times for any of the nine fixed parameter values, and the three other parameters take the eight or twelve possible combinations. Furthermore, the number of outliers is given, i.e. the number of runs whose running times fell below/exceeded a 50% radius around the average solution time. These solution times give a first hint that strong branching yields an enormous benefit. Furthermore, it can be observed that user cuts only help for rather long runs of CPLEX. Finally, a MIP emphasis on optimization seems to help.

Table 10 puts another perspective on the 600 computations, leading to another conclusion in particular concerning the last point: We consider the parameter settings which lead to the three shortest solution times for the five cycle bases. Here, one can see that only for MST span a MIP emphasis on optimization entered the best three runs. Rather, a combination of strong branching together with a MIP emphasis on integer feasibility provides one of the three shortest solution times for *any* of the five cycle bases which we consider here.

Although we do not put a focus on quickly finding (good) first feasible solutions, let us present the best results of the 120 computations. In only eight of them, the first feasible solution is found after less than one second of CPU time[7]. Both the quickest and the best first solution are attained using a minimal cycle basis: With user cuts ac-

---

[7] We multiply the total solution time with the ratio of the first feasible B&B node divided by the total number of B&B nodes investigated.

Table 9. Solution Times on ICE small for Various CPLEX Parameter Settings

Tree	MST span		UV one	UV nspan	Horton
Fund. improve	no	yes	yes	no	–
LP relax (%)	88.4	87.6	89.5	89.3	86.6
Additional valid inequalities as pure rows					
Min (s)	63	86	83	482	68
Average (s)	**249**	**907**	**3262**	**4184**	**365**
Max (s)	607	3849	21600	21600	987
# Outliers	5/3	7/3	8/3	6/1	4/3
Additional valid inequalities as user cuts					
Min (s)	36	83	106	578	82
Average (s)	**316**	**1097**	**2988**	**3626**	**1115**
Max (s)	972	3658	12652	10984	6620
# Outliers	4/2	7/4	6/3	4/4	8/3
Default variable selection strategy					
Min (s)	36	134	1126	1306	199
Average (s)	**353**	**1859**	**6009**	**5925**	**1340**
Max (s)	972	3849	21600	21600	6620
# Outliers	3/3	4/4	5/2	4/2	7/2
Strong branching variable selection strategy					
Min (s)	63	83	83	482	68
Average (s)	**211**	**144**	**241**	**1884**	**141**
Max (s)	471	303	437	5598	308
# Outliers	4/2	0/2	2/3	6/3	1/1
Default MIP cut generation					
Min (s)	36	83	83	482	68
Average (s)	**120**	**845**	**3305**	**1503**	**493**
Max (s)	272	3849	12652	4115	2612
# Outliers	1/3	8/3	7/4	6/3	7/2
Aggressive MIP cut generation					
Min (s)	199	94	151	985	82
Average (s)	**444**	**1159**	**2945**	**6306**	**987**
Max (s)	972	3658	21600	21600	6620
# Outliers	1/2	7/3	7/1	3/2	8/2
Default MIP emphasis					
Min (s)	90	93	139	482	84
Average (s)	**327**	**973**	**1883**	**5938**	**1363**
Max (s)	972	3658	7593	21600	6620
# Outliers	3/2	6/2	4/2	3/2	5/2
Integer feasibility MIP emphasis					
Min (s)	63	83	83	486	68
Average (s)	**271**	**1275**	**5466**	**2510**	**564**
Max (s)	837	3849	21600	6614	2000
# Outliers	2/1	4/2	5/2	3/2	3/2
Optimization MIP emphasis					
Min (s)	36	106	196	633	127
Average (s)	**249**	**758**	**2026**	**3266**	**292**
Max (s)	471	2861	6166	6132	786
# Outliers	3/3	4/1	4/2	2/3	2/1

tivated, and the other parameters as defaults, after 0.33s a solution of objective value 156% is found. With a MIP emphasis on optimization and an aggressive generation of MIP cuts, after 1.22s a solution with value 100.5% is constructed.

MIP Summary. The most definitive result of our study is that it is essential to add valid inequalities to the problem formulation. Here, one should consider both cycle inequalities and change cycle inequalities. It seems to be advantageous to add many valid inequalities in every iteration of the cutting plane algorithm, and then remove such inequalities which are no longer tight in subsequent iterations. Furthermore,

Table 10. Solution Times on ICE small for Best CPLEX Parameter Settings

Tree	MST span						UV one			UV nspan			Horton		
Fund. improve	no			yes			yes			no			–		
Solution time (s)	**36**	63	63	83	86	93	83	106	139	482	486	578	**68**	82	84
User cuts	1	–	–	1	–	1	–	1	1	–	–	1	–	1	1
Strong branching	–	–	1	1	1	1	1	1	1	1	1	1	1	1	1
Aggressive cuts	–	–	–	–	–	–	–	–	–	–	–	–	–	1	1
MIP Emphasis	2	2	1	1	1	–	1	1	–	–	1	1	1	1	–

–: default setting
1: feature activated / MIP emphasis on integer feasibility
2: MIP emphasis on optimality

our computations on ICE small suggest that about twice the number of rows of the initial MIP suffice as additional inequalities.

For the few CPLEX parameters we investigated, we suggest emphatically to use strong branching and to put the MIP emphasis on integer feasibility. Possibly, the positive effect of strong branching can even be intensified by modifying the related CPLEX parameters which control the strong branching limits (ILOG SA (2004)). In any case, aggressive MIP cut generation should only be activated if long running times are expected, in particular if the size of the branch and bound tree has to be limited.

For the choice of the integral cycle basis to use in problem formulation (6b), Table 6 indicates that shorter cycle bases allow shorter solution times, even after having added identical sets of valid inequalities. However, the cycle basis MST span does seem to have something *magic* in it, being resistant against our classification numbers "width of the cycle basis" and "objective value of the LP relaxation," but which has an extremely positive effect on the MIP solver of CPLEX.

Local Search Procedures: In contrast to U Berlin the genetic algorithm has no problems with satisfying constraints on ICE small. Typically within the first three to five generations a feasible solution is found and the solutions stay feasible in subsequent generations. Thus there is no need to use the median when comparing the cost function between test runs.

Fig. 2 shows test runs on five different settings for the population size and the mutation intensity. Every plot represents 20 runs on one parameter set. Within the given runtime of about 15 minutes, the test runs reach an average cost value of 34-38% above the optimum. On a longer test run (without plot) an average cost value of 26% above the optimum is reached after 75 minutes.

This time (again in contrast to U Berlin) no clear result about the best parameter set can be obtained. The apparently best runs are (pop 2, mut 49) and (pop 10, mut 9), while the run (pop 3, mut 33), whose parameter set is "between" the best ones, seems to be worst. But the difference between these plots is small in comparison with the associated standard deviation; see Fig. 3. Thus the plots of Fig. 2 should be considered as being identical.

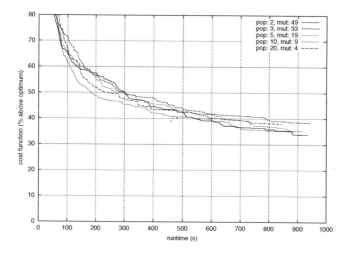

Fig. 2. Runtime Behaviour of the Genetic Algorithm on **ICE small** (Every Plot is Averaged over 20 Single Runs)

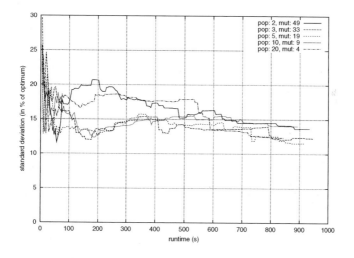

Fig. 3. Standard Deviation to Fig. 2

When dealing with large deviations on randomized algorithms, a promising idea is to start a couple of those algorithms and take the best result as the output of the whole process. In the special case of genetic algorithms, the selection of the best result can be done by collecting the individuals of all runs to a common population, on which a final collecting run is started. In doing so the genetic algorithm has the chance to combine different good solutions to a possibly better one.

We use two different strategies to test this approach:

- **Multiple long GA:** To stay close to the initial idea of just selecting the best solution, the collecting run is kept short. We start five runs of 35 generations each and use a collecting run of only 25 generations.
- **Multiple short GA:** To focus on the aspect of combining solutions in the collecting run, we extend it to 100 generations. In return, the initial runs have to be shortened to 20 generations each.

In both cases we use a large population size in the collecting run to avoid a fast domination of certain individuals while mutation is turned off. Fig. 4 shows the result of those two strategies in comparison with the simple genetic algorithm. For illustrative purpose we only use the best and the worst plot of Fig. 2.

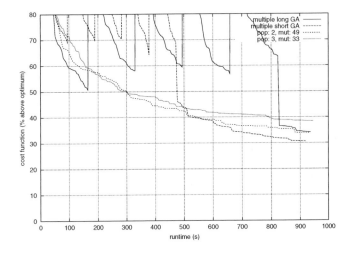

Fig. 4. Runtime Behaviour of Multiple Genetic Algorithms on ICE small (Every Plot is Averaged over 20 Single Runs)

Both plots of the multiple genetic algorithms show a sawtooth pattern in their first phase, when the five independent genetic algorithms are performed. In the second phase the cost function drops down to the minimum of the initial runs. Further improvement is made in the second phase by the collecting run. The results generated by the "multiple long GA" strategy are at 34% above the optimum and thus between the two given references, tending towards the better one. The "multiple short GA" reaches, with 31% above the optimum, a better result than the simple genetic algorithms. But due to the high variance on all these results, this should not be interpreted as a clear advantage of this strategy. The standard deviation of "multiple short GA" in the last phase is about 13% of the optimal value, while the standard deviation of "multiple long GA" is still about 8%.

These plots also show another interesting effect: The collecting run of the "multiple short GA" starts with cost values, that are comparable to or even a little worse than those of the reference plots at the same runtime. But in the remaining runtime it is able to improve much faster than the references. A possible explanation for the effect could be that the collecting run gains from combining a number of good but very different solutions. In a regular run of a genetic algorithm the individuals of a population tend to be more and more similar, due to the domination of the best.

On this data set we also try the simulated annealing algorithm. We use the geometric cooling schedule with an initial acceptance ratio of 40%, a stop acceptance ratio of 0.1%, and with at least six levels after the last improvement. But the results are rather bad. Within a runtime of 60 minutes we reach only an average cost value of 68.3% above the optimum (averaged over 30 single runs). The standard deviation is, with 23.7% of the optimum, even higher than at the genetic algorithms.

For this data set we conclude that all used approaches of the genetic algorithm can solve the problem to an average cost value of 30-40% above the optimum within a runtime of 15 minutes. Better values can be achieved on longer runs. Simulated annealing performs much worse than the genetic algorithm.

Constraint Programming: Unfortunately, on ICE small our heuristic of strengthening some constraints makes the problem infeasible. However, for the original formulation the first feasible solution is found in less than half a second on a SUN Ultra-SPARC-IIi with 333 MHz. After one minute, the best objective value is attained by standard variable selection combined with the look ahead propagation rule (202.1%). Six hours later, this value has only been reduced to 200.7%. Here, choosing the variable with minimal domain behaves slightly better (196.8%), although after 60s it has only an objective value of 226.2%. Summarizing, the time needed to construct a feasible solution – which is the original application of constraint programming – is indeed fully competitive to CPLEX, and much superior to local search procedures. Anyway, for minimizing a linear objective over a PESP instance, the constraint programming approach does not seem to help much.

Summary: On ICE small, only CPLEX is able to construct a solution of minimal objective value. With appropriate parameter settings, this can even be obtained in only 36 seconds. Nevertheless, with other parameter values, CPLEX does not find an optimum solution within six hours.

The genetic algorithm takes 10-15 minutes to find a solution with objective value about 30% above the optimum. Even after one hour, our simulated annealing algorithm is 68% above the optimum. Only constraint programming behaves worst: more than 90% above the optimum after six hours.

5.3 Solving ICE big

MIP Formulations: Since solving the much bigger instance ICE big will yield much longer solution times, we will concentrate on the smallest (generalized) fundamental cycle basis (Horton) for this instance, and on the smallest strictly fundamental cycle basis (MST span). Moreover, we are no longer able to vary the parameters of the cutting plane algorithm. Rather, based on our findings on ICE small, we will always add up to 2000 valid (change) cycle inequalities. Under these fixed settings, we will analyze the impact of the CPLEX parameters, where we omit the value "optimization" for the parameter MIP emphasis.

The first observation is that out of the eight following parameter combinations, only in one case CPLEX is able to construct a feasible solution with a minimal cycle basis chosen in the most promising problem formulation (6b):

- all parameters at their default values
- precisely one parameter at its non-default value; MIP emphasis on integer feasibility, strong branching (after 42888s, first feasible solution with value 1075421)
- precisely two parameters at their non-default values; MIP emphasis on integer feasibility and user cuts, MIP emphasis on integer feasibility and strong branching
- precisely one parameter at its default value; no aggressive MIP cut generation, no user cuts
- all four parameters at their non-default values.

Nevertheless, the best lower bounds are achieved with a minimal cycle basis. In runs where strong branching is not activated, the memory limit of 512 MB is reached after between six and sixteen hours. Otherwise, the time limit of 48 hours applied. With all four parameters at their non-default values, the value 735385 is proven as a lower bound. Notice that if the 741 user cuts are added as pure valid inequalities, then the LP relaxation has an optimal value of 654906, and if no cuts are added, the LP relaxation already yields 383074.

Fortunately, with the cycle basis MST span, CPLEX is able to construct feasible periodic timetables for ICE big very reliably. More precisely, in *any* of the eight parameter combinations we investigate, a feasible solution is found within a time limit of 24 hours, cf. Table 11. Complementing the analysis of CPLEX on ICE big, we give a more detailed impression of the solution process leading to the best timetable in Fig. 5. There it can be seen that the optimal value of the LP relaxation with cuts refined is 584692. What cannot be seen is that without the 1332 valid inequalities added, a lower bound of only 59432 can be achieved.

Table 11. Performance of CPLEX Computing 24 Hours on ICE big

User cuts	–	1	–	–	–	–	–	1
Strong branching	–	–	1	–	–	1	1	1
Aggressive cuts	–	–	–	1	–	–	1	1
MIP Emphasis	–	–	–	–	1	1	1	1
First solution (s)	295	515	230	3074	782	49	7743	455
First solution value	1342529	1583975	1142024	1630884	2021758	1030613	1480226	1567532
Best solution (s)	295	515	74817	16613	11769	65207	73658	35234
Best solution value	1342529	1583975	1057918	1445637	1317983	934630	**922262**	977034
% above best solution	45.6%	71.7%	14.7%	56.7%	42.9%	1.3%	0.0%	5.9%
Final Lower bound	667887	605373	700002	666708	604029	697970	696135	**708796**
% below best solution	27.6%	34.4%	24.1%	27.7%	34.5%	24.3%	24.5%	23.1%
–: default setting	1: feature activated							

Summarizing, even with the most promising parameter settings, CPLEX is not able to terminate with an optimality proof for ICE big. Although a minimal cycle

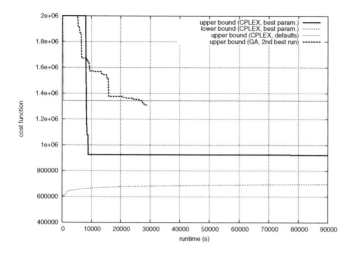

Fig. 5. Performance of CPLEX and of the Genetic Algorithm on ICE big

basis yields the second best solution times on ICE small, there are obviously only few parameter combinations for CPLEX to detect a feasible solution on ICE big with such cycle bases.

Rather, one should choose the MST span cycle basis. Moreover, it is very important to choose strong branching as the variable selection strategy, because otherwise the quality of the solution is much worse, in our examples by at least 25%. Similar to ICE small, the best behaviour can be seen when (at least) strong branching and an emphasis on integer feasibility are combined.

Local Search Procedures: On ICE big it seems to be difficult again to produce feasible solutions. On both test runs we start, one out of ten single runs is not able to find a feasible solution within the given runtime of about 8 hours. While most of the runs found their first feasible solution within the first 20 minutes, it took some others more than 2 hours. Hence, we use the median again for the analysis. Since we do not know the optimal cost value of ICE big, we measure the cost function in % above the upper bound, i.e., the best known solution.

Consider Fig. 5. It shows the median of the two test runs we made. Again we vary the population size and mutation intensity to get almost the same runtime per generation. Both plots reach a cost value of 60% above the upper bound within the first 50 minutes. During the remaining runtime both make further improvements and reach 33.97% (pop 30, mut 4) and 35.65% (pop 50, mut 2). If we ignore the infeasible run in every test run, the standard deviation is, with about 5% of the upper bound, much smaller than on ICE small. Over this background, we see a small advantage of (pop 30, mut 4), whose plot is below the one of (pop 50, mut 2) during the whole runtime. But this advantage vanishes towards the end of the runtime.

Constraint Programming: A really interesting fact about constraint programming is that even on the largest instance, it takes less than half a second to construct a first feasible solution. Only if we choose standard variable selection in combination with look ahead propagation, no solution is found even after six hours. Nevertheless, comparing these times to the ones achieved by CPLEX, recall that we do not really tune CPLEX in order to quickly find some first feasible solution. Furthermore, the quality of the solutions is rather poor. Selecting the variable with minimal domain and performing forward checking, after 60s a solution with objective value 2007630 is available. In the next six hours, this decreases only down to 1989110.

After all, our heuristic of strengthening some constraints in advance provides significantly better solutions for the same CP strategies: after one minute, we already obtain 1795830. But the improvements attained during the next six hours are again only marginal (1755060). In total, CP solutions are already considerably worse than feasible solutions obtained by both our genetic algorithm and CPLEX only with its standard parameter settings.

Summary: Also for ICE big, CPLEX computes the best solutions. But here, we were not able to terminate with a proof of optimality within one day. The best solutions were achieved with the cycle basis MST span and the parameter strong branching activated. But notice that depending on the values of the other parameters it may take more than two hours until CPLEX finds the first feasible solution.

Much similar to ICE small, the genetic algorithm misses the best solution of CPLEX by about 30%. Also, constraint programming keeps its gap of 90%. Notice that the similarity between the values on ICE big and on ICE small could be caused by the similar structure of these two data sets, cf. Section 5.2.

6 Conclusion

In Table 12 we provide a rough summary of our computational study. The entries are to be read as follows. The row "Quality" indicates the quality of the best solution that we obtained with a specific algorithm on a particular instance, having tried various parameter settings. The row "Time" represents the time that was necessary to obtain the best solution, where an entry ++ stands for the shortest solution times. Finally, if there exist (reasonable) parameter settings that cause an algorithm to produce solutions that are significantly worse than the best solution it is able to attain with other settings, this is indicated by a minus sign.

Due to the immense differences between the three data sets that were available to us, the entries do not follow general thresholds. Rather, they represent the performance relatively to the other algorithms on the very same data set. A minus entry in the row "Quality" is a knockout criterion for an algorithm. Also, a minus entry in the row "Time" prevented us from elaborating this algorithm on larger instances. Notice that there always exist parameter settings such that CPLEX computes the best solutions within a relatively small amount of time. Nevertheless, even on ICE small and ICE big the compositions of these optimal parameter sets do not coincide. Hence, we are not able to elect the best general purpose periodic railway timetabling algorithm.

Overall we can state that, given the current state of methods and machines, it is possible to calculate the timetable for the complete (long distance) network of one

Table 12. Overall Performance of Five Solution Techniques for PESP Instances

Algorithm	MIP (CPLEX)					Genetic Alg.			Sim. Ann.	CP (ILOG Solv.)		
	formul. (6b) + cuts			other								
Data	U Bln	ICE s.	ICE big	U Bln	ICE s.	U Bln	ICE s.	ICE big	ICE s.	U Bln	ICE s.	ICE big
Quality	++	++	++	++	++	++	+	+	−	++	−	−
Time	++	++	o	+	−	+	+	+	−	+	++	++
Independence of parameters	+	−	−−	−−	−−	+	+	+	+	−	+	+

of the largest railways in a very satisfying way, with respect to the production time and to the quality of the results. On the one hand, the comparison of various methods that we report in this paper was the basis for selecting the genetic algorithm as the method of choice for the Deutsche Bahn. The genetic algorithm turned out to be the most stable solution procedure, although the others are serious competitors. Depending on further developments this picture can change. On the other hand, we think that this comparison is an important and helpful step towards really understanding the timetabling problem. This is an ongoing process, so this is a report on work-in-progress.

Acknowledgement: This work has been supported by the DFG Research Center "Mathematics for key technologies" in Berlin.

References

Barták, R. (1999).   Constraint programming: A survey of solving technology. *AIRONews journal IV*, **4**, 7–11.

Berger, F. (2002).   Minimale Kreisbasen in Graphen. Technical report, Lecture on the annual meeting of the DMV, Halle.

Bixby, B. (2003). Personal communication. Rice University.

Bussieck, M. R., Winter, T., and Zimmermann, U. (1997).   Discrete optimization in public rail transport. *Mathematical Programming (Series B)*, **79**, 415–444.

Deo, N., Prabhu, M., and Krishnamoorthy, M. S. (1982).   Algorithms for generating fundamental cycles in a graph. *ACM Transactions on Mathematical Software*, **8**, 26–42.

Deo, N., Kumar, N., and Parsons, J. (1995).   Minimum-length fundamental-cycle set problem: A new heuristic and an SIMD implementation. Technical report CS-TR-95-04. University of Central Florida.

Emden-Weinert, T. and Proksch, M. (1999).   Best practice simulated annealing for the airline crew scheduling problem. *Journal of Heuristics*, **5**, 419–436.

Horton, J. D. (1987).   A polynomial-time algorithm to find the shortest cycle basis of a graph. *SIAM Journal on Computing*, **16**, 358–366.

ILOG SA (2004). CPLEX 8.1. http://www.ilog.com/products/cplex.

Laube, J. (2004). Taktfahrplanoptimierung mit Constraint Programming, diploma thesis, in German.

Liebchen, C. (2003). Finding short integral cycle bases for cyclic timetabling. In G. D. Battista and U. Zwick, editors, *Algorithms-ESA 2003, Lecture Notes in Computer Science 2832*, pages 715–726. Springer.

Liebchen, C. and Möhring, R. H. (2003). Information on MIPLIB's timetab- instances. Technical report 049/2003, TU Berlin.

Liebchen, C. and Möhring, R. H. (2007). The modeling power of the periodic event scheduling problem: Railway timetables – and beyond. This volume.

Liebchen, C. and Peeters, L. (2002a). On cyclic timetabling and cycles in graphs. Technical report 761/2002, TU Berlin.

Liebchen, C. and Peeters, L. (2002b). Some practical aspects of periodic timetabling. In P. Chamoni, R. Leisten, A. Martin, J. Minnemann, and H. Stadtler, editors, *Operations Research Proceedings 2001*, pages 25–32. Springer, Berlin.

Lindner, T. (2000). *Train Schedule Optimization in Public Transport*. Ph.D. thesis, TU Braunschweig.

Mühlenbein, H. (1997). Genetic algorithms. In E. H. L. Aarts and J. K. Lenstra, editors, *Local Search in Combinatorial Optimization*, pages 137–171. John Wiley & Sons.

Nachtigall, K. (1994). A branch and cut approach for periodic network program- ming. Hildesheimer Informatik-Berichte 29.

Nachtigall, K. (1996). Cutting planes for a polyhedron associated with a periodic network. Technical report, DLR Interner Bericht 17.

Nachtigall, K. (1998). Periodic network optimization and fixed interval timetables, habilitation thesis.

Nachtigall, K. and Voget, S. (1996). A genetic algorithm approach to periodic rail- way synchronization. *Computers & Operations Research*, **23**, 453–463.

Odijk, M. (1997). *Railway Timetable Generation*. Ph.D. thesis, TU Delft.

Proksch, M. (1997). Simulated Annealing und seine Anwendung auf das Crew- Scheduling-Problem, Diploma thesis, in German.

Schrijver, A. and Steenbeek, A. (1993). Dienstregelingontwikkeling voor Neder- landse Spoorwegen N.V. Rapport Fase 1, in Dutch. Technical report, Centrum voor Wiskunde en Informatica.

Serafini, P. and Ukovich, W. (1989). A mathematical model for periodic scheduling problems. *SIAM Journal on Discrete Mathematics*, **2**, 550–581.

Mixed-Fleet Ferry Routing and Scheduling

Z.W. Wang, Hong K. Lo, and M.F. Lai

Department of Civil Engineering, The Hong Kong University of Science and Technology, Clear Water Bay, Hong Kong, P.R.C, cehklo@ust.hk

Summary. This study formulates a mixed-fleet ferry routing and scheduling model while considering passengers' choices for differential services. Ferry services with different operation characteristics and passengers with different preferred arrival time-windows are considered in the model. The logit model is applied to determine passengers' service choices. The formulation then determines the best mixed-fleet operating strategy, including interlining schemes, so as to minimize the objective function that combines both the operator and passengers' performance measures. Mathematically, this mixed-fleet routing and scheduling problem is formulated as a mixed integer nonlinear program. This study then develops an iterative heuristic algorithm to solve this problem. The results show that the algorithm could improve the operations of the system given different initial points. Nevertheless, finding the global optimal solution could be difficult due to the inherent non-convex nature of the problem.

1 Introduction

Ferry services in Hong Kong are supplementary for cross-harbor traffic but essential for the outlying islands. The government plays an important role in the provision of these services by ensuring a financially viable environment to entice private sector participation and hence avoid subsidizing their operations. The current practice of the Hong Kong government is to bundle ferry services into packages, with each of them operated by a different company. It is then up to the operator of each package to determine the service schedules, interlining strategies, ferry types (fast and ordinary), and fleet size, so as to maximize their overall profit by providing services that are acceptable to passengers.

The problem addressed in this study can be considered as a service network design problem, which involves determining the service network and its passenger flows simultaneously, so as to achieve a certain objective. Magnanti and Wong (1984) first formulated this problem as a mixed integer linear program. Crainic and Laporte (1997) and Crainic (2000) presented state-of-the-art reviews on this topic. Indeed, this problem finds applications in many contexts. For example, Lai and Lo (2004) developed a single ferry fleet management model. Yan and Chen (2002) studied the scheduling of inter-city bus carriers, and Yan and Tseng (2002) developed a multi-fleet airline routing and scheduling model. However, none of the previous service network design problems considered passenger preferences for differential services.

Lai and Lo (2004) developed a ferry fleet management model and accompanied heuristic algorithm to optimize the fleet size, ferry routing, and service schedules. The model is formulated as a mixed integer multi-commodity network flow problem with a single ferry type. In reality, operators may offer different services, e.g., fast ferry with higher fare versus slow ferry with lower fare, to accommodate the different market segments. As a result, passenger preferences and choices on fare, service quality, and journey time become important factors for planning and coordinating service schedules, routings, and ferry type allocation. This paper aims at developing a multi-fleet ferry routing and scheduling model with mode choice integration. We apply the logit model to determine passengers' mode choice. For the context of ferry services to outlying islands in Hong Kong, in the absence of alternative ground transportation, we consider the total demand for each origin-destination pair as fixed. On the other hand, passenger demand for each particular type of ferry service is driven by its service disutility, including fare, journey and waiting times, subject to the service's capacity constraints. Furthermore, to more accurately reflect reality, travelers are segregated according to their preferred arrival time windows at destinations. Arrival before or after the preferred time windows will incur early or late arrival penalties.

The model developed in this study primarily considers the perspective of the operator, in terms of minimizing the operation costs or maximizing their profits. Nevertheless, due consideration must be given to its service performance according to the perspective of passengers. Poor service performance leads to long-term migration from the outlying islands, causing a drop in demand, or the possibility of losing the franchise of operating the services all together. Both consequences are undesirable to the operators. Therefore, the model combines the operator's as well as users' objectives, as is typically accomplished in transit network design studies (see, e.g., Ceder and Wilson (1986)).

The outline of this study is as follows. Section 2 depicts the model formulation. A heuristic algorithm is developed in Section 3. Section 4 presents the numerical study. Finally, Section 5 provides some concluding remarks.

2 Model Formulation

2.1 Assumptions

(a) *Passenger demand*: As this study focuses on ferry services to the outlying islands, which have limited alternative transportation modes, we assume captive demand. That is, the total demand for each origin-destination (OD) pair is given. In the long run, people might change their residences or job locations, rendering demand elastic. This is not considered in this study. However, even though the total demand for each

OD pair is fixed, the demand for a particular ferry service is elastic, to be determined by its service quality and passengers' choices.

(b) *Arrival time window and linear delay penalty*: Passengers are segregated according to their preferred arrival time windows at destinations. Linear delay penalties are imposed for early and late arrivals outside passengers' preferred arrival time windows.

(c) *Logit-modal split*: The logit model is applied to estimate passenger demands for different ferry services based on their fares, journey times, and delays.

(d) *Transfer is not allowed*: Transfer between different ferry services is not considered in the model as it rarely happens.

(e) *Overnight empty ferry repositioning*: At a specific pier, the number of ferries at berth at the start of the day is not necessarily equal to that at the end of the day. This assumption is justified by the fact that the cost of ferry repositioning at the end of the day is relatively insignificant as compared to the total operation cost.

2.2 Variable Definitions

This service network design problem involves determining both the ferry routing and service schedules for the planning horizon, which requires specifying the time dimension within the formulation. For this purpose, we draw upon the convenience of a time-space network structure, in which each node represents a specific location at a specific time, whereas each arc represents the temporal and spatial connection between the two corresponding nodes. The problem involves the determination of two types of arc variables: (i) ferry arc flows specify the ferry routes and departure schedules, and (ii) passenger arc flows depict the passenger movements given the ferry arc flows. The formulation, therefore, constitutes two types of time-space networks: the ferry flow and passenger flow networks. Each of these networks can be further divided into a group of sub-networks to handle different ferry types and OD demands. The detailed description of the ferry and passenger time-space networks refers to Lai and Lo (2004). The variable notations are defined as follows:

Sets

R	set of OD pairs
F	set of ferry service types
G	set of arrival time-windows
N^f, A^f	sets of nodes and arcs, respectively, in the f ferry flow network (for ferry service type f)
$N^{d,f}, A^{d,f}$	sets of nodes and arcs, respectively, in the d passenger flow network (for demand on OD pair d) associated with ferry service type f, notated as the d–f passenger flow network below
N_b^f, N_e^f	sets of nodes at the beginning and ending of the planning horizon, respectively, in the d–f ferry flow network; subsets of N^f
$N_b^{d,f}, N_e^{d,f}$	sets of nodes at the beginning and ending of the planning horizon, respectively, in the d–f passenger flow network; subsets of $N^{d,f}$

$S^f, S^{d,f}$ sets of service arcs in the f ferry and the d–f passenger flow network, respectively

$W^f, W^{d,f}$ sets of wait arcs in the f ferry and the d–f passenger flow network, respectively

$O^{d,f}$ set of origin arcs in the d–f passenger flow network

$D^{d,f}$ set of destination arcs in the d–f passenger flow network

$D_e^{d,f}, D_l^{d,f}$ sets of destination arcs for arrivals earlier and later than the arrival time-window g in the d–f passenger flow network, respectively; subsets of $D^{d,f}$

$M^{d,f}$ an artificial node in the d–f passenger flow network

Parameters

d an OD pair

f a ferry service type

g an arrival time-window

κ^f fixed cost associated with owning or hiring a ferry of type f for one day

V^f maximum fleet size of f type ferry

$B^{d,g}$ exogenous passenger demands with arrival time-window g on OD pair d

Q^f capacity of f type ferry

T_{ij} travel time between node i and j

$T^{d,f}$ travel time between OD pair d based on direct service of f type ferry

$\overline{T}^{d,f,g}$ average total travel time of passenger on OD pair d and arrival time-window g, utilizing f type ferry service

β_{ij}^g time duration between the time dimension of node i and the arrival time-window g for destination arc (i, j)

$\alpha^{d,f}$ fare of f type ferry service for OD pair d

C_{ij}^f operating cost per trip between node i and j of f type ferry service

$u^{d,f,g}$ utility function segregated into different OD pair d, ferry service type f and arrival time-window g

$\theta_1, \theta_2, \theta_f$ weights of fare, average total travel time and alternative specific constant in the utility function $u^{d,f,g}$, respectively

ω_e, ω_l weights of early arrival delay penalty and late arrival delay penalty

v_t value of travel time

v_w value of waiting time

ξ weight to capture the relative importance of total passenger disutility to operation costs

U_{ij}^f upper bound of the ferry flow between node i and j for f type ferry service (note that U_{ij}^f equals 1 for the service arcs in set S^f and is a positive integer value for wait arcs in set W^f)

Decision Variables

Y_{ij}^f ferry flow (i, j) in the f ferry flow time-space network

$X_{ij}^{d,f,g}$ passenger flow (i, j) in the d–f passenger flow time-space network with preferred arrival time window g

2.3 Network Description

Ferry time-space network: The time-space network of ferry flow, shown schematically in Fig. 1, is defined by a graph $G(N^f, A^f)$, in which f specifies the ferry service type. N^f is the set of nodes in the time-space network, A^f is the set of arcs representing ferry movements. A^f consists of two subsets: service arc set S^f and wait arc set W^f, such that $A^f = S^f \cup W^f$.

Each service arc describes a ferry trip, whose journey time, origin and destination are specified by the corresponding time-space nodes. Arc flow is represented by a binary variable, which equals 1 for a provided service; 0 otherwise. Arc costs encompass operating costs, including fuel, maintenance, and labor. The fixed cost of owning or hiring a ferry per day is imposed on arcs originating from the beginning of the planning horizon. Each wait arc, or vertical arc in the ferry network, indicates ferries idling at a pier without providing service. It is represented by a non-negative integer variable, denoting the number of ferries berthing at a pier. We assume that wait arcs have negligible operating costs.

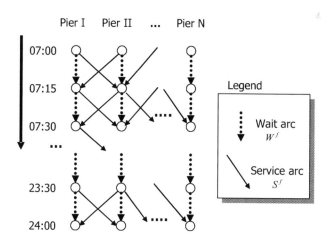

Fig. 1. The Ferry Time-space Network Schematic

Passenger time-space network: The time-space network of passenger flow is defined by a set of graphs $G(N^{d,f}, A^{d,f})$, where d refers to an OD pair, f the ferry service type, $N^{d,f}$ the set of nodes, and $A^{d,f}$ the set of arcs representing passenger movements. Similar to A^f, the set $A^{d,f}$ consists of two subsets, service arc set $S^{d,f}$

and wait arc set $W^{d,f}$ such that $A^{d,f} = S^{d,f} \cup W^{d,f}$. Moreover, associated with every graph $G(N^{d,f}, A^{d,f})$ are one artificial node, $M^{d,f}$, and two types of artificial arcs: origin arc $O^{d,f}$ and destination arcs $D^{d,f}$. Fig. 2 schematically illustrates the passenger time-space network.

Service arcs denote passenger trips between piers, whose journey times are specified by the corresponding nodes of the time-space network. Each arc flow represents the number of onboard passengers, which is constrained by the capacity of the ferry. The flow on the wait arc, on the other hand, describes the number of passengers waiting at the pier, which could be a result of either early arrivals at the pier, or insufficient capacity of the departed ferry to carry all the demand. Similar to wait arcs in the ferry network, passenger wait arcs are represented by vertical arcs.

The passenger flow network associated with each OD pair and each ferry service has an artificial node $M^{d,f}$. An origin arc is constructed to connect $M^{d,f}$ to the origin node of the last time interval, as illustrated in Fig. 2, whose arc flow represents the amount of passengers not served at the end of the planning horizon. One may interpret this flow as the unsatisfied or lost demand within the planning horizon. If serving all demand is an important consideration, one may set a large penalty for the origin arc, so that more frequent services are arranged to carry all the demand, at the expense of a higher operating cost. In addition, the passenger network includes a set of destination arcs that connect the destination nodes to the artificial node $M^{d,f}$. These destination arcs delineate the arrival times of passengers at their destinations, which are used to determine the schedule delay penalty for early or late arrivals outside the preferred arrival time-windows.

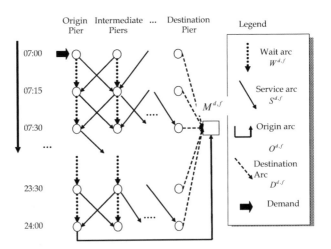

Fig. 2. The Passenger Time-space Network Schematic

2.4 Mathematical Formulation

Minimize: $Z =$

$$\left[\sum_{f\in F}\left(\sum_{i\in N_b^f}\sum_{j\in N^f\setminus N_b^f}Y_{ij}^f\kappa^f + \sum_{ij\in S^f}Y_{ij}^f C_{ij}^f - \sum_{g\in G}\sum_{d\in R}\sum_{ij\in D^{d,f}}X_{ij}^{d,f,g}\alpha^{d,f}\right)\right] +$$

$$\xi\left[\sum_{f\in F}\sum_{g\in G}\sum_{d\in R}\sum_{ij\in D^{d,f}}X_{ij}^{d,f,g}\beta_{ij}^g v_w\right.$$

$$\left.+\sum_{f\in F}\sum_{g\in G}\sum_{d\in R}\left(\sum_{ij\in S^{d,f}}X_{ij}^{d,f,g}T_{ij} - \sum_{ij\in D^{d,f}}X_{ij}^{d,f,g}T^{d,f}\right)v_t\right] \tag{1}$$

Subject to:

$$\sum_{j\in N^f}Y_{ij}^f - \sum_{k\in N^f}Y_{ki}^f = 0 \quad \forall i\in N^f\setminus(N_b^f\cup N_e^f), \forall f\in F \tag{2}$$

$$\sum_{i\in N_b^f}\sum_{j\in N^f\setminus N_b^f}Y_{ij}^f \le V^f \quad \forall f\in F \tag{3}$$

$$\sum_{j\in N^d}X_{ij}^{d,f,g} - \sum_{k\in N^d}X_{ki}^{d,f,g} = \begin{cases} \dfrac{e^{u^{d,f,g}}}{\sum\limits_{f'\in F}e^{u^{d,f',g}}}B^{d,g} & \begin{array}{l}\forall i\in N_b^{d,f},\forall d\in R,\\ \forall f\in F,\forall g\in G\end{array} \\ 0 & otherwise \end{cases} \tag{4}$$

$$\sum_{g\in G}\sum_{d\in R}X_{ij}^{d,f,g} \le Y_{ij}^f Q^f \quad \forall ij\in S^f, \forall f\in F \tag{5}$$

$$X_{ij}^{d,f,g} \ge 0 \quad \forall ij\in A^{d,f}, O^{d,f}, D^{d,f}, \forall d\in R, \forall f\in F, \forall g\in G \tag{6}$$

$$0 \le Y_{ij}^f \le U_{ij}^f \quad \forall ij\in A^f, \forall f\in F \tag{7}$$

$$Y_{ij}^f \in integer \quad \forall ij\in A^f, \forall f\in F \tag{8}$$

Where:

$$u^{d,f,g} = \theta_1\alpha^{d,f} + \theta_2\overline{T}^{d,f,g} + \theta_f \tag{9}$$

$$\overline{T}^{d,f,g} = \frac{\sum\limits_{ij\in D_e^{d,f}}X_{ij}^{d,f,g}\beta_{ij}^g\omega_e + \sum\limits_{ij\in D_l^{d,f}}X_{ij}^{d,f,g}\beta_{ij}^g\omega_l + \sum\limits_{ij\in S^{d,f}}X_{ij}^{d,f,g}T_{ij}}{\sum\limits_{ij\in D^{d,f}}X_{ij}^{d,f,g}} \tag{10}$$

$$\forall d \in R, \forall f \in F, \forall g \in G$$

This model primarily considers the perspective of the operator, aiming to minimize the operation costs, with the revenue from fare collection expressed in negative cost terms. Thus, a negative objective function value indicates a profit. On the other hand, the objective function incorporates passengers' delay and travel times as part of the "costs" to be considered. The objective function (1) seeks to minimize the total operating costs, comprising five terms: (i) fixed cost associated with owning or hiring a ferry for the service period; (ii) trip operating cost; (iii) revenue (i.e., expressed in negative terms to offset the costs); (iv) total arrival schedule delay penalty; (iv) total penalty cost of multi-stop trips. All the variable definitions are provided in Section 2.2.

Specifically, the objective function in (1) consists of two main brackets on the right hand side. The first bracket sums the operation costs; whereas the second bracket sums the passenger disutilities, with ξ being the relative weight between these two main brackets. The first term within the first bracket refers to the total fixed cost; the second term depicts the total trip operating cost; and the third term gives the total revenue, where $\alpha^{d,f}$ is the fare of type f ferry service on OD pair d. As for the second main bracket, the first term inside defines the total schedule delay penalty. The product $\beta_{ij}^g v_w$ refers to the cost of arrival delay for passengers on the destination arc, which incurs due to arrivals either earlier or later than their preferred arrival time windows. The second term represents the total multi-stop trip penalty, which is measured by the cost of additional travel time experienced by passengers on multi-stop trips, relative to the travel time on direct services. The term $\sum_{ij \in S^{d,f}} X_{ij}^{d,f,g} T_{ij}$ measures the total travel times for passengers of OD pair d, ferry type f and arrival time-window g. The summation of destination arc flows $\sum_{ij \in D^{d,f}} X_{ij}^{d,f,g}$ represents the total passengers reaching their destinations. The product $\sum_{ij \in D^{d,f}} X_{ij}^{d,f,g} T^{d,f}$ represents the total passenger travel time, had they been able to use direct services. Therefore, the difference between $\sum_{ij \in S^{d,f}} X_{ij}^{d,f,g} T_{ij}$ and $\sum_{ij \in D^{d,f}} X_{ij}^{d,f,g} T^{d,f}$ measures the total additional travel time due to multi-stop or indirect services. If there is no multi-stop trip, i.e. $T_{ij} = T^{d,f}, \{\forall ij \mid X_{ij}^{d,f,g} > 0$ and $ij \in S^{d,f}\}$, then this penalty cost is zero.

Constraint (2) denotes the conservation of ferry flows at each node i in each f ferry network. Constraint (3) requires that each type of ferry in operation be subject to the corresponding maximum fleet size. Constraint (4) states the passenger conservation condition at every node in the passenger flow network after considering the exogenous demand. Note that the logit demand splits for different ferry services are captured as part of (4). This introduces nonlinearity and in fact, non-convexity in the formulation. Constraint (5) combines the passenger flows of all OD pairs and arrival time-windows between (i, j) and requires that the total passenger volume be subject to the ferry capacity on each service arc (i, j). Constraints (6) and (7) provide the bounds of passenger flows and ferry flows between (i, j), respectively. Constraint (8) defines the ferry flow variables to be integer. Equation (9) defines the utility function, which comprises the attributes of ferry service, including fare $\alpha^{d,f}$ and average total travel time $\overline{T}^{d,f,g}$, and an alternative-specific constant. Equation (10) derives average total travel time by weighting the early arrival delay, late arrival delay and journey time. The early or late arrival delay applies if travelers' arrival times do not fall within their preferred arrival time-windows.

3 Heuristic Algorithm

An iterative heuristic algorithm is developed to solve this mixed integer nonlinear program (MINLP). This algorithm first relaxes and decomposes the original problem and then solves a series of mixed integer linear subproblems iteratively. Note that the nonlinear nature of the original problem comes from the logit modal-split function, which captures the interrelationship of service disutilities among the different ferry types. If only the demands for the different services are given and fixed, the original problem can be relaxed to a mixed integer linear program (MILP). In other words, given the initial (fixed) passenger demands for each ferry type, i.e. $B^{d,f_1,g}, B^{d,f_2,g}, \ldots$ ($f_1, f_2, \ldots$ refer to different ferry types), the original MINLP can be decomposed into a set of independent MILP subproblems, with each pertaining to a particular ferry type. Fig. 3 depicts the relaxation and decomposition processes. For the MILP subproblems, many existing algorithms can be applied to solve them. After solving these independent MILP subproblems, we obtain the passenger flows (i.e., $X_{ij}^{d,f_1,g}, X_{ij}^{d,f_2,g}, \ldots$) and ferry flow, (i.e., $Y_{ij}^{f_1}, Y_{ij}^{f_2}, \ldots$) for

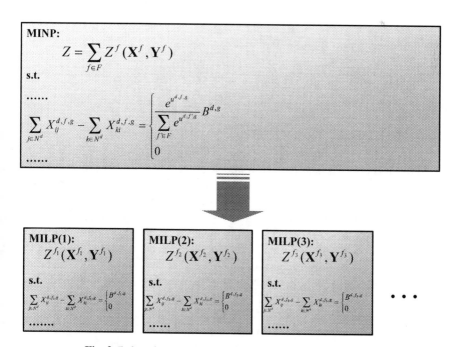

Fig. 3. Relaxation and Decomposition of the Original MINLP

different ferry service types. The service disutilities for different ferry types (i.e., $u^{d,f_1,g}, u^{d,f_2,g}, \ldots$) can be calculated according to (9) and (10). Then, according to the logit split function, we re-estimate the corresponding passenger demands for the different ferry service types. If the gap between the newly estimated passenger demands and the initial demands falls within a specified tolerance, consistency is achieved and the algorithm is stopped. The ferry flows obtained as such depict the "optimal" ferry scheduling and routing. In the case that the gap lies outside the specified tolerance, the newly obtained passenger demands are fed back into the MILP subproblems, which are solved again. This whole process is repeated, as schematically shown in Fig. 4, until convergence is achieved.

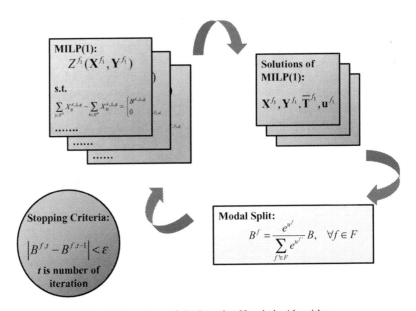

Fig. 4. Procedure of the Iterative Heuristic Algorithm

In this algorithm, it is important to initialize the passenger demands for the different ferry types, (i.e., $B^{d,f_1,g}, B^{d,f_2,g}, \ldots$) for the first iteration, or define the initial solution. In this study, we split the exogenous passenger demands $B^{d,g}$ arbitrarily to obtain an initial solution. Also, to ensure convergence of the algorithm, the method of successive averages (MSA) is used. Specifically, the service disutility defined by (9) and (10) is used to conduct the MSA procedure. In each iteration, we take the average of the service disutilities from the current as well as previous iterations, where each service disutility is derived from the solutions of the decomposed MILP subproblems. Let $u_k^{d,f,g}$ be the calculated disutility in the kth iteration; the average disutility is determined as: $\left(\frac{1}{k}\right)\sum_{n=1}^{k} u_k^{d,f,g}$.

Summarizing, the steps of the heuristic are as follows:

Step 0: Define the tolerance $\epsilon > 0$ and the initial solution $B_0^{d,f,g}$. Set $k = 1$.

Step 1: With given $B_k^{d,f,g}$, solve the decomposed independent MILP subproblems for each ferry service type, which yields $X_{ij}^{d,f,g}$ and $Y_{ij}^{d,f,g}$.

Step 2: Calculate $u_k^{d,f,g}$ based on $X_{ij}^{d,f,g}$ and $Y_{ij}^{d,f,g}$ determined in *Step 1*.

Step 3: Calculate $\tilde{u}_k^{d,f,g} = \left(\frac{1}{k}\right) \sum_{n=1}^{k} u_k^{d,f,g}$ based on the method of successive averages.

Step 4: Calculate $B_{k+1}^{d,f,g}$ based on the logit split function and $\tilde{u}_k^{d,f,g}$ determined in *Step 3*.

Step 5: If $\left| B_{k+1}^{d,f,g} - B_k +^{d,f,g} \right| < \epsilon$ then stop; otherwise set $k = k + 1$ and return to *Step 1*.

4 Numerical Studies

We implement the heuristic algorithm for a ferry route package in Hong Kong. The problem involves two ferry routes that share similar characteristics in terms of patronage, journey time and fare: CBD-Mui Wo (C-MW) and CBD-Peng Chau(C-PC). Both MW and PC are outlying islands. The details of the problem setting refer to Lai and Lo (2004).

We solve the problem for the two-hour morning peak (7:00a.m. - 9:00a.m.). The time interval in both the ferry and passenger flow time-space networks is set to be 15 minutes. Two types of ferry services, i.e. fast ferry with higher fare and ordinary ferry with lower fare, are available. Passengers are segregated into two different groups according to their preferred arrival time windows at destinations, 8:00a.m.-8:30 a.m. (the first time-window) and 8:45a.m.-9:15 a.m. (the second time-window). With the segregation ratio pre-set to be 7:3, we obtain the passenger demands for the different arrival time windows on different OD pairs, i.e., $B^{d,g}$.

For this problem scenario, each decomposed MILP subproblem involves 64 binary variables, 36 integer variables, 840 real variables, and a total of 450 constraints. We use the commercial optimization package CPLEX-6.0-MIP (ILOG (1998)) to solve the MILPs. The parameter x is set to be 1 and the stopping tolerance ϵ is 0.01.

Firstly, we apply the heuristic algorithm with two different initial solutions. In Case 1, the initial demand is estimated from the set of services that incur no delay to passengers (or the best scenario from passengers' perspective); whereas in Case 2, the initial demand is estimated from the existing service schedule. Fig. 5 illustrates how the objective value changes for both cases. Fig. 5 shows that the resultant solution depends on the choice of the initial solution, due to the non-convex nature of this problem. However, in both cases, the heuristic algorithm is able to drive down the objective function value. The drop or improvement for Case 1 is more pronounced due to the choice of an extreme initial solution. As for Case 2, using the existing schedule as a starting point, the result shows that one can still improve the performance of the system substantially, around 25% of the objective function value.

To demonstrate the non-convex nature of the problem, we also solve the heuristic algorithm with more than 200 randomly chosen initial solutions. The final result expressed in terms of objective function values is plotted in Fig. 6. From this figure,

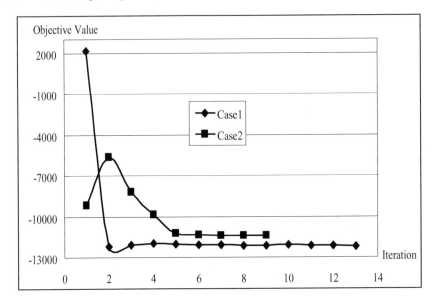

Fig. 5. Objective Function Value Against Iteration for Case I and Case 2

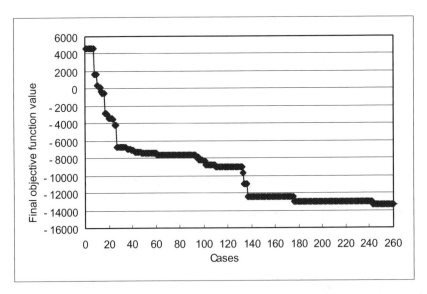

Fig. 6. Objective Function Values Obtained with Different Initial Solutions

we can see that different starting points can produce similar final solutions. Actually, each point in Fig. 6 represents a local minimum. Due to the non-convex nature of the network design problem, one needs to determine all the local minima in order to find the global minimum. In this sense, we cannot ascertain whether the solution obtained with a particular initial solution is globally optimal or not. In addition, the large gap between the best and worst solutions among these 200 plus trials shows that the choice of the initial solution has an important impact on the quality of the final solution. From this limited numerical experience, it is, however, not easy to determine *a priori* what is a good initial solution.

To examine the final solutions obtained from these 200 plus initial solutions, we compute their cumulative probability versus their objective function values. We find that there are big performance discrepancies among the solutions obtained. However, overall, the heuristic is more likely to lead to good results. For example, 87.2% of the solutions obtained have objective functions values that are lower (or better) than -6976.8, which is the best objective value if the operator only offers one ferry service type. Nevertheless, we note that how to construct good *a priori* initial solutions remains an important research question.

5 Concluding Remarks

This paper developed a multi-fleet ferry routing and scheduling model while considering passengers' mode choice preferences. Ferry services with different operation characteristics and passengers with different preferred arrival time windows are considered in the model. The logit model is applied to determine passengers' service choices. The formulation then determines the best mixed-fleet operating strategy, including interlining schemes, so as to minimize the objective function that combines both the operator and passengers' performance measures. Mathematically, this mixed-fleet routing and scheduling problem is formulated as a mixed integer nonlinear programming problem. This study then develops an iterative heuristic algorithm to solve this problem.

Case studies of ferry services in Hong Kong were examined to demonstrate the characteristics of the heuristic algorithm. The results showed that the solution produced by the heuristic was highly dependent on the choice of the initial solution, due to the non-convex nature of the network design problem. Actually, the problem stated in this study can be formulated as a bi-level programming problem, in which the upper level determines the ferry and passenger flows while the lower level models passengers' service choices. One may use the method of iterative balancing to solve this bi-level problem heuristically. However, it is known that this iterative method does not always yield the global optimal network design (Bell and Iida (1997)). Our current research focuses on exploring algorithmic improvements within the framework developed herein.

Acknowledgement: This study is sponsored by the Competitive Earmarked Research Grants, HKUST 6083/00E and HKUST 6161/02E, of the Hong Kong Research Grant Council.

References

Bell, M. G. and Iida, Y. (1997). *Transportation network analysis*. Wiley, Chichester.

Ceder, A. and Wilson, N. (1986). Bus network design. *Transportation Research*, **20B**, 331–344.

Crainic, T. G. (2000). Service network design in freight transportation. *European Journal of Operational Research*, **122**, 272–288.

Crainic, T. G. and Laporte, G. (1997). Planning models for freight transportation. *European Journal of Operational Research*, **97**, 409–438.

ILOG (1998). *CPLEX 6.0 User's Manual. ILOG, Inc.* Incline Village.

Lai, M. F. and Lo, H. K. (2004). Ferry service network design: Optimal fleet size, routing, and scheduling. *Transportation Research A*, **38**, 305–328.

Magnanti, T. L. and Wong, R. T. (1984). Network design and transportation planning: Models and algorithms. *Transportation Science*, **18**, 1–55.

Yan, S. and Chen, H. L. (2002). A scheduling model and a solution algorithm for inter-city bus carriers. *Transportation Research A*, **36**, 805–825.

Yan, S. and Tseng, C. H. (2002). A passenger demand model for airline flight scheduling and fleet routing. *Computers & Operations Research*, **29**, 1559–1581.

Generating Train Plans with Problem Space Search

Peter Pudney[1] and Alex Wardrop[2]

[1] University of South Australia, Peter.Pudney@unisa.edu.au
[2] WorleyParsons Rail, wardrop@ar.com.au

Summary. Planning train movements is difficult and time-consuming, particularly on long-haul rail networks, where many track segments are used by trains moving in opposite directions. A detailed train plan must specify the sequence of track segments to be used by each train, and when each track segment will be occupied. A good train plan will move trains through the network in a way that minimises the total cost associated with late arrivals at key intermediate and final destinations.

Traditionally, train plans are generated manually by drawing trains on a train graph. High priority trains are usually placed first, then the lower priority trains threaded around them. It can take many weeks to develop a train plan; the process usually stops as soon as a feasible train plan has been found, and the resulting plan can be far from optimal.

Researchers at the University of South Australia and WorleyParsons Rail have developed scheduling software that can generate optimised train plans automatically. The system takes a description of the way trains move through the network and a list of trains that are required to run, and quickly generates a train plan that is optimised against key performance indicators such as delays or lateness costs.

To find a good plan, we use a probabilistic search technique called Problem Space Search. A fast dispatch heuristic is used to move the trains through the network and generate a single train plan. By randomly perturbing the data used to make dispatch decisions, the Problem Space Search method quickly generates hundreds of different train plans, then selects the best. The automatic scheduling system can be used to support applications including general train planning, real-time dynamic rescheduling, integrated train, crew and maintenance planning, infrastructure planning and congestion studies.

One of the first applications of the system has been for an Australian mineral railway, to prepare efficient train plans to match mineral haulage requirements. The product is mined at six sites and transported by rail to a port. The numbers and sizes of train loads from each site are determined by grading requirements to meet the product specification for shipping. The train plan is then the orderly translation of these transportation requirements into an efficient timetable which resolves meets and crosses over a long single track railway. These train movements are thus part of an integrated mine-to-ship logistics chain.

1 Introduction

Most of Australia's long-haul rail network is single-line track that is shared by trains travelling in different directions, with occasional refuges or crossing loops. Trains are often delayed waiting for track to become available. Moving trains through the rail network without incurring significant delays requires careful planning. A detailed train plan must specify the sequence of track segments to be used by each train, and when each track segment will be occupied. Developing such train plans is difficult and time-consuming.

Traditionally, train plans are generated manually by drawing trains on a train graph. High priority trains are usually placed first, then the lower priority trains threaded around them. It can take many weeks to develop a train plan. The process usually stops as soon as a feasible train plan has been found, and the resulting plan can be far from optimal. Furthermore, train plans are modified many times between their first inception and the day of operation. Different planning stages often use different – and incompatible – tools.

Researchers at the University of South Australia and WorleyParsons Rail have developed scheduling software that can generate optimised train plans automatically. The system takes a description of the way trains move through the network and a list of trains that are required to run. Instead of plotting trains on a train graph, train planners specify when they want trains to depart and desired arrival times at key locations along the route. The system then automatically searches for a schedule that moves the trains through the network in a way that minimises the total cost associated with late arrivals at journey destinations and key intermediate points.

2 Problem Formulation

The problem of scheduling trains over a network of track segments is similar to the well-known job-shop problem of scheduling jobs on machines. A rail network comprises a set of track segments which cannot be occupied by opposing trains at any instant, just as machines in a job-shop can process only one job at a time.

Much of the previous work on automated scheduling has concentrated on simplified rail networks, such as single line track with crossing loops. We have been careful to develop a method that uses a very general description of the rail network – one that allows parallel tracks, alternative routes, complicated junctions, and realistic separation rules.

A rail network can be represented by a mathematical graph – that is, a set of vertices and a set of edges. Vertices correspond to locations on the rail network such as junctions, line ends, diamond crossings and timing points. Edges on the graph correspond to track segments on the rail network. There may be more than one edge between any pair of vertices, such as at crossing loops. Balloon loops may start and finish at the same vertex.

We represent a rail network using *track segments* that correspond to edges on the mathematical graph. Extra track segments are used to represent diamond crossings,

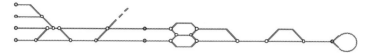

Fig. 1. A Rail Network can be Represented by a Mathematical Graph

stations without loops, or sets of points that form one-to-many or many-to-many junctions. We are also able to ignore many of the smaller edges, such as crossovers between parallel tracks. Fig. 2 shows the segments required to represent the network graph in Fig. 1.

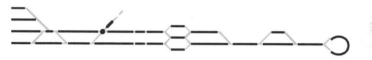

Fig. 2. Track Segments Used to Represent the Network in Fig. 1

Track segments have the following properties:

- a track segment may not be occupied by opposing trains;
- any point on the track at which an arrival time or departure time is required defines the end of a track segment;
- every valid train movement can be described as a sequence of track segments; and
- every pair of conflicting train movements shares at least one common track segment.

Track segment parameters include:

- the length of the segment;
- the directions in which the segment can be traversed (up, down, bidirectional);
- the segment type (mainline, loop, siding, diamond, junction);
- the separation required between the rear of one train and the front of a following train; and
- the time delay required between one train clearing a point on the segment and the next train arriving at that point

The motion of a train on the network is defined by a sequence of train *movements*. A movement describes how a train moves forward from its current track segment to another track segment on which it can stop without blocking opposing movements. A movement is a sequence of *movement segments*; each movement segment specifies:

- the track segment to be traversed;
- the direction in which the track segment will be traversed;
- the time taken for the front of the train to traverse the segment; and

- the entry and exit speeds.

  Fig. 3 shows a portion of a rail network.

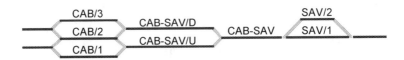

Fig. 3. Portion of a Network with a Station, Double Track, Single Track and Another Station

The possible movements on this portion of the network are:

- CAB/3, CAB-SAV/D
- CAB/2, CAB-SAV/D
- CAB-SAV/D, CAB-SAV, SAV/1
- CAB-SAV/D, CAB-SAV, SAV/2
- SAV/1, CAB-SAV, CAB-SAV/U
- SAV/2, CAB-SAV, CAB-SAV/U
- CAB-SAV/U, CAB/2
- CAB-SAV/U, CAB/1

For each movement we can also specify additional time taken if the movement starts from rest, additional time taken if the movement finishes at rest, and the dwell required at the end of the movement.

A *trip* is a set of possible movements that can be used by one or more trains; a template for a journey that can be made by a class of train. The trip movements specify all possible routes for a trip.

A *train* is an instance of a trip. Train parameters include:

- a list of track segments from which the train can start;
- the departure date and time;
- the length of the train; and
- a list of journey targets

A *target* is a point along the journey with a desired arrival time or where the train is required to dwell for a specified duration. A train must include at least one target – the final destination – but may also include intermediate targets where timing is important, such as at crew change locations. The parameters of each target are:

- a list of track segments that may be used by the train at the target;
- the desired arrival date and time;
- a lateness cost function;
- the dwell time; and
- the earliest departure time.

If you require the train to stop at a target for 20 minutes, but do not care what time it arrives, you can specify an arbitrary arrival time and a zero lateness cost function.

The total cost of a timetable is the sum of the lateness costs over all targets of all trains. If the true cost of lateness is not known, these cost functions can be set to form objective functions such as total delay (time spent waiting for track to become available), total weighted delay, or sum of delay squared.

The problem data specifies the network infrastructure, the way trains move on the network, and the train requirements. The train requirements specify the earliest time that a train may start, and desired arrival times at key locations along each train's journey. Our aim is to find a train plan that moves each of the trains across the network in accordance with its trip and target requirements (and normal railway operating constraints), and with minimum total lateness cost.

3 Problem Space Search

Realistic rail scheduling problems are often sufficiently large and complicated that formulating and solving the problem using mathematical programming techniques is intractable. Instead, we use a probabilistic search technique, Problem Space Search (Naphade *et al.* (1997)), to search for good solutions.

The principle of Problem Space Search is simple: a fast dispatch heuristic is used to generate a single solution to the problem, then random perturbations to the problem data cause the dispatcher to generate alternative solutions. We evaluate each of the generated solutions and retain the best.

We use a fast dispatch heuristic to generate a sequence of train movements that will move each train through the network to its destination. The dispatcher considers the trains on the network and the trains that are scheduled to move onto the network, chooses which train movement to make next, and iterates until all trains are at their destinations.

A *first-to-start* dispatcher chooses the next train to be moved as follows:

- For each train on the network, set the *dispatch decision time* to be the earliest time at which the train will be ready to start its next movement. A given train may have more than one possible next movement; in this case we select the earliest.
- Choose the train with the earliest possible dispatch decision time. If there is more than one, pick any one.

A *first-to-finish* dispatcher is similar, but chooses the train movement with the earliest finish time. Between *first-to-start* and *first-to-finish* are a range of dispatchers that choose the movement with the earliest $t = (1 - \alpha)t_0 + \alpha t_1$, where t_0 is the earliest movement start time, t_1 is the earliest movement finish time and $\alpha \in (0, 1)$. We have found that $\alpha = 0.5$ gives good results.

The possible movements for a class of trains is described in the trip data. However, the dispatcher also checks that movements for a particular train are feasible. For example, it will not move a long train onto a short crossing loop.

The result of applying the dispatcher to a scheduling problem is a single train plan – though not necessarily a good one. To find a good plan, the Problem Space Search method makes random perturbations to the problem data used by the dispatcher to decide which movement to make next. By perturbing the data used to make dispatch decisions, alternative decisions are made and alternative train plans are generated. The randomly perturbed data is used only to make the dispatch decision; the original, unperturbed data is still used to calculate the movements.

The desirable characteristics of the perturbations are:

- the probability of swapping the dispatch order of any two trains should be 0.5 if the trains have the same dispatch decision time;
- the probability of swapping the dispatch order of any two trains should decrease as the difference between their dispatch decision times increases; and
- the probability of swapping the dispatch order should be non-zero.

We use a normal distribution with zero mean and a standard deviation based on the mean movement duration for the trains on the network.

We can bias the dispatcher to favour trains with high priority, such as passenger trains, by reducing the dispatch time of these trains. We set the dispatch time for each train to

$$t_D = (1 - \alpha)t_0 + \alpha t_1 - N(0, \sigma) - \beta w$$

where t_0 is the segment start time, t_1 is the segment finish time, $\alpha \in [0, 1]$, $N(0, \sigma)$ is a random number drawn from a normal distribution with mean 0 and standard deviation σ, β is a constant, and w indicates the importance of the train; normal trains have $w = 1$, passenger trains might have $w = 2$. The constant β is chosen so that for two trains with the same times $(1 - \alpha)t_0 + \alpha t_1$, a train with $w = 2$ has a probability of about 0.8 of moving before a train with $w = 1$.

The 'goodness' of a train plan is calculated from the completed train plan. Each plan is evaluated, and the best plans are retained.

Some sequences of dispatch decisions may end in deadlock – a network config-uration from which it is not possible for all trains to reach their destinations. If only a small proportion of train plans end in deadlock, these can simply be discarded. Otherwise, it is possible to modify the dispatch heuristic to reduce the likelihood of deadlock.

The scheduler has been tested using data from real Australian rail networks in-cluding:

- New South Wales, North Coast, 780km, non-branching, 68 refuges or crossing loops, 42 trains per day;
- New South Wales, Illawarra, 210km, double and single track, non-branching, 35 refuges or crossing loops, 260 trains per day;
- Sydney – Melbourne, 900km, non-branching, 47 refuges or crossing loops, 118 trains per day;
- A mineral ore network, 300km main line, 5 branch lines, 26 refuges or crossing loops, 24 trains per day.

The scheduler generates train plans that are significantly better than the plans generated by the dispatcher using unperturbed data. On the 900km Sydney North Coast line, with 42 trains and 47 refuges or crossing loops, the search reduced the total train delay by 30%.

Fig. 4 shows a histogram of total delays from 835 train plans generated for the mineral railway test case discussed below. A smooth histogram usually indicates that the solution space has been searched adequately.

score range	tally	%	
20000 – 22000	1	0.1	-
22000 – 24000	11	1.3	–
24000 – 26000	43	5.1	——
26000 – 28000	88	10.5	————
28000 – 30000	115	13.8	—————
30000 – 32000	177	21.2	———————
32000 – 34000	160	19.2	——————
34000 – 36000	111	13.3	—————
36000 – 38000	71	8.5	————
38000 – 40000	36	4.3	——
40000 – 42000	19	2.3	—
42000 – 44000	3	0.4	-

Total 835 100.0

Fig. 4. Histogram of Scores for 835 Train Plans

There is a significant difference between the traditional train planning method and our method. Traditionally, entire train journeys are removed and added one-at-a-time from an existing train plan, and must be threaded around the existing trains. Decisions about which train should wait at a cross are made locally; but a sequence of local decisions that each appear to be reasonable do not necessarily lead to a good overall train plan.

We start with trains poised on the edge of an empty network and then move the trains forwards simultaneously. To add a new train to a plan, we simply put the new train into the train requirements and optimise again, starting from an empty network. The train planner is no longer able to directly place a train; instead, the paths of individual trains must be controlled via the train requirements, using targets and lateness costs.

This application of Problem Space Search frees timetable development from the tyranny of time and effort which bedevils manual timetable development. Our data description and dispatch heuristic apply to a general railway network so that timetable development can take place over a complete railway rather than an artificial portion. Our system can handle a range of railway track configurations between control points and refuging locations. It can also handle trains which might have specific network restrictions, such as long freight trains may be over-size for particular refuge locations. Most importantly, the speed of computation to obtain an efficient train plan allows the user to experiment and finesse the development of a timetable. Alternatively, this computation speed should open the way to providing real-time dispatch advice to train controllers, provided that they can receive timely information on train progress.

4 Applications

We are able to generate and evaluate hundreds of optimised train plans per minute. Potential uses for an automated train planning tool are described below.

Train Planning

Train plans are traditionally created by drawing trains one-at-a-time onto a train graph, either manually or using a computer. It can take many weeks to create a feasible train plan. As the day of operation approaches, the train plan is extensively revised to reflect changes in demand and in the network operating conditions. Train planners spend most of their time trying to maintain a feasible timetable, and have little time to look for better alternatives.

Given a system that can produce optimised train plans almost instantly, train planners can spend more time investigating the effects of alternative departure times, arrival times at key locations, and lateness costs. Adding and removing trains becomes simple – the system automatically recalculates an optimised train plan that meets the new train requirements.

Dynamic Rescheduling

In a control centre, an automated train planning system can be used in real-time, in the background, to revise train plans to take into account the actual state of trains on a network. One possible objective would be to recover, as much as possible, to the published timetable. Alternatively, the system could abandon the original train plan and instead calculate a new plan that meets, as closely as possible, given the new state of the network, the original train requirements.

Integrated Scheduling

Our scheduler can be used to generate many good train plans, each of which can be assessed against additional criteria such as track maintenance requirements and crew rostering requirements. We are also working to extend the system so that maintenance requirements are included in the problem specification; the system will be extended from a train planner to become a track possession planner.

Infrastructure Planning

Using an automated scheduler, the impact infrastructure changes on train plans can be assessed almost instantly. The system can also be used to quickly generate new train plans suited to new infrastructure.

Congestion Studies

The scheduler generates many good timetables. By analysing these timetables, we could construct a 'congestion map' that indicates where and when the network is congested. Congestion can be relieved by either changing the train requirements (e.g., shifting some trains into the less congested times of the day), or by adding infrastructure.

5 Case Study

One of the first applications of our system was for an Australian mineral railway that is currently shipping in excess of 50 million tonnes of product annually. However, it wants to increase production by 50% in response to increasing demand for high quality product. The mining and shipping operations have been integrated into a single logistics chain, of which the railway is an important part. In this environment, the railway operations have to fit into the production and shipping schedule rather than the other way around.

Thus, the company determines what the flow from the different mines should be to meet both the product specification and the forthcoming shipping schedule. This translates into mining plans and transportation plans. From the railway perspective, it is required to haul minerals in varying quantities from the different mines up to the physical capacity of either the available wagon and locomotive fleet or the railway network. In the short term, the company is constrained by its rolling stock resources. However, it is ordering more wagons and locomotives in anticipation of increased production. In the longer term, it may be constrained by its current railway infrastructure. While it is able to increase single track line capacity by dividing long sections with new crossing loops there are limits to how far this process can be taken. In the meantime, the company needs to be able to plan for increased mineral transportation over a long single track railway (over 300 kilometres of main line plus more than 100 kilometres of branch lines). Fig. 5 schematically displays the current railway network. The bottom line is the main line. Each of the other five horizontal lines represents a branch line. The labelled points are timing points, crossing loops, junctions or yards.

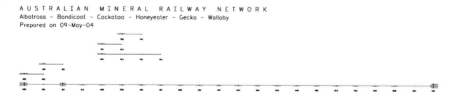

AUSTRALIAN MINERAL RAILWAY NETWORK
Albatross - Bandicoot - Cockatoo - Honeyeater - Gecko - Wallaby
Prepared on 09-May-04

Fig. 5. Schematic Diagram of the Mineral Rail Network

Trains are ordered daily to meet weekly (and longer) production schedules. To make best use of the train unloaders at the port, train round trips need to be dispatched in such a way that there is a relatively even flow of laden returns to the port. At the same time, trains are being dispatched over a single track railway which inherently must delay most trains somewhere in their travels. The train operations challenge is to meet the production schedules with the minimum of rolling stock and the minimum of en-route delays.

Table 1. Line Capacity and Corrected Usage on the Mineral Rail Network

section	capacity	usage (%)
Grevillea – Hovea	114	9.3 -
Gecko – Hakea	74	14.4 -
Honeyeater – Hakea	32	33.6 —
Hakea – Hovea	105	20.4 ————
Hovea – Heron	39	95.3 ————
Cassowary – Cockatoo	31	33.9 —
Bandicoot – Bilby	109	14.7 -
Albatross – Cockatoo	37	84.3 ———
Cockatoo – Dingo	152	21.1 –
Dingo – Emu	62	51.7 ——
Emu – Finch	87	36.8 —
Finch – Goanna	43	62.1 ———
Goanna – Heron	50	42.4 —
Heron – Ibis	116	41.2 —
Ibis – Jacana	81	52.5 ——
Jacana – Kangaroo	78	54.8 ——
Kangaroo – Lyrebird	56	85.3 ———
Lyrebird – Malleefowl	106	50.1 ——
Malleefowl – Numbat	68	62.6 ——
Numbat – Oyster	119	44.9 —
Oyster – Possum	38	124.9 ————
Possum – Quokka	39	123.4 ————
Quokka – Rosella	39	109.6 ————
Rosella – Shearwater	65	81.5 ———
Shearwater – Thylacine	119	44.9 —
Thylacine – Wallaby	86	61.8 ——

We can statically estimate sectional line capacity from what we know of the physical layout of the railway and the sectional running times of the empty and laden trains. We can deduce sectional usage from an input list of pre-resolution train requirements – in this case a hypothetical schedule with twelve round trips dispatched each day. However, input train requirements (and output train plans) are rarely uniformly distributed throughout a working day. Therefore, these train requirements need to be corrected for their non-uniformity. The modified usage can then be compared to the previously calculated line capacity and the level of sectional usage calculated. Table 1 shows line capacity and the corrected usage for the rail network with twelve round trips each day. The table indicates that the railway between Oyster and Rosella would be severely stressed by the proposed train requirements.

Our scheduler was then applied to the input train requirements to flow over the railway network. The objective was to minimise the total delay experienced by all the input trains. No distinction was made between delays to empty trains and delays to laden trains. Nevertheless, it would be quite straightforward to differentially weight empty and laden train delays. However, differential weighting, or any other form of objective function, will not change the way in which Problem Space Search produces feasible train plans. Instead, the choice of objective function will change the ranking of feasible solutions so that different types of solutions will be favoured by different objective functions. Fig. 4 presents a frequency distribution of the total delays generated from 835 feasible solutions to this train planning problem in 28 seconds. The problem was run over a 36 hour period to cover the lead-in and lead-out from a full working day, and included sixty-five long distance (port–mine) and short distance (junction–mine) trains. Fig. 6 displays a train diagram (time versus distance) of the best train plan. Delays averaged roughly 14% of the total travel time and favoured empty trains over laden trains.

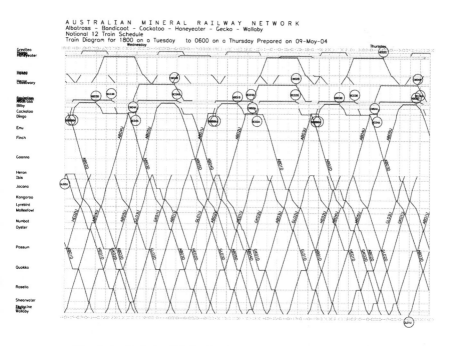

Fig. 6. An Optimised Train Plan for the Mineral Railway Network

Because the static capacity analysis flagged an incipient lack of capacity in a key section of the network we also looked at the impact that increasing the numbers of trains would have on the use of line capacity. The infrastructure was held constant but different numbers of mainline return trips were run – 8, 10 and 12 round trips per day. Table 2 summarises the results of these train plan resolution trials. It is clear that increasing the numbers of trains while keeping the current infrastructure fixed will increase the average delay experienced by each train. Delay time increased non-linearly, as a proportion of total time, as the number of trains in the system increased. The question for the company is how much this increase in train delay may cost it in lost production as against the cost of relieving line capacity in three single track sections.

Table 2. Scheduling Results from 1000 Trials with Varying Numbers of Trains per Day on the Mineral Railway Network (The Number of Trains is the Number of Different Main Line and Branch Line Trains in the 36-Hour Scheduling Period.)

	Trains per day		
	8	10	12
Number of trains	51	58	65
Number of feasible timetables (/1000)	606	744	835
Best delay (min)	789.8	1279.8	2104.3
Time to complete 1000 trials (sec)	13	20	28
Total travel time (min)	10077	12631	15125
Accumulated delay (min)	790	1280	2104
Delay percentage of total time	7.8	10.1	13.9
Total distance travelled (km)	9583	12090	14333
Average travel time (min)	197.6	217.8	232.7
Average delay time (min)	15.5	22.1	32.4
Average distance (km)	187.9	208.5	220.5
Average speed (km/h)	52.9	52.1	49.9

6 Conclusion

Problem Space Search has proved itself to be a powerful tool for the development of effective train plans over a general railway network. It offers the user good results within a short computation time.

The key to our scheduling system is our representation of the problem. We are able to represent train movements on general railway networks, with branching and looping and different sectional track configurations. Trains are progressed, one movement at a time, through the network under the control of a suitable dispatch heuristic. Problem Space Search is invoked to randomise the decision process to produce different feasible train plans. These train plans are then scored according to a user-specified objective function of arbitrary sophistication. The user is then free to select the best solutions for further examination.

Our scheduling system is currently being used by a mining company to plan train movements from its mines to the port. It has been applied to current operations and for planning future operations using increasing numbers of physical trains. The process is not limited to varying the numbers of trains in the input train requirements. It has also been designed to allow for changes in railway infrastructure, the opening of more mines, and the introduction of additional rolling stock.

References

Naphade, K. S., Wu, S. D., and Storer, R. H. (1997). Problem space search algorithms for resource-constrained project scheduling. *Annals of Operations Research*, **70**, 307–326.

School Bus Routing in Rural School Districts

Sam R. Thangiah[1], Adel Fergany[1], Bryan Wilson[1], Anthony Pitluga[1], and William Mennell[2]

[1] Artificial Intelligence and Robotics Laboratory, Computer Science Department, Slippery Rock University, Slippery Rock, Pennsylvania, USA sam.thangiah@sru.edu
[2] Robert H. Smith School of Business, University of Maryland, College Park, Maryland, USA

Summary. The Commonwealth of Pennsylvania has the nation's largest rural population and the Commonwealth plays an important role in providing transportation for students to travel to their respective schools. State and local governments reimburse school districts for student transportation costs in Pennsylvania. Effective policies for governing the transportation of students can result in large cost savings for the respective governments and reduced travel time for the students. This paper presents heuristics to solve a complex rural school bus routing problem using digitized road networks that can lead to cost savings for both State and local governments. The school bus routing problem addressed and solved in this paper is a mixed-fleet, multi-depot, site-dependent, split-delivery problem with side constraints. Computation of real road distances for the rural school district between pickup points, depots and schools, consisting of 4200 road segments, was done using digitized road networks obtained from the U. S. Census Bureau. Heuristic algorithms were designed and implemented to solve a school bus routing problem with real life data obtained from a rural school district. Feasible solutions to the complex rural school bus routing problem, consisting of 13 depots, 5 schools, 71 pickup points and 583 students, were obtained in less than 10 minutes of CPU time.

1 Introduction

The routing of school buses in rural areas is similar to a classical vehicle routing problem (VRP) (Christofides and Eilon (1969)). A classical VRP consists of a set of vehicles that start from a central depot and either pickup or deliver goods to a set of customers. The objective of the classical VRP is to minimize the total number of vehicles and distance traveled without exceeding the capacity of the vehicles. School bus routing for a rural school district is a complex VRP. In its simplest form, a school bus routing problem consists of a finite number of students at known pickup locations that are to be routed to a single school while reducing the overall routing cost. In a classical VRP an unlimited number of homogenous vehicles are available to service customers from a central depot with each vehicle constrained by capacity and the total distance traveled. The distance between customers is calculated in Euclidean space and the capacity is measured in uniform units. The last few decades have seen the outgrowth of powerful algorithms for solving the VRP using exact and heuristic methods. Surveys on classifications and applications of the VRP can be found in (Bodin *et al.* (1983), Laporte (1992), Fisher (1995), Laporte and Osman (1995), Cordeau *et al.* (2002))

A rural school district consists of a collection of elementary, middle and high schools that require students to be picked up from their homes and dropped off at their respective schools. The elementary, middle and high schools can start at different times. Due to the multiplicity of elementary, middle or high schools in a rural school district, students end up going to different schools. School buses can start at the bus depot, a warehouse or a bus driver's home and pick up all the students going to one or more school(s). The concept of a central starting and ending location does not exist in real-life school bus routing problems as each school bus can have multiple starting and ending locations.

In the Commonwealth of Pennsylvania, the cost of transporting students is borne by the taxpayers at the local and State level. As such, contractors of school buses are required to bid competitively to transport students. The school district has to consider multiple contractors, mix fleet, multiple depots and heterogeneous vehicles to service a rural school district. School buses vary in capacity, length, equipment available for special needs of students and fixed and variable costs. The responsibility of a rural school district is to select the number and type of school buses required to transport students while minimizing the cost of transportation. The mix of students present at each pickup point must be taken into account. Special needs of students, such as those in wheelchairs, would require a school bus with a wheelchair lift in comparison to a regular bus. A pickup point with a regular and a wheelchair student may require service of multiple buses of different types. That is, more than one vehicle is required to service the same pickup point. The vehicle selection process has to consider road constraints imposed on school buses. A large capacity bus may not be able to negotiate narrow roads or make sharp turns on locations with limited visibility to on-coming traffic. In addition, due to sparse roads in a rural school district, combined with natural obstacles such as streams, hills and pedestrian roads, Euclidean distance is often not the right measure of the actual distances between pickup points (Thangiah and Nygaard (1992)). Thus, unlike densely populated regions, real road network distances between pickup points need to be used to get feasible and useable solutions.

This paper presents a heuristic algorithm to solve a complex rural school bus routing problem using digitized road networks obtained from the U. S. Census Bureau. The road network for the rural school district consisted of 4200 road segments, and it was used to calculate real road distances between depots, schools and student pickup points. Heuristic algorithms were implemented to solve a real life school bus routing problem with data obtained from a rural school district consisting of five schools, 583 students, 71 pickup points and 13 depots. The implemented heuristic, for the school bus routing problem, solves a mixed-fleet, multi-depot, site-dependent, split-delivery problem with side constraints. Solutions to the problem were obtained in less than 10 minutes of CPU time on a 3.05GHz Pentium IV computer system.

The next section of this paper explains the school bus routing problem and its associated complexities in more detail. Section 3 describes the digitized road network used in calculating distances and travel times. Section 4 presents the conceptual and mathematical formulation for the complex rural school bus routing problem. Section 5 develops the cost analysis functions of the heuristic algorithm for solving the problem. Insertion heuristics and local optimization methods for improving the solution are described in Section 6. Computational results on a data set obtained from a school district are detailed in Section 7, with concluding remarks and future work given in Section 8.

2 The School Bus Routing Problem

In this section we discuss the school bus routing problem, with special emphasis on the complexities involved in solving it.

2.1 Simple School Bus Problem

The simple school bus routing problem (SSBRP) can be considered to have a collection of heterogeneous vehicles starting from multiple depots and serving students located at different pickup points. This simplification–namely, removal of site-dependent, split-delivery options–allows us to solve the problem using a multi-depot, mixed-fleet formulation, or a variant of it for which there are implemented heuristics from the literature.

In solving the SSBRP we ensure that the total number of students transported by a bus does not exceed the capacity of the bus and the total travel time of the bus does not exceed the maximum allowable travel time for a student. The travel time of a student is the sum total of the distance traveled by the school bus and service time incurred at each of the student's pickup points, from the student's pickup point to the corresponding school. Service time is the sum of time spent in stopping, student boarding and departing from a student pickup location.

The mathematical model for finding optimal route assignments for the SSBRP belongs to the class of $\mathcal{NP}$-complete problems as it has components of the VRP and the traveling salesman problem (TSP) in it. For problems in the $\mathcal{NP}$-complete class, the time taken to obtain an optimal solution increases exponentially with respect to the size of the problem. Due to the intrinsic difficulty of the problem, search methods based on heuristics are most promising for solving practical size problems. Real-life school bus routing problems have a much richer set of constraints than the SSBRP and can therefore be expected to have a much higher computational complexity.

2.2 The Complexity of Routing School Buses

The significance of the school bus routing problem is attributed to its impact on economic and social objectives, in addition to its monetary objectives (Serna and

Bonrostro (2001)). Pennsylvania, with 23% of the state population living in rural areas, has the nation's largest rural population based on the census conducted in 2000. State and local governments in Pennsylvania reimburse the cost of transportation for students to travel to and from their respective public schools. The State and individual school districts bear the cost of transporting students in rural areas. Since each school district is responsible for developing its own school bus routes, most school districts have analysts who use manual methods or commercial systems to generate school bus routes. In theory, either the analyst or the commercial programs have to consider many of the following constraints when routing school buses in rural areas:

- One-way roads
- Hazardous roads or roads without walkways
- Speed zones
- Multiple origination points of buses
- Student pickup and drop-off points
- Students having to cross multi-lane roads to get to a student pickup point
- Deadhaul distance (the distance from the origination point of an empty school bus to the first student pickup point)
- Linehaul distance (distance traveled by a bus with at least one student onboard)
- Presence of student pickup points on inclined roads during winter
- Transportation of handicapped students on school buses equipped with wheelchair lifts or special-restraint seats
- Railroad crossings

In addition to the above constraints in routing school buses, there are objective functions that should be minimized; in particular, the number of school buses and the travel time of the students. Commercial school bus routing systems do not support all the factors that need to be considered when routing school buses. Therefore analysts rely on manual methods to route the school buses or manually change the routes generated by commercial systems to conform to the constraints.

Manual methods for routing school buses have their limitations as the human mind overloads rapidly when working with complex combinatorial problems. Analysts who deal routinely with combinatorial problems tend to rely on simplifying assumptions in order to lessen the degree of complexity. It has been observed that manual solutions for complex combinatorial problems are 5-30% short of optimal solutions (measured in vehicles and/or total miles traveled) (Bodin and Berman (1979)).

The average annual student transportation cost for a rural school district, using either its own buses or contracted buses, is approximately 40% of the annual school district budget. A school district that manually routes buses designs routes with little attention to the quality or "goodness" of the resulting routes. Since there are no alternate school bus routes which may serve as a point of reference for the quality of the analyst's manually created bus routes, the first feasible set of routes obtained become s the final set of routes.

Instead of the above complex routing constraints and objective functions, the school district takes into consideration essentially the following three important factors that affect the routing process:

1. Local/State regulations governing the transportation of students
2. Reimbursements obtained by school districts
3. Travel time of students

When routing school buses, the first priority is to ensure that local and State regulations governing the transportation of students are observed. The next step in the process is to route the school buses such that one can obtain the maximum reimbursement from the State. The reimbursements received by the school district is positively correlated to the total linehaul, rather than on the efficiency of the routes, such as the reduction in the number of school buses used or travel time of the students.

We now consider the above three factors and discuss how each one of them influences the routing process.

2.3 Local and State Regulations

Local and State governments have rules and regulations governing the transportation of students. These rules and regulations are for the safety of the students. The most important regulations that govern the transportation of students are:

- Students are assigned to pickup locations such that the path they have to take from their home to the location should not be hazardous.
- Students within one mile of school are required to walk to school unless the path to the school is deemed hazardous.

These rules are very subjective and cannot be easily automated. As such, the district transportation officer's knowledge is used for determining the assignment of students to pickup locations.

2.4 Reimbursement for School Districts

In Pennsylvania, the State and local governments reimburse the cost of transporting students to public schools. A high percentage of the student transportation cost is reimbursed by the State using a complex reimbursement formula. The percentage of transportation cost not reimbursed by the State is covered by the local government using income from school taxes levied on the local residents of that school district. The complex reimbursement formula used by the State is based on factors such as:

- Total number of school buses
- Year of manufacture of the bus chassis
- Capacity of each school bus
- Average number of miles traveled by the bus for the school year
- Average number of miles traveled by the bus on a single day

- Total number of students traveling on the bus each day
- Cost Price Index (CPI) for the year. The CPI is used to determine the rate of inflation
- An aid ratio which computes the total taxes that are collected from the residents of the district

The formula involving the above factors, we believe, has evolved over time and consists of incremental additions appended to the original formula over the years. In further studying the formula using linear programming models, the primary factor having the largest impact on the cost was the total mileage traveled by the bus. The secondary factor was the total number of students in a bus. Transportation cost can be minimized by maximizing the number of students in a bus and the total travel time of the bus. As a school bus has limited capacity and needs to minimize the maximum travel time of a student, the objective is to find a set of routes that minimizes the total distance traveled by the buses, with each student seated comfortably in the bus.

2.5 Travel Time of Students

Fig. 1 shows the bus route for four students that are to be transported to a school. The strategy is to pickup the student that is furthest away from the school, Student 4, and then design a route that picks up the other students as the bus winds its way towards the school. That is, Student 4 would be picked up first followed by Student 3, then Student 2 and then Student 1, where Student 1 is closest to the school.

This would be the most efficient route from the students point of view, as the student closest to the school has to travel the minimum distance and no student travels any further than Student 4 who is furthest away from the school. The deadhaul distance, or the distance for which the bus travels without any students, is the distance from the school to Student 4.

When school districts route school buses, inefficient routing principles are used in order to increase the reimbursement. Fig. 2 shows the type of routes used by school buses to maximize reimbursements, resulting in students traveling a greater distance.

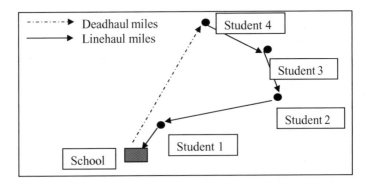

Fig. 1. An Efficient School Bus Route to Minimize Student Travel Time

The deadhaul distance for Fig. 2 is the distance from the school to Student 1. Most school districts use a routing strategy similar to Fig. 2, even though such strategy increases the travel time for most students. In order to increase reimbursement, school bus routes are designed to minimize deadhaul miles at the cost of increasing student travel time. A student who is closest to the school is usually picked up first, to minimize deadhaul, followed by other students. Another factor contributing to the adoption of inefficient routing strategies is that the State does not reimburse the school district for the deadhaul miles traveled that exceed the linehaul distance. This is counter-productive to the principle of reimbursement, resulting in the State and local governments, as well as the students, incurring higher costs.

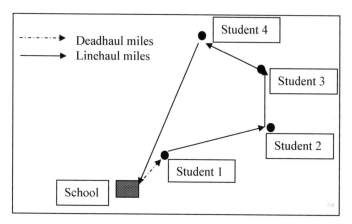

Fig. 2. A School Bus Route that Maximizes Reimbursement

The policy on reimbursement should be correlated to the efficiency of the travel time of the students. This would result in efficient routes that minimize the total distance traveled by the students.

3 Digitized Road Network Map

Pennsylvania is comprised of counties, townships and boroughs. That is, the State is divided into counties, which are further divided into townships and the boroughs exist within the townships. Rural school districts are comprised of multiple boroughs and townships. The distance between two student locations in a rural school district may be geographically short, but the traveled distance may be far off based on the available road network and the conditions of such roads. For example, in Fig. 3 the Euclidean or Manhattan distance between Student-1 and Student-2 is smaller than the road network distance, which involves traversing road segments <J, I>, <I, H>, <H, K> and <K, D>. The use of Euclidean or Manhattan distance is not a good measure of the travel distance between two student locations especially in rural areas.

Unlike a road network in an urban setting, the majority of rural areas do not have grid-like road networks. Rural roads wind around natural barriers such as rivers, streams or hills. In addition, rural areas have low density road networks with man-made barriers such as railroads and farmlands. In order to use realistic distances between locations one has to use the actual digitized road networks to calculate the distance.

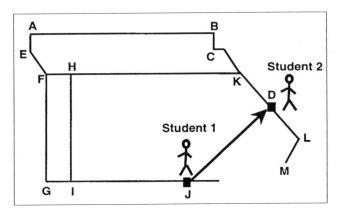

Fig. 3. Euclidean Distance Versus Road Network Distance for Traveling from Pickup Location J to Pickup Location D

The cost of obtaining digitized road network data can be prohibitive. A more value-based solution is to obtain a free copy of the Tiger maps from the U.S. Department of Census. Most commercial companies use the digitized maps obtained from the Census Bureau as the base and refine it using satellite imagery and physical road surveys. For the purpose of this research the Tiger maps from the U.S. Census Bureau proved more than adequate. The road networks in the Tiger files are a collection of road segments. Each road segment is a sequence of road links that define the shape of the road and do not have any intersections except at the starting and ending points of the road segments. For this research, special data structures were implemented to extract the data from the Tiger files in order to use such data for computing road network distances, which were used to compute shortest-path distances between various points on the map.

Each student has a residence and a pickup location. Depending on the location of the student's residence, either the residence itself could be the pickup point or the student would have to walk to a pickup point. The transportation officer determines the assignment of pickup points on the digitized road segments. The digitized road network was used to compute the shortest path between student pickup points, locations of the schools, contractor depots and bus driver homes. The shortest path between two locations on the map was computed using Dijkstra's algorithm (Horowitz and Sahni (1988)). The shortest path distances obtained from the digitized networks were used for solving the rural school bus routing problem.

4 Rural School Bus Routing Problem Formulation

In this section we discuss the various facets of the rural school bus routing problem (RSBRP), provide the conceptual and mathematical formulation of the problem, and address the constraints of a potential solution.

4.1 Multi-Depot, Mixed-Fleet, Site-Dependent, Split-Delivery VRP

When routing school buses, heuristics based on VRP's can be used to solve the problem. However, this reliance on VRP heuristics usually leads to an oversimplification of the problem, as there are a number of factors that make the RSBRP more complex than the classical VRP. Those factors that contribute to the complexity of RSBRP are discussed in this section.

In designing routes for a rural school district, the transportation officer is required to send out bids for the school bus companies. Multiple contractors bid to route the students for the school year. Each contractor has a set of multiple-capacity, multiple-cost buses to transport students. Each bus can start from a depot, warehouse, bus driver's home or a school. As the school buses can start from multiple locations, RS-BRP introduces the added complexity of multiple depots into the problem. As buses used for transporting students are mixed-fleet, i.e., vehicles of varying capacities and sizes, the RSBRP becomes a multi-depot, mixed-fleet VRP.

School buses servicing students are constrained by the roads, turns and the needs of students to be serviced. For example, smaller buses would be used to pickup students on narrow roads, since the smaller buses are more capable of negotiating the narrow turns. Students in wheelchairs would need a bus with a wheelchair lift. Students with different needs at an assigned pickup point have to be serviced by a compatible bus, thus introducing site-dependency to RSBRP. That is, the type of buses servicing a pickup point is dependent on the mix of students at that pickup point. In addition, the students at one pickup location can be serviced by more than one bus, even if the students have the same needs. As multiple-buses might be required to service the same pickup location, the RSBRP has a component of split-delivery.

The combination and presence of the different components discussed above makes RSBRP a multi-depot, mixed-fleet, site-dependent, split-delivery VRP, which is more complex than the classical VRP.

The multi-depot VRP (MDVRP) has one more layer of complexity than a standard VRP. As the name suggests, multi-depot problems contain more than one depot. Since any bus can be assigned to any depot, and multiple buses can start from a single depot, each depot with its set of buses resembles a VRP (Laporte *et al.* (1988), Chao *et al.* (1993), Renaud *et al.* (1996), Cordeau *et al.* (1997), Thangiah and Salhi (2001)).

In a mixed-fleet VRP (MFVRP), the set of trucks used to solve the problem are heterogeneous. Each type of truck has a limited capacity in addition to fixed and variable costs. Each truck must still service a series of customers without exceeding its capacity. The total cost, which is a function of the fixed cost, variable cost and distance, is to be minimized. The goal is to find the best fleet composition that can service all available customers, while minimizing the cost (Gheysens *et al.* (1984), Golden *et al.* (1984), Desrochers and Verhoog (1991), Chao *et al.* (1993), Salhi and Rand (1993), Cordeau *et al.* (1997), Gendreau *et al.* (1999)). The multi-depot, mixed-fleet VRP is a combination of MDVRP and MFVRP (Salhi and Sari (1997)).

In a site-dependent VRP the set of trucks is not only heterogeneous, but so is the set of customers. A one-to-one relationship exists between customers and trucks. As each type of truck can only visit one type of customer, it can be characterized as a multilevel routing problem. At the first level, customers are mapped to trucks. At the second level, a VRP is solved for each type of vehicle. At the final level, a TSP is solved for each route (Nag *et al.* (1998), Chao *et al.* (1999), Cordeau and Laporte (2001), Chao *et al.* (2004)).

In a split-delivery VRP a customer can be serviced by more than one vehicle. For example, if a customer has to ship products that cannot fit into a single truck, the products are split so that it can be distributed between two or more trucks (Dror and Trudeau (1989), Dror *et al.* (1994)). Similarly, in the RSBRP, the students assigned to a pickup point may be serviced by more than one bus.

In the RSBRP each bus starts from a depot (whose location varies from one bus-contractor to another) or a school and terminates at one of the schools. The starting and ending points of the school buses can be different. In addition, depending on the starting times of the schools, a school bus route may have more than one school as terminating points. That is, both elementary and secondary school students may be transported on the same school bus.

4.2 Mathematical Notation

Parameters:

$$
\begin{aligned}
P &= \{1, \ldots, p_{max}\} \text{ set of pickup points} \\
N &= \{1, \ldots, n_{max}\} \text{ set of students} \\
T &= \{1, \ldots, t_{max}\} \text{ set of available school buses} \\
U &= \{1, \ldots, u_{max}\} \text{ set of source or depot locations} \\
V &= \{1, \ldots, v_{max}\} \text{ set of destinations or schools} \\
Y &= \{1, \ldots, y_{max}\} \text{ set of vehicle types}
\end{aligned}
$$

Decision Variables:

$R_{i,j}$ = shortest road network distance between pickup points i and j $(i, j \in P)$

P_i^u = shortest road network distance from depot u to pickup point i $(i \in P, u \in U)$

Q_i^v = shortest road network distance from pickup point i to destination v $(i \in P, v \in V)$

R_u^v = set of shortest paths from depot u $(u \in U)$ to school(s) v $(v \in V)$

r_{iy}^{uv} = route i, which is served by vehicle of type y $(y \in Y)$, starting from depot u $(u \in U)$ to school v $(v \in V)$

C_i^{uv} = least extra cost of servicing student i $(i \in N)$ from depot u $(u \in U)$ to school v $(v \in V)$

p_i^k = pickup point i $(i \in P)$ serviced by the vehicle which services route k $(k \in \{r_{jy}^{uv}, \forall j\})$

q_i = demand of the ith $(i \in N)$ student, assumed to be of unit value

S_{ip}^y = student i $(i \in N)$ requiring vehicle type y $(y \in Y)$ assigned to pickup point p $(p \in P)$

Functions:

$N(S_{ip}^y)$ = total number of students at pickup point p

CAP_y = capacity of vehicle type y $(y \in Y)$

VC_y = variable cost of vehicle type y $(y \in Y)$

FC_y = fixed cost of vehicle type y $(y \in Y)$

$Q(r_{iy}^{uv})$ = sum of demands of students in route r_{iy}^{uv}

$L(r_{iy}^{uv})$ = length, in miles, of route r_{iy}^{uv}

$TT(r_{iy}^{uv})$ = total travel time of route r_{iy}^{uv}

$TC(r_{iy}^{uv})$ = total cost for servicing route r_{iy}^{uv}

AC_r^n = cost of appending student n to an existing route r

NC_r^n = cost of inserting student n to a new route r

4.3 School Bus Assignment to Source (Depot)

School buses and drivers are associated with a contractor and can start from either the contractor depot, a warehouse belonging to the contractor, the driver's home or from a school. Initially all school buses serving a particular route are assumed to start from the same depot (a school). That is, for each route, r_{iy}^{uv}, $y \in Y, U = V$ and $|U| = 1$. During the local optimization of the feasible solution obtained for the RSBRP, the school buses are assigned to different starting depots to evaluate the cost of buses starting from locations other than schools.

4.4 School Bus Assignment to a Destination (School)

Each school bus is capable of having either one single destination or multiple destinations. An example of a single destination is a bus that services all students going to an elementary school or a middle school, but not both. An example of a bus with multiple destinations is a bus that services all students going to either an elementary school or a middle school. In the latter case, the bus will first drop off all students going to an elementary school followed by dropping off all students going to the middle school. The implemented heuristics for the RSBRP can handle either of these cases. Thus in each route, r_{iy}^{uv}, $v \in V$ and $|V| \geq 1$.

4.5 Student Assignment to Pickup Points

School buses pickup students from a designated pickup point. The pickup point to which a student is assigned is based on safety regulations, which are enacted by State and local governments. At the current time, there is no algorithmic formulation that can assign students to pickup points. Such assignment is left to the director of transportation for the school. The assignment S_{ip}^{y} of student i $(i \in N)$ to pickup point p $(p \in P)$ is done manually such that $N(S_{ip}^{y}) \geq 1$. Assignments of students to a pickup point are constrained by:

- The location of the pickup point on a road segment. That is, the pickup point might be located on a road segment that is too narrow for a bus with a long body to take turns. Road constraints also include sections where a road is too narrow or winding for a bus to traverse safely.
- The type of students at the pickup point. Students with or without additional needs might be assigned to the same pickup point. A bus that does not have such resources will not service students with additional needs, such as wheelchair lifts or monitors on bus.

Though students are assigned to a pickup point on a road network, there is no guarantee that all students at that pickup point will be serviced by one vehicle type. That is, a student may be served by any vehicle type that satisfies the student's minimum needs.

4.6 School Bus Turn Constraints

A turn constraint represents an instance where a bus cannot travel from one road segment to another. Blind turns before intersections and places where the crest of a hill obstructs another driver's view are examples of turn constraints. School buses are much longer than cars; therefore, turns that would be safe for a car to make may be dangerous for a school bus. School bus turn constraints, based on vehicle type, are integrated into the routing process.

The next section details cost functions for routing school buses used in heuristics implemented to solve the RSBRP.

5 Cost of Routing School Buses

Designing heuristics to solve the RSBRP requires a metric to measure the cost of routing school buses. The metric has to take into consideration locations of the pickup points, assignment of students to pickup points, type of students, total number of school buses with fixed and variable costs, type of vehicles, and the shortest road network distances between all points such as the depots, schools and pickup points, in addition to road and turn constraints.

5.1 Fixed and Variable Costs in School Bus Routing

Each academic year the school district accepts bids from multiple contractors for transporting students. Each contractor provides the total number of buses available with the maximum and minimum capacities of each bus, the type of equipment available on the bus, such as wheelchair lifts, and the fixed and variable cost for the bus.

The total number of school buses tendered by contractors is usually greater than the total required by the school district. Therefore, if $Q(r_{iy}^{uv}) = 0$, then it is assumed that the vehicle was not selected in the bidding process. The cost of using vehicle i in the routing process can be computed as: $FC_i + (VC_i \times L(r_{iy}^{uv}))$

The total cost of transporting students is computed as follows:

$$\sum_{\forall i \in T} FC_i + (VC_i \times L(r_{iy}^{uv})), \ \forall Q(r_{iy}^{uv}) \neq 0 \tag{1}$$

In Equation 1, reduction in the total number of vehicles is done implicitly. The primary objective is to schedule all the students while the secondary objective is to minimize the total distance and the number of vehicles. Reduction of cost in Equation 1 will implicitly lead to reduction in vehicles and distance.

5.2 Cost of Inserting a Student into a Route

When inserting a student S_{jp}^y into a route r_{iy}^{uv}, the least cost of insertion, C_j^{uv} is computed in the following manner. A student can be inserted into a non-empty route, with at least one student in it (Insertion Type I), or into an empty route (Insertion Type II). When assigning a student to a vehicle on the route, the type of vehicle required by the student and the vehicle type servicing the student must be compatible. The implemented heuristics uses either Insertion Type I or Insertion Type II.

Inserting Students into an Existing Route: Insertion Type I

- Compute the total cost of serving before and after inserting student n

$$TC(r_{iy}^{uv}) = (VC_y \times L(r_{iy}^{uv})) + FC_y$$

$$TC(r_{iy}^{uv} \cup \{n\}) = (VC_y \times \left[L(r_{iy}^{uv}) + [R_{pre-p,p} + R_{p,post-p} - R_{pre-p,post-p}]\right]) + FC_y \tag{2}$$

- Compute added cost of inserting student i into the existing route $r = r_{iy}^{uv}$

$$AC_n^r = TC(r_{iy}^{uv}) - TC(r_{iy}^{uv} \cup \{n\}) \tag{3}$$

Student n is inserted into route r_{iy}^{uv} between two successive pickup point's *pre-p* and *post-p* in the route with the least cost computed using Equation 3. The two points, *pre-p* and *post-p*, can be a depot and a student pickup point, two student pickup points or a student pickup point and a school, respectively. The vehicle type for route r_{iy}^{uv} must be compatible with the vehicle type requested by student S_{jp}^y. In addition, for insertion to take place into the route r_{iy}^{uv}, the constraint $CAP_y \geq Q(r_{iy}^{uv}) + q_n$ must be satisfied.

Inserting Students into an Empty Route: Insertion Type II

Insertion Type II inserts a student into a new bus that is empty. The new cost of adding a bus route $r = r_{iy}^{uv}$ to service a student $n = S_{jp}^{y}$ is calculated as:

$$NC_r^n = (VC_y \times (R_{u,p} + R_{p,v})) + FC_y \tag{4}$$

The student n is inserted in route r_{iy}^{uv} between a depot and a school at a cost obtained using Equation 4 with a vehicle type capable of serving the student. That is, the vehicle type must match the requested student type. In addition, for insertion to take place in route r_{iy}^{uv}, the constraint $CAP_y \geq q_n$ must be satisfied.

6 School Bus Routing Heuristics

A solution to the RSBRP is obtained using cost Equations 3 and 4 by first obtaining an initial feasible solution and then improving the solution by minimizing Equation 1. The improvement of a route is achieved using intra-route and inter-route local optimization methods.

6.1 Obtaining an Initial Solution to the Problem

In order to obtain an initial feasible solution, the following algorithm is used:

> *Sort all available school buses in increasing order of capacity*
> **for** *each available bus* $t := 1$ *to* t_{max} **loop**
>   **for** *each* $S_{ip}^{x} \in N$ $(i := 1, \ldots, |n_{max}|)$ **loop**
>   **if** $(x = y$ *in* r_{iy}^{uv} *and* $S_{ip}^{x})$ **and** $(q_i + Q(r_{iy}^{uv}) \geq CAP_y)$ **then**
>   *Insert* S_{ip}^{x} *into route* r_{iy}^{uv} *using Eq. 3*
>   **else**
>   *Insert* S_{ip}^{x} *into empty route* r_{iy}^{uv} *using Eq. 4*
>   **end**
>   *Execute intra-route optimization*
>   *Increment* $Q(r_{iy}^{uv})$ *by* q_i
>   *Tag* S_{ip}^{x} *as assigned*
>   **end**
> **end**

The above algorithm gives us an initial feasible solution. Though each student is being inserted independently into a route, students are clustered by the pickup points to which they have been assigned. That is, the bus will visit a pickup point only once in its route as all students belonging to that pickup point are clustered together.

Once the initial solution is obtained, both intra- and inter-route improvement heuristics are applied to improve the solution.

6.2 Intra-Route Local Optimization Methods

The intra-route heuristics locally optimize a single route using methods such as 1-opt and 2-opt. Local optimization methods 1-opt and 2-opt (Lin (1965), Lin and Kling-man (1973)) operate on a single route in order to reduce the distance traveled along a bus route. The local optimization methods move pickup points to a different location within a route, if the move leads to a reduction in Equation 1. The local optimization starts with an arbitrary Hamiltonian Cycle, in this case the route under consideration. Assuming each pickup point on the route is a node and the path between the pickup points is an edge, the local optimizations removes links, and creates new ones. After each switch, the feasibility of the route is checked and the cost is calculated. All possible combinations are checked and the combination that leads to the maximum savings is retained.

6.3 Inter-Route Improvement Heuristics

The inter-route improvement heuristic moves students between routes, relocates the starting point of a bus and reduces the total number of buses required to transport students in order to minimize transportation cost. These heuristics are similar to the ones implemented by Salhi and Rand (1993) for solving the MFVRP. The Salhi-Rand heuristics had unlimited trucks available for selection and did not have to consider site-dependent and split-delivery of customers. The heuristic methods implemented for the RSBRP are the Student-Interchange, Sharing, Reduction, Combine, and Swap, which are discussed in the following sections.

Student-Interchange Heuristic Method

The student-interchange method is based on the interchange of customers between sets of routes. This technique has also been successfully applied to solve complex VRPs (Osman and Christofides (1994), Thangiah *et al.* (1993), Thangiah (1996), Thangiah *et al.* (1996), Thangiah and Petrovic (1998)).

Given a solution to the problem represented by a set of routes $S = \{R_1, \ldots, R_p, \ldots, R_q, \ldots, R_K\}$, where each route is the sequence of students serviced on this route, a student-interchange between a pair of routes R_p and R_q is defined as a replacement of a sequence of students $S_1 \subseteq R_p$ of size $|S_1| \leq \Theta$ by another sequence $S_2 \subseteq R_q$ of size $|S_2| \leq \Theta$ to get two new routes $R'_p = (R_p - S_1) \cup S_2$, $R'_q = (R_q - S_2) \cup S_1$ and a new neighboring solution $S' = \{R_1, \ldots, R'_p, \ldots, R'_q, \ldots, R'_K\}$. More specifically, if one of the sequences is empty, then the students of one route are simply moved to the other route (all possible insertion places being considered). If both sequences contain at least one student, then these sequences are swapped (i.e., each sequence takes the place of the other sequence in each route). The neighborhood $N_\Theta(S)$ of a given solution S is the set of all neighbors S' generated in this way for a given value of Θ. The order in which the neighbors are searched is specified as follows for a given solution $S = \{R_1, \ldots, R_p, \ldots, R_q, \ldots, R_K\}$:

$$(R_1, R_2), (R_1, R_3), \ldots, (R_1, R_K), (R_2, R_3), \ldots, (R_2, R_K), \ldots, (R_{K-1}, R_K)$$

Hence, all possible pairs of routes (R_p, R_q) are examined to define a cycle of search. For a given pair of routes (R_p, R_q), the order of application of the student-interchange operators must also be defined. Here we consider the case $\Theta = 2$ that results in one or two students being shifted from one route to another or exchanged between two routes. The search in the neighborhood of the current solution applies the operators in the following order on each pair of routes: (0,1), (1,0), (1,1), (0,2), (2,0), (2,1), (1,2) and (2,2). The operators (0,1), (1,0), (2,0) and (0,2) on routes (R_p, R_q) indicate a shift of one or two students from one route to another. The operator (1,1) indicates an exchange of one student between the two routes. The operators (1,2), (2,1) and (2,2) are defined similarly and indicate an exchange of students between the two routes.

For a given operator and a given pair of routes, the students are considered sequentially and systematically along the routes in order to find a better solution. Once the generation of the neighborhood is established, the first improvement strategy selects the first solution found in the neighborhood of the current solution. The strategy accepts the first neighboring solution that decreases the cost of the current solution.

Sharing Heuristic Method

The sharing heuristic removes all pickup points from a bus and allocates them to other non-empty buses. All student movements consist of moving the pickup points between buses. When a pickup point is moved, all the students that are associated with that pickup point are moved as a block. If all the removed students cannot be allocated into other non-empty buses, they are placed into an empty bus. After all the pickup points from the initial bus are placed in other buses, the cost is calculated. If the new cost is less than the initial cost, the routes are retained. If not, the original routes are restored and the next non-empty bus is selected for sharing. The heuristic implementation is as follows:

> Let $Q(r_{iy}^{uv}) \geq 1, i = 1, \ldots, m$
> Let $Q(r_{jy}^{uv}) = 0, j = m + 1, \ldots, t_{max}$
> **for** $(i := 1 \text{ to } m)$ **loop**
> $\quad C1 := \sum_{h=1}^{m} TC(r_{hy}^{uv})$
> $\quad$ **for** *each* $x \in r_{iy}^{uv}$ *where* $x \in P$ **loop**
> $\quad\quad$ *Transfer* x *to* r_{ky}^{uv} *where* $k = 1, \ldots, m$ *and* $k \neq i$ *using Eq. 3*
> $\quad\quad$ **if** *x was not transferred* **then**
> $\quad\quad\quad$ *Transfer* x *to* r_{hy}^{uv} *using Eq. 4*
> $\quad\quad$ **end**
> $\quad$ **end**
> $\quad$ **if** $(Q(r_{iy}^{uv}) = 0)$ **then**
> $\quad\quad C2 := \sum_{h=1}^{m} TC(r_{hy}^{uv})$ *where* $h \neq i$
> $\quad\quad$ **if** $(C2 < C1)$ **then**
> $\quad\quad\quad$ *Keep the changes*
> $\quad\quad$ **else**
> $\quad\quad\quad$ *Restore old routes*

>   **end**
>   **else**
>   *Restore old routes*
>   **end**
> **end**

Reduction Heuristic Method

The Reduction optimization removes all pickup points from a bus and moves them to other non-empty buses. The Reduction optimization will not use new buses. Once a bus has emptied, the new cost is calculated. If the new cost is less than the initial cost, the new routes will be retained; otherwise, the original routes are restored. The heuristic implementation is as follows:

> Let $Q(r_{iy}^{uv}) \geq 1, i = 1, \ldots, m$
> Let $Q(r_{jy}^{uv}) = 0, j = m + 1, \ldots, t_{max}$
>   **for** $(i := 1$ to $m)$ **loop**
> $C1 := \sum_{h=1}^{m} TC(r_{hy}^{uv})$
>   **for** *each* $x \in r_{iy}^{uv}$ *where* $x \in P$ **loop**
>   *Transfer* x *to* r_{ky}^{uv} *where* $k = 1, \ldots, m$ *and* $k \neq i$
>   **if** $(Q(r_{iy}^{uv}) = 0)$ **then**
> $C2 := \sum_{h=1}^{m} TC(r_{hy}^{uv})$ *where* $h \neq i$
>   **if** $(C2 < C1)$ **then**
>   *Keep the new routes*
>   **else**
>   *Restore old routes*
>   **end**
>   **else**
>   *Restore old routes*
>   **end**
>   **end**
>   **end**

Combine Heuristic Method

The combine heuristic removes all the students from two buses and assigns them into one empty bus. This heuristic tries to reduce the total cost by trading fixed and variable costs of two bus routes for the fixed and variable cost of one larger bus. In addition, the newly created bus route is relocated to all compatible depots in search of a starting location for the bus that reduces the total travel time. If the newly created route has a lower cost than the previous two routes, the new route is retained; otherwise the two old routes are restored. The heuristic implementation is as follows:

> Let $Q(r_{iy}^{uv}) \geq 1, i = 1, \ldots, m$
> Let $Q(r_{jy}^{uv}) = 0, j = m + 1, \ldots, t_{max}$

for $(i := 1$ to $m - 1)$ **loop**
$\qquad C1 := \sum_{h=1}^{m} TC(r_{hy}^{uv})$
$\qquad$ **for** $(j := m + 1$ to $t_{max})$ **loop**
$\qquad\qquad$ **for** *each* $l \in r_{iy}^{uv}$ **and** $n \in r_{i+1,y}^{uv}$ where $l, n \in P$ **loop**
$\qquad\qquad\qquad$ *Transfer* l, n to r_{jy}^{uv}
$\qquad\qquad$ **end**
$\qquad$ **end**
$\qquad$ **if** $(Q(r_{iy}^{uv}) = 0)$ **and** $(Q(r_{i+1,y}^{uv}) = 0)$ **then**
$\qquad\qquad C2 := \sum_{k=1}^{m} TC(r_{ky}^{uv})$
$\qquad\qquad$ **if** $(C2 < C1)$ **then**
$\qquad\qquad\qquad$ *Keep the new routes*
$\qquad\qquad$ **else**
$\qquad\qquad\qquad$ *Restore old routes*
$\qquad\qquad$ **end**
$\qquad$ **else**
$\qquad\qquad$ *Restore old routes*
$\qquad$ **end**
end

Swap Buses Heuristic Method

The swap buses heuristic relocates the starting points of buses to find new routes with reduced travel times. Each route has a starting and ending depot. The ending depot is the school where the student is being dropped off. The starting depot is relocated in search of solutions that reduce the total route cost. In this heuristic, each bus is assigned to each of the compatible depots. If the travel time and cost is reduced after a bus is relocated to a different depot, the bus with the new depot is retained. If the relocation leads to an increase in cost or travel time, the bus is restored to its old starting depot. This is done for all buses. The heuristic implementation is as follows:

$\qquad$ Let $Q(r_{iy}^{uv}) \geq 1, i = 1, \ldots, m$
$\qquad$ Let $Q(r_{jy}^{uv}) = 0, j = m + 1, \ldots, t_{max}$
$\qquad$ **for** $(i := 1$ to $m)$ **loop**
$\qquad\qquad C1 := \sum_{h=1}^{m} TC(r_{hy}^{uv})$
$\qquad\qquad$ **for** $(x := 1$ to $u_{max})$ **loop**
$\qquad\qquad\qquad$ **if** x *is compatible with* y **then**
$\qquad\qquad\qquad\qquad C2 := \sum_{h=1}^{m} TC(r_{hy}^{uv})$
$\qquad\qquad\qquad\qquad$ **if** $(C2 < C1)$ **then**
$\qquad\qquad\qquad\qquad\qquad$ *Keep the new routes*
$\qquad\qquad\qquad\qquad$ **else**
$\qquad\qquad\qquad\qquad\qquad$ *Restore old depot*
$\qquad\qquad\qquad\qquad$ **end**
$\qquad\qquad\qquad$ **end**
$\qquad\qquad$ **end**
$\qquad$ **end**

6.4 Heuristic for the RSBRP

The RSBRP implementation utilizes the above defined heuristics to solve the problem in the following sequence:

Step 1: *Obtain initial solution in Results*
Step 2: *Perform local 1-opt and 2-opt for each of the routes in Results*
Step 3: *count := 0; FoundCostImprovement :=* True
Step 4: **while** (*FoundCostImprovement* **and** *count* < *10*) **loop**

      Comment: Incrementally accumulate routes' improvements in *Results*
      FoundCostImprovement := False
      Apply student-interchange heuristic with $\Theta = 2$
      Perform local 1-opt and 2-opt for each of the routes
      Apply Sharing heuristic
      Perform local 1-opt and 2-opt for each of the routes
      Apply Reduction heuristic
      Perform local 1-opt and 2-opt for each of the routes
      Apply Combine heuristic
      Perform local 1-opt and 2-opt for each of the routes
      Apply Swap-Buses heuristic
      Perform local 1-opt and 2-opt for each of the routes
      Increment count by 1
      comment: Optimization heuristics set *FoundCostImprovement*
   end
Step 5: *Write out Results*

The above heuristic algorithm for the RSBRP was used to solve a real life problem from a rural school district.

7 Computational Results

The generalized heuristic algorithm described above was used to solve a RSBRP with data obtained from a local school district. The problem consisted of 583 students, 71 pickup points and 13 depots. The breakdown of the depots was three contractor's depots, four warehouses, and six driver's home depots. A total of five schools were used as destinations. A total of 18 school buses were made available through bids from contractors. The maximum travel time for a bus was set to 70 minutes as determined by the school district.

The type of students that were to be serviced for the local school district consisted of regular students, students needing wheel chair assistance, students requiring buses with wheelchair lifts and students who have to be monitored while on the bus. The 583 students consisted of 540 regular students (93%), 15 monitored students (3%), 18 wheelchair students (3%) and 10 wheelchair/lift students (1%). The heuristic algorithm was implemented in Java and executed on a 3.05 GHz Pentium IV machine with 1GB of RAM on a Windows 2000 operating system.

The solutions obtained by the implemented heuristics reduced the distance and the total number of buses in comparison to the manual solutions obtained by the school district. The solution available from the school district is not comparable to the solution obtained by the implemented heuristics due to a gulf between the cost function used by the school district to determine the efficiency of a bus route in comparison to actual cost efficiency of a route.

School districts tend to maximize the reimbursement that can be obtained from the State and local tax base. A five-minute increase in the route travel time of a school bus may lead to an approximate savings of 0.1% in transportation cost for each school bus, as this may help avoid adding new buses to the routing process. Similarly, decreasing the route travel time of a school bus by five minutes may result in an approximate increase of 0.1% in reimbursement to the school district for each school bus. As school districts are not reimbursed for deadhaul distances that exceed the linehaul distances, manual routes tend to either minimize or eliminate the deadhaul distances entirely for a school bus. Reduction in deadhaul distances lead to an increase in the travel time of a student as a school bus would pick the student closest to the starting depot at the start of the journey.

The objective of the implemented heuristic algorithm was to reduce the transportation cost. Reduction in transportation cost results in maximizing deadhaul distance, minimizing the total travel time of the students and minimizing the total number of school buses and distances traveled by the school buses. The implemented heuristics were tested using different methods for obtaining initial solutions. The two main methods of obtaining an initial solution were by assigning the selected student to the first available school bus or the best available school bus in terms of cost. For each of these assignments, the students were picked up according to three different strategies: the order of the furthest away from the depot, the order of the closest to the depot or in a random order.

Table 1 details solutions obtained by placing students in the first school bus in terms of minimal cost. Table 2 details solutions obtained by placing students in the best feasible school bus. All the school buses start from a single depot initially. The local heuristics search for alternate starting depots for the school buses during the implementation. In all the solutions, the results indicate that it is advantageous to start from multiple depots when servicing the students. This is not practiced currently by the school district.

Assignment of students to the first bus leads to feasible solutions irrespective of how the students are selected as detailed in Table 1. All solutions in Table 2 start from one depot but fan out to multiple depots. Selection of students, either randomly or in the order of the furthest away from the school, leads to buses starting from six depots compared to buses starting from five depots when selecting students in the order of the closest to the school.

When students are assigned to the best possible bus, as in Table 2, irrespective of how the students are chosen for placement, all solutions obtained terminate with some of the students not assigned to any of the school buses. Assignment of students to the best fitting school bus, initially, leads quickly to local optimization. The rapid convergence into a locally optimal solution, initiated by placing students in the best fitting bus, prevents the inter- or intra-route heuristics from improving the solution.

In addition, selection of students randomly and in the order of the furthest away also leads to a reduction in the total number of school buses used in transporting students. Reduction in school buses does not necessarily lead to cost savings as in a mixed-fleet problem, where two smaller buses could have been traded for a larger bus leading to an increase in cost. Both Tables 1 and 2[3] list the total cost of the buses in the Cost column, which gives the sum of the fixed and variable costs for all school buses used for transporting students. The best solution found by the heuristic is when students initially selected are furthest away from the school. All of the solutions allow deadhaul to be integrated into the routes as they lead to reduction in the travel time of students. The school buses that are used in the routing process also fill the buses to approximately 90% of their capacity.

Table 1. Details of Solutions Obtained by Placing Students Initially into the First Bus

First Bus	D	B	S	!S	PC	SDPC	TD	TT	DH	MaxTT	AvgTT	Cost	CPU
Closest Student													
initial solution	1	12	583	0	88.4%	0.02	257.29	366.09	32.22	53.31	6.44	2949.54	3.75
final solution	5	12	583	0	88.4%	0.03	235.29	353.25	23.20	49.01	5.04	2805.86	405.93
Furthest Student													
initial solution	1	12	583	0	79.3%	0.01	253.35	363.78	27.93	50.78	6.59	2918.97	4.20
final solution	6	11	583	0	89.4%	0.03	223.41	346.37	20.43	50.86	5.21	2629.45	466.90
Random Student													
initial solution	1	12	583	0	79.3%	0.02	327.52	407.05	30.92	54.36	6.72	2968.86	4.29
final solution	6	11	583	0	89.4%	0.03	228.56	349.32	17.52	46.51	4.69	2651.74	344.73

The solutions for the RSBRP obtained by the implemented heuristics indicate a number of factors that must be considered when routing school buses, namely:

- Deadhaul distances should be integrated into school bus routes to reduce the maximum travel time of students. Decreasing or eliminating deadhaul distances increases the cost of the bus routes and the maximum travel time of students.

---

[3] Legend for Tables 1 and 2:

D	Depots
B	Buses
S	Students Serviced
!S	Students Not Serviced
PC	Percent Capacity of Buses Filled
SDPC	Std. Dev. of Percent Capacity of Buses Filled
TD	Total Distance
TT	Total Travel Time
DH	Total Deadhaul Distance
MaxTT	Maximum Travel Time
AvgTT	Average Travel Time
Cost	Total Route Cost of Buses
CPU	CPU time in seconds on a 3.2 GHz Machine

Table 2. Details of Solutions Obtained by Placing Students Initially into the Best Bus

First Bus	D	B	S	!S	PC	SDPC	TD	TT	DH	MaxTT	AvgTT	Cost	CPU
Closest Student													
initial solution	1	14	563	20	84.6%	0.03	285.29	377.42	36.83	53.31	5.97	3148.75	12.58
final solution	8	12	563	20	88.3%	0.03	224.77	342.11	16.63	49.01	3.61	2637.91	437.33
Furthest Student													
initial solution	1	14	572	11	86.0%	0.03	281.95	377.72	38.80	50.78	5.89	3065.77	16.11
final solution	8	12	572	11	87.5%	0.04	233.22	349.29	21.84	50.86	3.05	2695.74	355.55
Random Student													
initial solution	1	14	577	6	86.7%	0.03	337.56	411.41	37.32	54.36	5.78	3068.25	15.36
final solution	6	12	577	6	89.1%	0.32	224.35	345.37	17.43	46.51	4.69	2663.25	491.42

- Buses starting from one central depot result in an increase in transportation cost. School buses should be routed from multiple depots to reduce the cost of transportation and maximum travel time of the students.
- Buses should be filled to 90% of its capacity. Increasing the number of students in a school bus to its full capacity will lead to an increase in the maximum travel time for students. Buses filled to full capacity can lead to uncomfortable rides for the students. Building a 10% redundancy in school bus capacity will allow for students to be added to a route without having to reroute school buses during the school semester.

State and local governments should look into the formula being used to reimburse transportation cost for school districts; as such formula is not an effective measure of minimizing either the cost of transportation or the travel time of students. State and local reimbursement formulas should have incentives for school districts that integrate deadhauls into their routes and reduce maximum travel time of students.

8 Conclusions and Future Directions

In this paper we have described the implementation of heuristics to solve a rural school bus routing problem that has multi-depot, mixed-fleet, site-dependent and split-delivery characteristics with side constraints. The heuristics were tested on real life data obtained from a rural school district. The implemented heuristics obtain cost effective solutions in under 10 minutes of CPU time on a Pentium IV machine running at 3.05 GHz.

There are a number of research avenues that deserve further investigations. At the current time all school buses start from a single depot, but it would seem that starting buses from multiple-depots might be advantageous. That is, assign each school bus to start from a different depot and then assign students to the bus. This would depend on how the students are distributed in the school district. In addition, the solutions can be further improved by using meta-heuristic search strategies such as genetic algorithms or tabu search. Even though adding a layer of meta-heuristics to the current set of heuristics will lead to an increase in the computational time, it would be a worthwhile effort as these routes are computed each semester and tend to stay the same for the entire semester. Therefore, expending extra processing time on the computational effort to get better solutions would result in efficient routes for the school district.

References

Bodin, L. and Berman, L. (1979). Routing and scheduling of school buses by computer. *Transportation Science*, **13**, 113–129.

Bodin, L., Golden, B. L., Assad, A. A., and Ball, M. O. (1983). Routing and scheduling of vehicles and crews. *Computers & Operations Research*, **10**, 63–211.

Chao, I.-M., Golden, B. L., and Wasil, E. (1993). A new heuristic for the multi-depot vehicle routing problem that improves upon best-known solutions. *American Journal of Mathematical & Management Sciences*, **13**, 371–401.

Chao, I.-M., Golden, B. L., and Wasil, E. (1999). A computational study of a new heuristic for the site-dependent vehicle routing problem. *INFORMS Journal on Computing*, **37**, 319–336.

Chao, I.-M., Golden, B., and Wasil, E. (2004). A computational study of a new heuristic for the site-dependent vehicle routing problem. *INFOR*, **37**(3), 319–336.

Christofides, N. and Eilon, S. (1969). An algorithm for the vehicle-dispatching problem. *Operational Research Quarterly*, **20**, 309–318.

Cordeau, J., Gendreau, M., and Laporte, G. (1997). A tabu search heuristic for periodic and multi-depot vehicle routing problems. *Networks*, **30**, 105–119.

Cordeau, J.-F. and Laporte, G. (2001). A tabu search algorithm for the site dependent vehicle routing problem with time windows. *INFOR*, **39**, 292–298.

Cordeau, J.-F., Gendreau, M., Laporte, G., Potvin, J.-Y., and Semet, F. (2002). A guide to vehicle routing heuristics. *Journal of the Operational Research Society*, **53**, 512–522.

Desrochers, M. and Verhoog, T. W. (1991). A new heuristic for the fleet size and mix vehicle routing problem. *Computers & Operations Research*, **3**, 263–274.

Dror, M. and Trudeau, P. (1989). Savings by split delivery. *Transportation Science*, **23**, 141–145.

Dror, M., Laporte, G., and Trudeau, P. (1994). Vehicle routing with split deliveries. *Discrete Applied Mathematics*, **50**, 239–254.

Fisher, M. L. (1995). Vehicle routing. In M. Ball, T. Magnanti, C. Monma, and G. Nemhauser, editors, *Network Routing. Handbooks on Operations Research and Management Science*, pages 1–33. North-Holland, Amsterdam.

Gendreau, M., Laporte, G., Musaraganyi, C., and Taillard, E. (1999). A tabu search heuristic for the heterogeneous fleet mix vehicle routing problem. *Computers & Operations Research*, **26**, 1153–1173.

Gheysens, F., Golden, B. L., and Assad, A. A. (1984). A comparison of techniques for solving the fleet size and mix vehicle routing problems. *OR Spektrum*, **6**, 207–216.

Golden, B., Assad, A. A., Levy, L., and Gheysens, F. (1984). The fleet size and mix vehicle routing problem. *Computers & Operations Research*, **11**, 49–66.

Horowitz, E. and Sahni, S. (1988). *Fundamentals of Computer Algorithms*. Computer Science Press, Maryland.

Laporte, G. (1992). The vehicle routing problem: An overview of exact and approximate algorithms. *European Journal of Operational Research*, **59**, 345–358.

Laporte, G. and Osman, I. H. (1995). Routing problems: A bibliography. *Annals of Operations Research*, **61**, 227–262.

Laporte, G., Nobert, Y., and Arpin, A. (1988). Optimal solutions to capacitated multi-depot vehicle routing problems. *Congressus Numerantium*, **44**, 283–292.

Lin, S. (1965). Computer solutions of the traveling salesman problem. *Bell Systems Technical Journal*, **44**, 2245–2269.

Lin, S. and Klingman, D. (1973). An effective solution to the traveling salesman problem. *Operations Research*, **20**, 498–516.

Nag, B., Golden, B. L., and Assad, A. A. (1998). Vehicle routing with site dependencies. In B. Golden and A. Assad, editors, *Vehicle Routing: Methods and Studies*, pages 149–159. North-Holland, Amsterdam.

Osman, I. and Christofides, N. (1994). Capacitated clustering problems by hybrid simulated annealing and tabu search. *International Transactions in Operational Research*, **1**, 317–336.

Renaud, J., Laporte, G., and Boctor, F. (1996). A tabu search heuristic for the multi-depot vehicle routing problem. *Computers & Operations Research*, **23**, 229–235.

Salhi, S. and Rand, G. K. (1993). Incorporating vehicle routing into the vehicle fleet composition problem. *European Journal of Operational Research*, **66**, 313–330.

Salhi, S. and Sari, M. (1997). A multi-level composite heuristic for the multi-depot vehicle fleet mix problem. *European Journal of Operational Research*, **103**, 95–112.

Serna, C. and Bonrostro, J. (2001). Minimax vehicle routing problems: Application to school transport in the province of Burgos. In S. Voss and J. Daduna, editors, *Computer-Aided Scheduling of Public Transport*, pages 297–317. Springer, Berlin.

Thangiah, S. R. (1996). Genetic algorithms for vehicle routing problems with time windows. In L. Chambers, editor, *Applications Handbook of Genetic Algorithms*, pages 253–277. CRC Press, Boca Raton.

Thangiah, S. R. and Nygaard, K. (1992). School bus routing using genetic algorithms. In G. Biswas, editor, *Proceedings of the Applications of Artificial Intelligence X: Knowledge-Based Systems*, pages 387–398. IEEE Press.

Thangiah, S. R. and Petrovic, P. (1998). Introduction to genetic heuristics and vehicle routing problems with complex constraints. In D. Woodruff, editor, *Advances in Computational and Stochastic Optimization, Logic Programming, and Heuristic Search*, pages 253–286. Kluwer Academic.

Thangiah, S. R. and Salhi, S. (2001). Genetic clustering: An adaptive heuristic for the multi-depot vehicle routing problem. *Applied Artificial Intelligence*, **15**, 361–383.

Thangiah, S. R., Osman, I. H., and Vinayagamoorthy, R. (1993). Algorithms for vehicle routing problems with time deadlines. *American Journal of Mathematical & Management Sciences*, **13**, 322–355.

Thangiah, S. R., Potvin, J.-Y., and Sun, T. (1996). Heuristic approaches to vehicle routing with backhauls and time windows. *Computers & Operations Research*, **23**, 1043–1057.

## Part III

Service Monitoring, Operations, and Dispatching

A Metaheuristic Approach to Aircraft Departure Scheduling at London Heathrow Airport

Jason A. D. Atkin[1], Edmund K. Burke[1], John S. Greenwood[2], and Dale Reeson[3]

[1] Automated Scheduling, Optimisation and Planning Research Group, School of Computer
Science and Information Technology, University of Nottingham, Jubilee Campus,
Wollaton Road, Nottingham, NG8 1BB, UK {jaa,ekb}@cs.nott.ac.uk
[2] National Air Traffic Services Ltd, NATS CTC, 4000 Parkway, Whiteley, Fareham,
Hampshire, PO15 7FL, UK
[3] National Air Traffic Services Ltd, Heathrow Airport, Hounslow, Middlesex, TW6 1JJ, UK

Summary. London Heathrow airport is one of the busiest airports in the world. Moreover, it
is unusual among the world's leading airports in that it only has two runways. At many airports the runway throughput is the bottleneck to the departure process and, as such, it is vital
to schedule departures effectively and efficiently. For reasons of safety, separations need to be
enforced between departing aircraft. The minimum separation between any pair of departing
aircraft is determined not only by those aircraft but also by the flight paths and speeds of aircraft that have previously departed. Departures from London Heathrow are subject to physical
constraints that are not usually addressed in departure runway scheduling models. There are
many constraints which impact upon the orders of aircraft that are possible and we will show
how these constraints either have already been included in the model we present or can be
included in the future. The runway controllers are responsible for the sequencing of the aircraft for the departure runway. This is currently carried out manually. In this paper we propose
a metaheuristic-based solution for determining good sequences of aircraft in order to aid the
runway controller in this difficult and demanding task. Finally some results are given to show
the effectiveness of this system and we evaluate those results against manually produced real
world schedules.

1 Introduction

London Heathrow is a busy two-runway airport which, due to its popularity with both
airlines and passengers, suffers severe aircraft congestion at certain times. Traffic in
airports is not evenly spread, for obvious reasons which pertain to airline and passenger preferences. There are, inevitably, times when the departure process is congested
but the arrivals are sparse. There are also times when the situation is reversed, and
times when both are congested. London Heathrow airport is actually situated on an
extremely small plot of land in comparison to other airports around the world and
with respect to how busy the airport is.

The airport capacity problem is concerned with estimating the capacity of an airport in terms of arrivals and departures. It has been examined for a number of years. Newell (1979) provided a model and showed that the capacity of the airport is increased when arrivals and departures can be alternated on both runways. Although mixed mode, where arrivals and departures are intermixed on a runway, is preferable for increasing the throughput, this is not currently possible at Heathrow due to the proximity of the surrounding residences. However, there is the future possibility of it being considered for peak times.

The departure flow at Logan airport was analysed in Idris *et al.* (1998a), Idris *et al.* (1998b), and Logan airport was compared to other major airports. Runway scheduling was seen to be a bottleneck upon the departure process and the authors concluded that it is vital to increase the throughput of the departure runway.

There are some similarities between the arrival and departure processes for the runways at an airport. Both processes are subject to sequence-dependent separation times between aircraft. Previous research has looked at the arrivals problem with the goal being to order arriving aircraft for a single runway so as to either minimise the total completion time or to minimise the total deviation from an ideal arrival time for each aircraft. Mixed integer zero-one formulations were presented in Beasley *et al.* (2000) and genetic algorithms were shown to be effective in Beasley *et al.* (2001).

Abela *et al.* (1993) looked at the arrivals problem for a set of aircraft with landing time windows. They presented a genetic algorithm to give an approximate solution and a branch and bound algorithm for solving the problem when formulated as a 0-1 mixed integer programming problem to give an exact solution. A heuristic approach for an upper bound and a branch and bound algorithm for the arrivals problem were given in Ernst *et al.* (1999). A network simplex method was used to assign arrival times given any partial ordering of aircraft. The arrivals problem, as it is presented in the literature, however, does not address the major constraints upon the departures problem at London Heathrow airport.

A constraint satisfaction based model for the departure problem was presented in van Leeuwen *et al.* (2002) for solution by ILOG Solver and Scheduler. A 15 minute time slot was assigned to each aircraft and separations were allocated based upon the size and speed of the aircraft and upon the exit point that the departing aircraft were going to use.

The departure process was analysed and a departure planner proposed in Anagnostakis *et al.* (2000), Anagnostakis and Clarke (2002) and Anagnostakis and Clarke (2003). A search tree was described and branch and bound techniques or an A* algorithm were recommended for solving the departure problem in Anagnostakis *et al.* (2001). A dynamic program was suggested in Trivizas (1998) to solve the departure order problem by limiting the possible number of aircraft that are considered for any place in the schedule, reducing the search space dramatically.

If only considering separations between adjacent aircraft and ignoring the physical constraints from the holding points, the departure problem can be seen to be a variant of the single machine job sequencing problem where jobs have sequence-dependent processing or set-up times. Substantial research has been undertaken into this problem. For example, Bianco *et al.* (1999) looked at the generalised problem with release dates as well as sequence-dependent processing times, showing the equivalence to the cumulative asymmetric travelling salesman problem with release dates. To ensure safety in the departure process, however, it is not possible to only consider adjacent pairs of aircraft and it is easy to produce schedules where all adjacent pairs have the required separations but other aircraft pairs do not.

Craig *et al.* (2001) did look at the effects of one holding point structure and gave a dynamic programming solution for scheduling take-offs. In practice, however, the holding point structures are more flexible than the one described here and a more general solution needs to be developed.

There are important constraints at London Heathrow airport that are not normally considered in the departure problem as it is presented in the current scientific literature. These are identified in the problem description below.

2 Problem Description

The objective of this paper is to increase the throughput of the departure runway subject to various constraints, with safety being paramount. There are currently only two runways in normal use at Heathrow; however, if environmental targets are met, there may be a possibility to add a third, parallel runway in the future. At any time of the day, only one runway can currently be used for departures.

The direction of the wind determines the direction in which the runways are used. The runways are labelled according to the direction in which they are employed and whether they are on the right or the left when facing that direction. The four runway configurations have been labelled in Fig. 1. For example, when arriving or departing heading west, the northern runway is referred to as 27R as it has a direction of 270 degrees and is the runway on the right.

There is actually a third runway already but this is only ever used for arrivals. It is shorter than the other two and not long enough for many Heathrow departures. It is used no more than twice per year. It also intersects both of the other runways so it is not practical to use it if either of the other two runways is in use. Indeed, it is usually used as a taxiway.

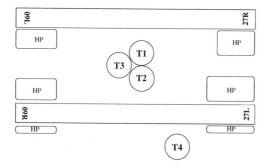

Fig. 1. The Layout of London Heathrow Airport

There are currently four terminals at London Heathrow, labelled T1 to T4 in Fig. 1. Three terminals are situated between the runways but the fourth is to the south of the southern runway.

When a flight is ready to depart a delivery controller has to give permission for engine start up. A ground controller then instructs the pilot in order to control the movement of the aircraft around the taxiways. Once an aircraft approaches the runway end and is no longer in conflict with any other aircraft the ground controller will relinquish control of the aircraft to the runway controller.

In this paper, we are concerned only with the operations of the runway controller. We assume that the ground controller and delivery controller are currently outside of the system and merely feed aircraft into the start of the system. Later research will look to include these roles into the model.

There are holding points, labelled HP in Fig. 1 at each end of each of the runways, and both north and south of the southern runway. Within these physical holding point structures the runway controller can reorder the aircraft before they reach the runway.

2.1 Holding Point Constraints

Aircraft go through holding points to get to the runways. Holding points can be considered to be one or more entrance queues to some maneuvering space where a final take-off order is produced for the runway. Where there are different entrance queues available, the ground controller will usually send an aircraft into the most convenient queue. The runway controller can request aircraft to be sent to specific queues but in practice, as the runway controller is very busy with the aircraft already in the holding points, there is rarely sufficient time to also consider the aircraft the ground controller has.

As mentioned before, Heathrow has very limited space so the holding point and taxi space is limited. Given the initial order of aircraft in the input queues to the holding points, the runway controller has to decide how to sequence the take-offs in order to maximise the throughput at the runway. This can be a very difficult task at times. Only limited amounts of reordering are possible at these holding points. The configuration of the holding points varies greatly between runway ends and will determine what reordering operations can take place and the costs involved in each operation.

2.2 Minimum Separations

To ensure safety, minimum separation times are imposed between aircraft taking off. The order of the aircraft for take-off can make a significant difference to the total delay that needs to be imposed upon the aircraft.

The minimum separation between aircraft is determined by:

- Wake Vortex: Large aircraft leave a stronger wake vortex than smaller, lighter aircraft and are also less affected by wake vortex. Every aircraft has a weight category and the wake vortex separation for any pair of aircraft can be determined by comparing their weight categories.

- Departure Routes: Aircraft will usually have a Standard Instrument Departure (SID) route assigned to them, giving a pilot a known departure route to follow. The relative SID routes of any two aircraft will impose a minimum departure interval between them. This ensures that safe minimum separation distances are kept while in flight. At times of congestion in the airspace, a larger than normal separation may be required between certain SID routes in order to increase the separation between flights heading into the congestion. These separations differ depending upon the runway in use at the time.
- Speed Group: The relative flight speeds of the aircraft can also make a difference to the separations which must be imposed upon aircraft flying the same or similar routes. The relative speed groups of the two aircraft modify the separation required for the relative SID routes. If the following aircraft will close the distance, then a larger initial separation is necessary. Conversely, if the following aircraft is slower then a lower separation can sometimes be applied.

The runway controller will aim for minimum separations between aircraft wherever possible. It should be noted here that a controller has some discretion as far as some separations are concerned. In particular some of the SID route based separations can be reduced in good visibility.

2.3 Other Constraints

The departure process is a dynamic system where aircraft are added to, and removed from, the system over time. The runway controller will have only limited knowledge about the aircraft that are not currently at the holding points.

The runway controller has a lot of information that is very hard to capture as hard data. In many cases a controller will be weighing the effects of contradictory constraints such as maximising throughput while minimising overtaking, to ensure fairness and minimising maneuvering, to reduce workload.

2.4 Overall Objective

The objective is to find candidate solutions for which the runway throughput is maximised and all constraints are met. We were told by one air traffic controller that the best figure obtained for Heathrow was 54 aircraft in an hour and that this figure is so good that it is extremely unusual.

For our research, we use a reduction in the holding point delay as a surrogate objective. Holding point delay is measured as the amount of time the aircraft spend in the holding point. Any objective to minimise this will have the effect of reducing the number of large separations and also of moving larger separations later in the take off order, so that they delay less aircraft. Moving larger separations to a later position in the schedule means that there is more opportunity to deal with them using new aircraft entering the system later. So a delay based objective for the problem at any instant in time is a good surrogate for a throughput based approach for the overall schedule. As the holding point arrival times are constant, the sum of take-off times could be used as an equivalent, but less meaningful, objective function.

3 Model Description

In this model we aim to maximise the throughput of the runway by minimising the total delay, D, suffered by the aircraft at the holding points. Let h_i be the arrival time for aircraft i at the holding point, where i is an integer ≥ 1. The integer i represents the position of the aircraft in the take-off order. If d_i is the take-off time for aircraft i from the runway, then we can calculate the total delay at the holding points using Equation (1) where n is the total number of aircraft departing.

We define a function $S(j, i)$ to give the minimum separation necessary between leading aircraft j and (not necessarily immediately) following aircraft i to meet all separation requirements. Function $S(j, i)$ incorporates all separation rules for weight classes, SID routes and speed groups.

If we assign each aircraft a route through the holding point structure then, given a holding point entry time, h_i, and a suitable function, $T(t_i)$, for the traversal time through the holding points along a traversal path t_i for aircraft i, the earliest time the aircraft can reach the runway can be calculated as $h_i + T(t_i)$.

For the model, we assume that all aircraft take off as early as possible, so for any aircraft, i, the take-off time, d_i, can be predicted as the earliest point that both allows sufficient time to reach the runway and complies with all of the required separation rules, Equation (2).

Function $S(j, i)$ can be taken to be the maximum of two functions: $W(w_j, w_i)$ which will calculate the required wake vortex separation from the weight categories w_i and w_j of aircraft i and j; and, $R(r_j, s_j, r_i, s_i)$ which will calculate the required separation based upon the SID routes, r_i and r_j, and the speed groups, s_i and s_j, of the aircraft i and j (see Equation (3)). The separations for SID routes differ depending on which runway the aircraft are departing from, so $R(r_j, s_j, r_i, s_i)$, like $T(t_i)$, is runway specific.

Both functions $W(w_j, w_i)$ and $R(r_j, s_j, r_i, s_i)$ are defined to return standard separation values in accordance with current regulations. It should be noted that the runway controller has some flexibility in good weather to reduce the separations given by $R(r_j, s_j, r_i, s_i)$ and a fully operational decision support system would allow the controller to do just that.

We can express this model as follows:

Minimize

$$D = \sum_{i=1}^{n} (d_i - h_i) \tag{1}$$

where

$$d_i = \max(h_i + T(t_i), \max_{j \neq i} (d_j + S(j, i))) \tag{2}$$

$$S(j, i) = \max(W(w_j, w_i), R(r_j, s_j, r_i, s_i)) \tag{3}$$

3.1 Holding Point Constraints

Any practical model must incorporate the holding point constraints. There is no point in presenting candidate solutions to a runway controller if he/she cannot actually achieve the order due to the physical constraints.

An example of a holding point structure can be seen in Fig. 2. The nodes are the valid positions for aircraft and the arcs show moves that aircraft could make. This network is more restrictive than the actual network at the associated holding point at Heathrow and is deliberately so. Any solution which is feasible for this network should be both feasible and sensible for the real network.

We investigate metaheuristic local search, as specified in Section 4. This means that the search will move from one solution to the next. A solution could consist of just a final take-off order or it could give details about all of the taxi movements within the holding points and a take-off order could be derived from this.

If a solution consisted of the order in which individual moves were made within the holding point, specifying details of how aircraft attain the reordering as well as the final take-off order achieved, the search space would be extremely large. Many solutions would give the same take-off order but differ in the paths used to traverse the holding point or in the order in which moves were made. The relative order in which many actions take place often does not matter. So, many apparently different solutions may, in fact, be identical. Some paths take longer to traverse than others, so some solutions will be much better than others that have the same take-off order. This manoeuvring cost would have to be considered within the objective function.

Rather than modelling the movement within the holding points, the selected model instead has solutions which specify only a take-off order rather than how the order is achieved. Not all potential take-off orders will be achievable, however, so this must be verified. The method, in which the reordering is attained, does have an impact and some ways are obviously better than others. We use a heuristic to assign holding point traversal paths to aircraft, then perform a feasibility check to verify that the solution is achievable, given the holding point structure.

3.2 Path Assignment Heuristic

The heuristic to assign paths through the holding point to aircraft is holding point specific. The first stage in the design is to identify the good paths through the holding point. This is performed by asking the runway controllers about the ease and feasibility of using possible paths and eliminating from consideration any which are difficult to use, leaving only good paths. Given each entrance point, multiple paths are available.

Some paths are faster than others, but all paths are easy to use even though some will be longer than others. The allocation heuristic allocates slower paths to aircraft that are overtaken and faster paths to aircraft that overtake. This ensures that all aircraft on longer, slower paths are being overtaken in the holding point and therefore have much more time available to traverse the holding point.

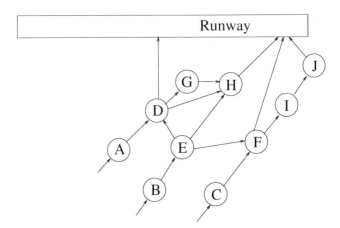

Fig. 2. An Example Holding Point Network Structure

For example, if two aircraft arriving at entrance A in Fig. 2 needed to reverse their order before take-off, the first would be assigned path ADGH and the second path ADH. The first would then hold at G while the second overtook it.

Once an aircraft is in the holding point the heuristic does not allow the assigned path to be changed so it is important to attempt to maintain flexibility when assigning paths to aircraft close to the holding point.

3.3 Directed Graph Model of the Holding Point

Once paths have been assigned to aircraft, the feasibility of the schedule is checked by feeding aircraft into the start nodes of the directed graph for the holding point, in the order they will arrive at the holding point. Fig. 2 shows the graph used for the 27R holding point. Rules are used to determine which aircraft to move next and whether moving a specific aircraft could block another aircraft. If the aircraft can exit the graph onto the runway in the desired take-off order then the schedule is deemed to be feasible.

Two levels of pre-processing are used. The first is based purely upon the holding point structure and the possible paths that could be employed. This stage is performed for each holding point graph prior to the start of the tests and can be performed off-line. It caches information about the later structure of the holding point beyond each node, recording for each of the paths entering the node, details of which other paths converge with it and how many nodes are not shared between them. The second pre-processing stage requires knowledge of the desired take-off order, so it is performed before each feasibility check. This stage calculates partial take off orders at each node, for sets of converging paths, ensuring that, for any pair of aircraft for which there is no possibility of changing order beyond this node, the aircraft enter the node in the correct order. Together, the pre-processing results provide knowledge about whether any aircraft can move without blocking another aircraft, ensuring that the feasibility check can be made both deterministically and quickly.

4 Departure Scheduling Algorithms

All of the search heuristics that we investigated had the same basic format but differed in the details. They are described below.

First descent: The first descent algorithm is the most simplistic algorithm and has the following structure.

1. Obtain initial current solution. An initial current solution will usually be a solution where the aircraft are in the order at which they arrived at the holding points. This solution has the advantage that it will always be feasible as no reordering is necessary within the holding points.
2. Evaluate the solution as described in Section 4.2, using the default holding point paths as no reordering is necessary so feasibility is guaranteed.
3. Generate a new candidate solution by selecting a solution from the neighbourhood of the current solution, as described in Section 4.1.
4. Heuristically assign holding point paths to aircraft, as in Section 3.2.
5. Check the feasibility at the holding point structure to ensure that the order of take-off is possible, as described in Section 3.3.
6. Evaluate the cost of the solution, as shown in Section 4.2.
7. If the candidate solution has a lower cost than the current solution then accept it as the new current solution.
8. If the given number of evaluations have been completed, then stop the algorithm and report the best result so far, otherwise return to Step 3.

Simulated annealing: The simulated annealing algorithm has the same structure as the first descent algorithm except in Step 7. In Step 7, rather than only accepting better solutions, the simulated annealing algorithm will sometimes accept moves to worse solutions, allowing it to escape from local optima. If the cost of the new solution is less than the cost of the current solution, then the new solution will always be accepted. If the cost of the new solution is more than the cost of the current solution then there is a small chance to still accept the new solution.

Let D_{curr} be the cost of the current solution and D_{cand} be the cost of the candidate solution.

The candidate solution will be accepted in Step 7 if:

$$D_{cand} < D_{curr} \tag{4}$$

or

$$R < e^{-\delta/T} \tag{5}$$

where $\delta = D_{cand} - D_{curr}$ is the difference between the current and candidate solutions, R represents a uniform random variable in the range [0..1] and T is a temperature which is initially large, so that many bad solutions are accepted, but decreases over time so that the simulated annealing algorithm slowly converges towards the first descent over time.

Steeper descent: The steeper descent and tabu search algorithms are similar to the first descent algorithm but both generate fifty candidate solutions at a time in Step 3 rather than just one. All of the fifty candidates are evaluated simultaneously in Steps 4, 5 and 6. In Step 7 the best of the feasible candidate solutions is adopted as the new current solution in Step 7. The best candidate is adopted even if it is worse than the current solution, which means this is more than a strict descent algorithm. This gives the algorithm a limited ability to move out of local optima but no method to avoid it moving straight back to the local optimum it just left.

Evaluations of candidates are expensive so, for comparison, the searches are limited to a number of evaluations rather than a number of iterations. This means that the first descent and simulated annealing algorithms run for fifty times as many iterations as the steeper descent and tabu search algorithms.

Tabu search: The tabu search algorithm is similar to the steeper descent algorithm except that it maintains a list of tabu moves. When a move is made, details of the move are stored on a tabu list. The tabu list stores details of which aircraft were moved and the absolute positions they were moved from, for the last ten moves made. If a future move attempts to place all of these aircraft back at the position from which they were moved then it will be declared tabu and rejected.

Like the steeper descent algorithm, the tabu search evaluates fifty candidate solutions at once. The only difference between the two algorithms is that, in Step 7, each candidate is evaluated and tested to see if it matches a move on the tabu list. The best of the feasible, non-tabu candidates is adopted and the details of the move made are stored on the tabu list. Again, the best candidate is adopted even if it is worse than the current solution, allowing the search to escape local optima. The tabu list ensures the search cannot quickly return to a local optimum from which it has escaped.

4.1 Neighbourhood Design

These algorithms all rely upon the selection of neighbouring solutions. Choosing a neighbouring solution is a matter of first randomly determining the move to use then randomly determining the details of that move. A large number of moves are available to the search methods.

Swap single aircraft: The *swap single aircraft* move takes two aircraft from the schedule and swaps the positions of the aircraft in the final take-off order. There is a 30% chance that this move will be used, selecting two aircraft at random.

Shift aircraft: The *shift multiple aircraft* move selects a consecutive group of one to five aircraft and moves them to a new random position in the schedule, either forwards or backwards. There is a 50% chance that this move will be made. Moving multiple aircraft is especially useful once the aircraft are in a north/south alternating pattern as, in this case, moving a single aircraft would usually make the schedule worse.

Randomise a set of aircraft: This move selects a consecutive set of aircraft as the target. Each aircraft within this set is then moved to a random position in the set. This move may emulate a shift, swap or a reversal in the order in some cases but some of the schedules attainable through this move are not attainable otherwise. There is a 20% chance that this move will be used. In experimental results, this move has shown a valuable contribution in finding good schedules, when not overused.

4.2 Objective Function

It is advisable to limit the amount of deviation from the holding point arrival order as well as to limit the delay. Reducing the number of 'swaps' of aircraft in the take-off order will help to reduce the workload for the pilots and controllers and it will also make it easier for the next iteration to build a feasible schedule.

With this goal in mind, the following objective function is used by the search algorithms:

$$D = \alpha \sum_{i=1}^{n} (A_i - i)^2 + \beta \sum_{i=1}^{n} (d_i - h_i) \qquad (6)$$

where n is the number of aircraft in the take-off schedule, d_i is the take-off time and h_i is the holding point arrival time of the ith aircraft in the take-off queue. A_i is the position, $1, 2...n$, in the initial holding point arrival order, of the ith aircraft in the take-off queue.

With the delay measured in seconds and separation rules specifying a minimum number of minutes separation, the constants α and β were chosen to be 1 and 5, respectively, to ensure that reducing the delay was the primary objective and reducing the reordering was only secondary.

4.3 Testing the Search Algorithms

We aim to determine the feasibility of a metaheuristic based approach to the real-time scheduling of aircraft at Heathrow given the holding point constraints that must be considered. We therefore test our algorithms by providing them with static problems of a type that may occur in a real system, where there is limited visibility of future aircraft and some constraints upon what can be done with the aircraft already in the holding point. We form a series of these problems by applying a rolling window of 25 aircraft at a time to each input dataset and applying the results of each search to the input for the next search. In a real system, not all suggested reorderings will be accepted, as the controller has a number of other objectives to keep in mind. Here, we are assuming that the metaheuristic order will always be accepted. It is important to attempt to automate the system, so that it can be tested in an objective rather than subjective manner, even though this is not how it would be used in practice.

An initial schedule was first built for the first 25 aircraft by employing the following procedure.

1. Add the first 20 aircraft to the system.

2. Run the search algorithms for 10000 evaluations. Keep the best result found.
3. Fix the take-off order, take-off time and traversal paths of the first 5 aircraft to take off. Traversal paths for aircraft overtaken by these aircraft were also fixed.
4. Add the next 5 aircraft to the system.
5. Run the algorithms for 5000 evaluations. Keep the best result found.

A second, iterated stage is then employed. This is the stage that more closely emulates what will happen in practice, with some aircraft having take-off slots or traversal paths already assigned. Each iteration took between 0.4 and 0.8 seconds. The second stage can be outlined as follows:

1. Fix the take-off order, take-off times and traversal paths of the first 10 aircraft to take off. Again, this also fixes the traversal paths of all of the aircraft they overtake.
2. Add the next aircraft to the system.
3. Remove the first aircraft from the system.
4. Run the search algorithms for 5000 evaluations. Keep the best result.
5. If there are no more aircraft to add then stop, otherwise return to Step 1.

As aircraft are removed from the system, the take-off order is recorded and at the end, the combined schedule of all of the departures is built and evaluated. This test was applied ten times to each dataset for each of the algorithms.

We have two main concerns in our testing. Firstly, we must verify whether our algorithms can find good results for the sub-problems within a very short search time, to verify their feasibility for use in a real-time system. Secondly, although the searches are considering only a subset of aircraft at once, it is the value of the entire schedule as a whole which actually matters. We would like to verify that solving the sub-problems will give good results for the entire schedule, validating the approach for a real system. To answer both of these questions, we evaluate the final schedule as a whole, predicting take-off times and calculating a total delay for all of the aircraft in the dataset.

5 Results

5.1 Input Data and Assumptions

Historical recorded data was used for the evaluation. Three datasets were used with different numbers of aircraft (123, 189 and 299, respectively).

The most convenient holding point entrance for the allocated stand was assigned to each aircraft. The real holding point arrival times from the historic data were used. In a real system, precise arrival times would not be known until the aircraft actually arrived at the holding points and estimated arrival times would have to be used until then.

Recorded data shows that it takes a minimum of just over a minute for an aircraft to traverse the holding point structure and get airborne but this time can vary widely. For this paper, all holding point traversal times were assumed to be equal and independent of the route taken, as only good paths were used. Two values for this time were tested: one and two minutes. A traversal time of one minute has the advantage of allowing aircraft to arrive, enter the runway and take-off very quickly, which is what often happens in practice at quiet periods. A two-minute traversal time, although no longer allowing fast entry at times when this is possible, seems better suited for the model in many ways as it can be assumed to account for some of the uncertainty in arrival time or traversal time that occurs in real life.

The real situation would have some aircraft already in the holding point. We simplify these tests by always starting aircraft at the holding point entrances to avoid having to make predictions for the positions of aircraft within the holding point. The danger of not predicting holding point positions for aircraft already in the holding point is that the reordering of earlier aircraft that have already taken off may have enforced certain manoeuvring upon the aircraft that have not taken off yet. To ensure that restarting aircraft at the holding point entrances does not increase the flexibility of later take-offs, we leave earlier aircraft in the system until they can no longer have any effect on the aircraft that have not yet taken off, thus re-enforcing the manoeuvring on the later aircraft.

Our model can easily consider aircraft already in the holding point by modifying the earliest take-off time appropriately and starting the feasibility check with the aircraft already in the intermediate nodes rather than at the holding point entrance. This would considerably reduce the complexity of the feasibility check in the holding point graph, but it would introduce a great deal of complexity into the test simulation with the need for a position prediction system.

5.2 Total Delay on Aircraft

The test schedule was executed ten times for each of the search approaches, on each set of data, for both one and two minute holding point traversal times. The mean values of the total delay in seconds for the ten runs are shown in the Tables 1 and 2. The best figures are presented in bold.

Table 1. Comparison of Mean Delays – 1 Minute Traversal Time

Algorithm	Dataset 1	Dataset 2	Dataset 3
Manual schedule	55140	136168	103692
First Descent	23548	49966	51438
Steeper Descent	**23511**	49158	50977
Simulated Annealing	**23511**	**48613**	50788
Tabu Search	23516	48767	**50661**

Table 2. Comparison of Mean Delays – 2 Minute Traversal Time

Algorithm	Dataset 1	Dataset 2	Dataset 3
Manual	62244	142828	121632
First Descent	**30831**	59170	69377
Steeper Descent	**30831**	58275	68916
Simulated Annealing	**30831**	57815	68728
Tabu Search	**30831**	**57504**	**68601**

5.3 Search Times

We aim to verify the feasibility of implementing a metaheuristic based system to provide real-time advice to a runway controller. One of the key objectives for this research is that results must be returned extremely quickly from each individual search. Although the important consideration for our research is the search time for a single iteration, the total test time is useful for evaluating the relative speeds of the algorithms. Tables 3 and 4 give the mean execution time, in seconds, for the tests performed with each of the four algorithms.

Table 3. Comparison of Total Search Time – 1 Minute Traversal Time

Algorithm	Dataset 1	Dataset 2	Dataset 3
First Descent	69.4	114.0	193.1
Steeper Descent	67.3	110.1	187.5
Simulated Annealing	71.6	116.5	197.8
Tabu Search	80.9	132.4	225.4

Table 4. Comparison of Total Search Time – 2 Minute Traversal Time

Algorithm	Dataset 1	Dataset 2	Dataset 3
First Descent	69.2	114.7	194.5
Steeper Descent	67.7	110.9	199.1
Simulated Annealing	72.0	117.2	238.2
Tabu Search	80.9	132.0	225.4

5.4 Evaluation of the Results

The metaheuristic solutions provide much lower total delays than the manual solution and this provides significant evidence for the high value of such approaches.

However, there are a number of reasons why our automated solutions are so superior (in terms of delay). In fact, the manual solutions are very good, with very few separations above the minimum. These reasons are outlined below.

1. Maximising throughput is not the same as minimising delay. The controller is trying to maximise throughput and is not directly attempting to minimise total delay. Minimising delay will have the effect of moving larger separations as late as possible in the schedule. Minimising the delay will maximise the throughput but the converse is not true. For example, assume a six minute period with only three aircraft available to take off. Two minute separations would give the same throughput as one minute separations but a lot larger delay. Where larger separations will be necessary, a runway controller may sometimes wish to have them earlier to avoid delaying aircraft which take advantage of these to cross the runway.

2. Some aircraft have a Calculated Time of Take-off (CTOT) which effectively designates a fifteen minute take-off time slot. It is important that such aircraft take off within this window. For the results in this paper, we have no CTOT information so we assumed no CTOT limitations.

3. In bad weather, a Minimum Departure Interval (MDI) could be applied to some routes. This temporarily increases the minimum separation allowed between aircraft using certain routes and so can increase delay. We have no data for whether any MDIs were present on the specified days so were forced to exclude MDIs from the evaluation.

4. This is a multi-objective problem and minimising delay only looks at one objective. Many conflicting objectives need to be satisfied and this is one reason why an automated solution can only ever be advisory.

5. Taxi times are not actually identical or predictable. We have no way of knowing whether certain aircraft were exceptionally slow or fast in practice.

6. The metaheuristics have more knowledge about the future than the runway controller did. Sometimes a good order from the metaheuristics has been a result of knowing which aircraft are going to be arriving later. Reducing the load on the runway controller via an advisory system should allow runway controllers to take account of these later arrivals themselves; something they do not currently have the time to do.

Minimising the delay is a good way to try to ensure maximal throughput of the runway as it makes it easier to reschedule as new aircraft enter the system.

The fact that the metaheuristics give better delays than the manual solution means that they hold significant promise for forming the basis of an advisory system. By reducing the work load of the runway controller and allowing more aircraft to be considered than are currently in the holding point structure, it should be possible to reduce the delay and increase throughput in practice.

Dataset 1 was from a less busy time of the day than the other two datasets. There were less possibilities to reorder aircraft as there were less aircraft in the holding points at any time. All but the first descent algorithm found the same good schedule for the aircraft in this dataset. The mean values of 23511 and 30831 were also the minimum values found for this dataset, by any of the algorithms. The tabu search failed on one execution to find this good schedule hence the slightly higher mean for the tabu search with one minute traversal time.

Datasets 2 and 3 were from busier times of the day. For both traversal times, for both Datasets 2 and 3, student t-tests showed that tabu search performed significantly better than the steeper descent algorithm and that both simulated annealing and tabu search performed significantly better than the first descent algorithm, with a confidence level of 99% in each case.

The simulated annealing algorithm gave good results across the datasets. It got the best results for Dataset 2 in Table 1 and equal best on Dataset 1 in both tables. Student t-tests performed on the results, however, failed to show a significance in the difference between the results for simulated annealing and tabu search, for either of the traversal times for Dataset 2, despite the difference in the mean values of the results.

With ten executions of the algorithms on each dataset for each traversal time, there are forty executions that can be compared for these datasets. Tabu search gave better results than the steeper descent algorithm on 39 of the executions and the same result on the other execution. The only difference between the two approaches is the presence of the tabu list so we conclude that the tabu list is contributing to the success of the search.

Tabu search produced the best result for Dataset 3 in Table 1 and the best results for all three datasets on Table 2, although all of the automated methods got equal best results for Dataset 1. Student t-tests showed that tabu search performed significantly better than simulated annealing for both traversal times for Dataset 3, with a confidence level of 99%.

However, there is a significant cost to maintaining and checking the tabu list, this being shown in the greater time that the tabu search takes to perform the search.

We aimed to determine whether a metaheuristic approach could solve the scheduling problem fast enough to be of use to a real time system and whether an approach which solves a number of sub-problems could attain a good overall delay for the entire schedule. The good overall delay for the schedule obtained when applying either the tabu search or simulated annealing algorithms to the problem assures us that the metaheuristic approach is a promising approach for a real-time decision support system for a runway controller as it can, with a very short search time, provide very good results for the sub-problems with which a real controller would have to deal, leading to very good overall delay figures.

6 Conclusions

The departure problem is a complicated one due to the many constraints upon the schedule and the sequence-dependent separations between aircraft. Most of the existing research has looked at the arrivals problem rather than the departure problem where the separations are based on the wake vortex categories of aircraft. In that case it is only necessary to check the separations between adjacent aircraft. However, the route and speed based separations at Heathrow are not only asymmetric, but also do not obey the triangle inequality, so it is not sufficient merely to look at adjacent pairs of aircraft. A schedule that provides safe separations for all adjacent pairs of aircraft will not necessarily provide safe separations for other aircraft pairs.

Many different techniques have previously been applied to this problem yet none account for the physical constraints upon reordering that exist at an airport like London Heathrow. There are many constraints upon a departure system that are not normally modelled and any solution should also aim to minimise other aspects such as controller and pilot workload and fairness.

This paper has presented a model for the system that can take account of the real life constraints. The initial results presented here include some of the constraints that are particularly important at Heathrow. The results show that it is feasible to check the effects of the holding points after schedules have been generated and that the metaheuristics will still perform well in the limited time that they have.

From the experiments carried out here we can conclude that tabu search obtained the best delays overall, although it was the worst performer on Dataset 1 in Table 1 and it did take the longest to run due to the overheads associated with the tabu list. Simulated annealing performed well across all the experiments but not always as well as tabu search. Further research will include much more experimentation to see whether these results apply in general for the Heathrow problem.

Both the tabu search and simulated annealing algorithms perform well in the very short search time permitted. We can determine from the results that the metaheuristic searches form a promising basis for an advisory system for a controller as they are suggesting schedules which improve on the delay in the schedules the controllers are currently implementing.

Further research will add to this model and evaluate the effects of the constraints that have not yet been included. Implementation using genetic algorithms and hybridised metaheuristics are also planned.

Acknowledgements: This work was supported by EPSRC (The Engineering and Physical Sciences Research Council) and NATS (National Air Traffic Services) Ltd. from a grant awarded via the Smith Institute for Industrial Mathematics and Systems Engineering.

References

Abela, J., Abramson, D., Krishnamoorthy, M., de Silva, A., and Mills, G. (1993). Computing optimal schedules for landing aircraft. In *Proceedings of the 12th National Conference of the Australian Society for Operations Research, Adelaide*, pages 71–90. Available at: http://www.csse.monash.edu.au/~davida/papers/asorpaper.pdf [30 March 2004].

Anagnostakis, I. and Clarke, J.-P. (2002). Runway operations planning, a two-stage heuristic algorithm. In *AIAA Aircraft, Technology, Integration and Operations Forum, Los Angeles, CA*. Available at: http://icat-server.mit.edu/Library/Download/167_paper0024.pdf [30 March 2004].

Anagnostakis, I. and Clarke, J.-P. (2003). Runway operations planning, a two-stage methodology. In *Proceedings of the 36th Hawaii International Conference on System Sciences (HICSS-36), Hawaii.*

Anagnostakis, I., Clarke, J.-P., Böhme, D., and Völckers, U. (2001). Runway operations planning and control, sequencing and scheduling. In *Proceedings of the 34th Hawaii International Conference on System Sciences (HICSS-34), Hawaii.*

Anagnostakis, I., Idris, H. R., Clarke, J.-P., Feron, E., Hansman, R. J., Odoni, A. R., and Hall, W. D. (2000). A conceptual design of a departure planner decision aid. In *3rd FAA/Eurocontrol International Air Traffic Management R & D Seminar, ATM-2000, Naples, Italy.* Available at: http://atm-seminar-2000. eurocontrol.fr/acceptedpapers/pdf/paper68.pdf [30 March 2004].

Beasley, J. E., Krishnamoorthy, M., Sharaiha, Y. M., and Abramson, D. (2000). Scheduling aircraft landings – the static case. *Transportation Science,* **34,** 180–197.

Beasley, J. E., Sonander, J., and Havelock, P. (2001). Scheduling aircraft landings at London Heathrow using a population heuristic. *Journal of the Operational Research Society,* **52,** 483–493.

Bianco, L., Dell'Olmo, P., and Giordani, S. (1999). Minimizing total completion time subject to release dates and sequence-dependent processing times. *Annals of Operations Research,* **86,** 393–416.

Craig, A., Ketzscer, R., Leese, R. A., Noble, S. D., Parrott, K., Preater, J., Wilson, R. E., and Wood, D. A. (2001). The sequencing of aircraft departures. In *40th European Study Group with Industry, Keele.* Available at: http://www.smithinst.ac.uk/Projects/ESGI40/ ESGI40-NATS/Report/AircraftSequencing.pdf [30 March 2004].

Ernst, A. T., Krishnamoorthy, M., and Storer, R. H. (1999). Heuristic and exact algorithms for scheduling aircraft landings. *Networks,* **34**(3), 229–241.

Idris, H. R., Delcaire, B., Anagnostakis, I., Hall, W. D., Pujet, N., Feron, E., Hansman, R. J., Clarke, J. P., and Odoni, A. (1998a). Identification of flow constraint and control points in departure operations at airport systems. In *Proceedings of the AIAA Guidance, Navigation and Control conference, Boston, MA.*

Idris, H. R., Delcaire, B., Anagnostakis, I., Hall, W. D., Clarke, J. P., Hansman, R. J., Feron, E., and Odoni, A. R. (1998b). Observations of departure processes at Logan airport to support the development of departure planning tools. Presented at the 2nd USA/Europe Air Traffic Management R&D Seminar ATM-98, Orlando, Florida. Available at: http://atm-seminar-98.eurocontrol. fr/finalpapers/track2/idris1.pdf [15 December 2003].

Newell, G. F. (1979). Airport capacity and delays. *Transportation Science,* **13,** 201–241.

Trivizas, D. A. (1998). Optimal scheduling with maximum position shift (MPS) constraints: A runway scheduling application. *Journal of Navigation,* **51,** 250–266.

van Leeuwen, P., Hesselink, H., and Rohling, J. (2002). Scheduling aircraft using constraint satisfaction. *Electronic Notes in Theoretical Computer Science,* **76.**

Improving Scheduling Through Performance Monitoring

Thomas J. Kimpel[1], James G. Strathman[1], and Steve Callas[2]

[1] Center for Urban Studies, Portland State University, 506 SW Mill St., Room 350, Portland, OR 97201, USA, E-mail: {kimpelt,strathmanj}@pdx.edu
[2] TriMet, 4012 SE 17th Ave., Portland, OR 97202, USA, E-mail: callasc@trimet.org

Summary. Historically, schedulers and operations management personnel have made decisions with limited information about various states of the transit system. The present study highlights innovative uses of data collected via automatic vehicle location and automatic passenger count technologies in the areas of scheduling and operations management at TriMet, the transit provider for the Portland, Oregon metropolitan region. Two main topics are addressed in this paper. First, we look at efforts at TriMet involving the use of archived operations data to improve bus schedules. Second, we look at the role of operator behavior in relation to service reliability and steps the agency is taking to reduce run time variability and maintain vehicle headways through better management of operators. The quality, quantity, and disaggregate nature of data at TriMet has greatly enhanced the agency's ability to generate performance reports as well as undertake special purpose studies targeting specific operational issues, providing essential feedback into the scheduling process.

1 Introduction

It is important for transit agencies to identify the causes of unreliable service in order to be able to provide high quality service to passengers in the most economical manner. Bus routes may exhibit poor performance due to operational problems or simply because schedules are poorly written (Guenthner and Hamat (1983)). If buses are consistently early or late, then this would indicate a scheduling problem and not an operational one. Ideally, schedules should be related to the measuring and monitoring of service performance which typically involves a comparison of actual to scheduled service. The scheduling process requires a number of inputs including information on passenger loads, running times, and various constraints imposed by labor rules and timed transfers in addition to clock frequency and policy headway considerations. The ability to measure and monitor operational performance has historically been limited by data availability (Benn and Barton-Aschman Associates (1995), Levinson (1980)). Data availability varies widely among transit agencies including the type, amount, quality, level of aggregation, and frequency of data collection and often hampers service planning and scheduling (Boyle (1998), Casey (1999), Furth (2000), Furth *et al.* (2003), Wilson *et al.* (1984)).

The present study highlights innovative uses of data recovered by the TriMet automated Bus Dispatch System (BDS). A detailed description of TriMet's experience with implementation of the BDS is presented in Appendix A of TRCP Project H-28 (Furth *et al.* (2003)). Two of the main components of the BDS are automatic vehicle location (AVL) and automatic passenger counter (APC) technologies. At TriMet, 100% of the bus fleet is equipped with AVL technology while approximately 72% of the vehicles are APC equipped. Like most agencies that have AVL systems in place, TriMet polls bus location at regular time intervals and transmits this information to dispatch centers in-real time. TriMet is somewhat unique among North American transit properties in that its AVL system was designed to collect stop-level information on bus operations. The data collection component of the BDS records information each time a bus passes a stop, regardless of whether any passenger activity occurs. The disaggregate nature of the BDS data provides unique opportunities for measuring and monitoring service performance at multiple summary levels. The types of data collected by the BDS at each stop include arrival and departure times, dwell times, door openings, lift operations, and maximum speed since the previous stop as well as boardings and alightings on APC equipped vehicles, providing the agency with a complete picture of bus operations for each bus in the system on a continual basis. This information is subsequently analyzed and used as inputs into various service planning, scheduling, and operations management functions.

Historically, schedulers and operations management personnel have made decisions with limited information about various states of the transit system. It has taken approximately two decades for TriMet to fully transition from a data poor to a data rich environment, beginning with the agency's initial testing of APC technology in the early 1980s. The core AVL and APC components in place today were initially implemented as part of a major upgrade of the agency's computerized dispatching system which began in 1993 and became fully operational in 1998. Interestingly, fiscal constraints related to the costs of manual data collection provided much of the impetus for change. The design of the BDS greatly benefited from a number of factors including 1) a dedicated project manager who was well rounded having previously served as an operator, trainer, and Section 15 data analyst, 2) the agency's past experience with difficulties associated with referencing APC data to trips and time points based on time stamps and odometer readings, and 3) the identification of the need for and potential benefits of using automatically collected operations data for scheduling purposes (Furth *et al.* (2003)). The decision to collect detailed operations information at the level of the bus stop and to archive the data can be attributed to having a diverse project team which included dispatchers as well as schedulers, service planners, operations analysts, and maintenance personnel. By not limiting the AVL system to real-time uses (poll-based data collection primarily benefiting dispatchers), the agency effectively increased the number of potential users of the data as well as the number of potential applications by several orders of magnitude. While certain benefits such as improved performance monitoring capabilities were foreseen prior to implementation of the BDS, there is little doubt that the present capabilities of the system have far exceeded initial expectations. The data collected via the BDS have greatly increased communication between various parts of the agency including, but not limited, to operations, dispatch, scheduling, maintenance, marketing, customer service, training, and upper management. Each group has specific data needs and the disaggregate nature of the data allow for summarization at multiple spatial and temporal levels. While it is not the purpose of this report to discuss all of the uses of AVL and APC data within the agency, it is important to provide the reader with an understanding of the full range of potential benefits of the technologies.

Two main topics are addressed in this paper. First, we discuss efforts at TriMet utilizing archived operations data to improve schedules. These efforts are notable because of the quality, quantity, and disaggregate nature of the data available at TriMet. Second, we look at the role of operator behavior in relation to service reliability and steps the agency is taking to reduce run time variability and maintain headways through better management of bus operators.

Scheduling is a complex process that consists of matching transit service to passenger demand subject to various constraints including timed transfers, policy headways, clock frequency headways, and work rules. At TriMet scheduling is the implementation of the service design which is based on an urban grid system in the more densely populated areas and a timed transfer system in lower density suburban areas. One of the aims of service planning is to match passenger demand, which varies over time and space, with the service design through scheduling. Schedules are written to accommodate the "typical operator," meaning that schedule can be maintained by most operators under normal conditions. Schedulers are careful not to add too much scheduled run time between time points. Too much run time adversely affects passengers in two principal ways: 1) if operators hold buses to maintain schedule adherence, then persons on board vehicles are delayed; and, 2) if operators do not hold, then buses will tend to run early, impacting passengers that arrive at their origin stops on time. Schedulers can be somewhat more generous when setting recovery times at the ends of trips to make up for any shortages in running times. Schedulers have to confront the fact that passenger demand is subject to certain amount of random variation and that the capabilities of individual operators vary considerably. Fig. 1 shows the theoretical relationship between passenger demand, operator behavior, traffic levels, transit service reliability, and scheduling. Passenger activity influences the amount of scheduled running time either directly through increases or decreases in demand over time or indirectly through service reliability impacts. An increase in demand over time necessitates the need for either additional scheduled run time or perhaps the addition of a new trip. Highly variable demand results in increased run time variation causing schedulers to add additional recovery time into schedules. Operator behavior influences the amount of scheduled recovery time largely through impacts on service reliability, or indirectly through service reliability impacts on passenger demand. Background traffic levels are accounted for when setting scheduled running time, although excess traffic congestion influences recovery time through impacts on service reliability.

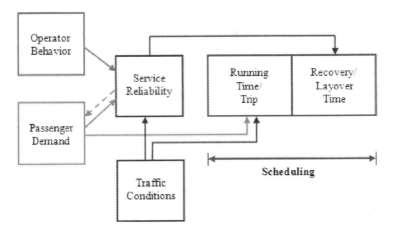

Fig. 1. Theoretical Relationships

The TriMet BDS has transformed the way the agency collects, analyzes, and uses data. Prior to implementation of the BDS, data collection proved to be an arduous task resulting in data of limited quantity and quality. Historically, the scheduling department at TriMet was responsible for developing and implementing procedures to systematically measure on-time performance (OTP) at regular time intervals. The use of AVL technology has largely replaced the need to send agency personnel into the field to collect departure time information at specific locations for service monitoring applications. The agency also regularly collects information on passenger activity including data necessary for annual National Transit Database (NTD) reporting, biennial cordon counts, and a comprehensive passenger census conducted every five years. These activities have benefited greatly from automated data collection. It has been shown that data collected by APCs are more accurate and subject to less bias than data collected by manual means (Strathman *et al.* (2001), Kimpel *et al.* (2004)) although the amount of error was found to vary by bus type. Widespread deployment of AVL and APC technologies on TriMet's vehicle fleet has effectively eliminated the need to assign vehicles to specific trips for data collection purposes.

2 Scheduling and Operations Research at TriMet

The quality and quantity of AVL and APC data has greatly enhanced the agency's ability to conduct special purpose studies that target specific operational issues and address specific research questions. The results of many of these studies, either undertaken by TriMet or in conjunction with the Center for Urban Studies at Portland State University, are presented here. The studies can be grouped into two broad categories: 1) studies that focus on the relationship between run time variation and schedule efficiency; and, 2) studies that address the relationship between operator behavior and transit service reliability. Both researchers and practitioners are aware of the influence of operators on transit service reliability (Abkowitz (1978), Levinson (1991), Strathman *et al.* (2002b), Woodhull (1987)). Since operators have differing levels of experience and behaviors, this translates into run time variability which adversely impacts schedules by necessitating additional run time between time points, extra layover/recovery time, or additional resources in the form of extra bus trips to address passenger loading problems.

Schedule efficiency is related to the amount of excess slack time in schedules. Levinson (1991) argues that scheduled run times should be set at a value slightly less than the mean or median run time in order to ensure that the majority of operators do not have to kill time in order to maintain schedule adherence. Levinson (1991) also contends that the optimal amount of layover/recovery time for a given bus trip is the 95th percentile run time minus the mean or median running time. This notion is depicted graphically in Fig. 2. A study by Portland State University and TriMet analyzed schedule efficiency from the perspective of run time variation (Strathman *et al.* (2002b)). A run time distribution for Route 14- Hawthorne Blvd. is presented in Fig. 3 to illustrate the Levinson optimal standard. The run time distribution is based on 1,026 trip level observations from the spring 2000 signup (booking). The data show that actual run times range from 38 to 69 minutes. The median actual run time is 50.4 minutes which is 17.2% greater than the mean scheduled run time of 43.0 minutes. The ideal recovery time of 11.1 minutes is 9.8 minutes less (46.9%) than the average scheduled recovery time of 20.9 minutes. The graph indicates that scheduled run time should be increased by 7.4 minutes and recovery time decreased by 9.8 minutes, resulting in an efficiency improvement of 2.4 minutes per trip. The graph is somewhat idealized, as work rules, headway synchronization, and maintenance of clock headways may necessitate additional layover/recovery time. Furthermore, the analysis is based on data which is aggregated over all trips whereas the scheduling process would benefit from analysis of multiple days of observations for individual trips.

A separate component of the study by Strathman *et al.* (2002b) estimated the annualized costs of schedule inefficiencies at the system level using information derived from run time distributions developed for each individual trip in the system. The analysis employed 281,305 trip-level observations encompassing 65 weekdays of service. Three alternative layover/recovery time scenarios were addressed in the analysis: 1) the Levinson optimal recovery consisting of the 95th percentile run time minus the median, 2) 10% of the median run time which is the minimum amount specified under the labor contract, and 3) 18% of the median run time which is rule-of-thumb standard used by TriMet schedulers. At the system level, the study found excess schedule time (run time plus layover/recovery time) of 7.3, 7.9, and 3.8 minutes per trip for the Levinson optimal, the contract minimum, and the rule-of-thumb standards, respectively. Estimated annual costs associated with excess schedule times ranged from a low of $5.7M for the rule-of-thumb standard to $7.1M for the Levinson optimal to a high of $7.7M for the contract minimum. When aggregated to the level of the individual route, the authors found that 81 of the 104 bus routes (77.9%) contained excess schedule time and that 23 routes (22.1%) had too little schedule time. These findings suggest that 1) schedule adjustments are necessary, and 2) ef-

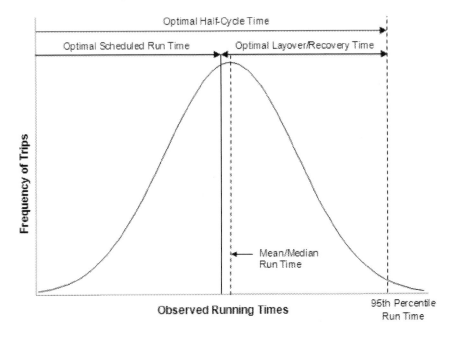

Fig. 2. Optimal Running and Layover/Recovery Times

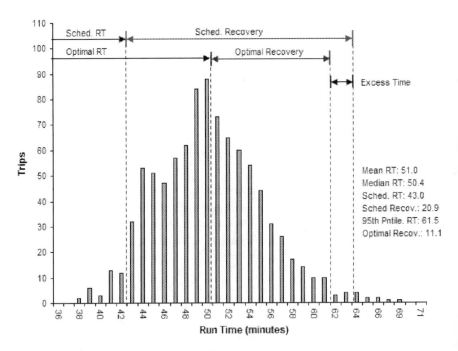

Fig. 3. Run Time Distribution

forts to reduce the amounts of excess schedule time in schedules would likely yield substantial costs savings.

The relationship between passenger loads and headway irregularity was addressed in another study conducted by Portland State University and TriMet (Strathman *et al.* (2002a)). A two stage least squares regression model was employed to address simultaneity between headway delay and passenger loads at peak load points. The data consisted of 12,593 observations representing ten bus routes operating during peak time periods during the peak hour of service. Controlling for the effects of passenger activity on headway delay, the study found that headway delays were a primary cause of passenger overloads. Sensitivity analysis showed that small reductions in headway delay would yield large reductions in overloads. Furthermore, the amount of headway delay at the peak load point was found to be largely determined by the amount of headway delay at the origin. Efforts to address origin delays center on better field supervision at the beginning of lines or, in the case of poor schedules, corrective action.

An analysis of the effects of individual operators on bus running times was included in the report by Strathman *et al.* (2002b). A fixed effects regression model containing dummy variables for each individual operator was employed. The data set consisted of 10,743 weekday bus trips associated with TriMet's 15 frequent service bus routes during the summer and fall 2000 signup periods. The study found wide variation in the parameter estimates for the individual operators and that the operator fixed effects were normally distributed. Sensitivity analysis based upon the 18% rule-of-thumb standard indicated that nearly 70% of the amount of recovery/layover time was needed to address differences in operator behavior, rather than variable operating conditions. The authors also found that operators accounted for 17% of observed run time variation. An additional regression model was used to test the effects of certain operator characteristics on bus running times. Each additional year of operator experience was estimated to result in a 6.8 second reduction in run time per trip. These findings highlight the fact that run times are affected not only by differences in operator behavior but also by operator experience.

With respect to the efficient utilization of vehicles and service hours, TriMet has been developing methods to test whether vehicle loading problems are due to uneven headways. The relationship between vehicle spacing and passenger loads is depicted graphically in Figs. 4 and 5. In Fig. 4, the Y-axis displays the actual passenger load for each bus trip in the p.m. peak period outbound direction for all of TriMet's frequent service bus routes for a three month period. The X-axis displays the *headway ratio* which is actual headway divided by scheduled headway at the peak load point. A headway ratio value greater than one indicates that bus spacing is increasing relative to schedule and a value less than one indicates that buses are too closely spaced. The areas to the left and right of the vertical bars at 50% and 150% of the headway ratio represent extreme headway variation. The horizontal lines at 20 and 55 passengers represents a somewhat arbitrary range of acceptable passenger loads. The diagonal line is a least squares regression line showing the effect of headway deviation on passenger loads. Observations with loads greater than 55 with a headway ratio greater than 150% of the scheduled headway (shaded- upper right) indicate overloaded trips

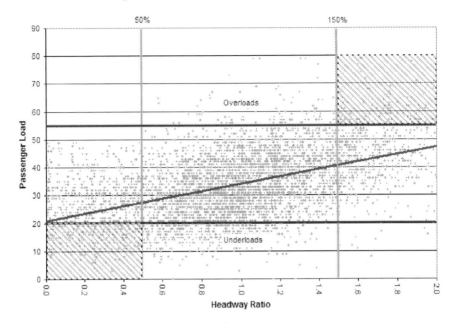

Fig. 4. Headway to Load Relationship

with vehicle spacing problems. Observations with loads less than 20 with a headway ratio less than 50% of the scheduled headway (shaded- lower left) represent trips with underutilized capacity due to irregular headways. Note that even when the headway ratio is 1.0, there still exist some over- and underloads. A more sophisticated way of determining to what extent vehicle spacing problems are responsible for overcrowding was developed using trip-level regression models to determine the percentage of trips that are overcapacity due to uneven headways. Fig. 5 shows the results for Route 14- Hawthorne Blvd. for the a.m. peak hour of service. The model estimates the maximum load on a trip as a function of the amount of headway deviation in minutes. In this example, a maximum load of 48.4 passengers is estimated at zero minutes of headway deviation. Approximately 31% of the variation in passenger loading is explained by the model. The arrows on the graph depict, based on the slope of the regression line, what the load would have been for the specific trip if headways had been evenly spaced. Using this methodology for all trips, the percent of "unavoidable" overloads and underloads can be estimated using the assumption of evenly spaced headways.

TriMet recently undertook a study comparing headway variability to passenger loads at the maximum load point to better understand the relationship between vehicle spacing and loads. Impetus for the study stemmed from a renewed interest at the agency to better manage headways and departure delays to address operational inefficiencies. Table 1 contains various statistics compiled from data collected on TriMet's 15 "frequent service" bus routes (route headways of 15 minutes or less seven days a

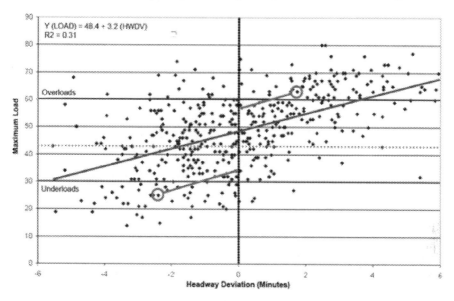

Fig. 5. Percent Overloads Due to Headway Deviation

week) during the fall 2003 signup. The statistics were derived from 5,183 trips passing peak load points. The data show that approximately 66% of all trips fall within
a headway deviation range of 0.5 to 1.5. For an eight minute scheduled headway,
this translates to an actual headway between four and 12 minutes. If a more stringent headway deviation standard of 0.75 to 1.25 were applied, only 44% of the trips
would fall within the acceptable range. For heavily loaded trips of 55 passengers or
greater, 55% of the trips were found to have a headway deviation greater than 1.5.
This value increases to 72% under the more stringent headway deviation standard of
1.25. Heavily load trips represent 7% of all trips in the analysis. For trips experienc-
ing underloads of 20 passengers or less, 35% fall below the lower headway deviation
value of 0.5. If the headway deviation standard were tightened to 0.75, the percentage of underloaded trips increases to 57%. Lightly loaded trips represent 11% of all
trips in the analysis. These results indicate that efforts to improve bus spacing will
yield positive efficiency benefits with respect to vehicle capacity utilization. Since
unreliable service often necessitates the need for additional bus trips due to capacity
issues, an effective headway management program has the potential to yield substantial cost savings in the short term in the form of reduced trips necessary to serve the
same level of demand.

Table 1 also contains information related to late origin departures and operator
years of experience which are believed to be two of the major causes of uneven headways. With respect to late departures, defined as more than three minutes late, 19%
of trips operated by drivers with three or more years of experience are leaving the
beginning of the line late. Operators with less than three years of experience are departing late 25% of the time. A number of factors may be responsible for such a high

Table 1. Frequent Bus Headway Deviation and Late Departure Analysis

Measure	Criteria	%Trips
Headway adherence	Ratio = 0.50 *to* 1.50	66%
	Ratio = 0.75 *to* 1.25	44%
Heavily loaded trips (>= 55)	Ratio >= 1.50	55%
	Ratio >= 1.25	72%
Lightly loaded trips (<= 20)	Ratio >= 0.50	35%
	Ratio >= 0.75	57%
Late trip departures	Operator experience >= 3 years	19%
	Operator experience < 3 years	25%

number of late departures including operator experience and behavior, inadequate field supervision or training, and poorly written schedules. If too much run time is built into schedules, operators may simply be basing their departure times on knowledge of actual operating conditions likely to be encountered on a given trip. Some operators may be departing trip origins late and then speeding to make up for lost time. Late departures may also be due to scheduling problems. If there is inadequate run time or too little recovery time built into schedules, operators may be departing trips late due to conditions encountered on the previous trip. Inadequate field supervision and training may be at fault as certain operators simply not be getting the message about the importance of departing trip origins on time.

The results of the previous studies make clear a number of points. First, there exists excess slack time in schedules. This time represents substantial costs to the agency. Second, there is wide variation in operator behavior which impacts both schedules and service reliability. The majority of recovery/layover time needed in schedules is due to operator variability. Operator experience has been shown to impact bus running times and terminal departure times. Uneven headways are largely responsible for passenger loading problems with much of the blame attributable to operators departing trips late. The next two sections describe performance monitoring efforts at TriMet and how BDS data is used to address scheduling inefficiencies through analysis of bus operations, service reliability, and operator behavior.

3 Monitoring Bus Operations Through Performance Reporting

Efficient schedules require ongoing monitoring and adjustment. The quality and quantity of data collected by the TriMet BDS have greatly expanded the agency's ability to undertake regular performance monitoring. One of the main advantages of the TriMet BDS is that the disaggregate nature of the data allows for the generation of performance reports at multiple summary levels (route, trip, stop direction, time of day, etc.) serving a number of different purposes. An excellent summary of data needs by level of detail and agency function is presented in Furth *et al.* (2003). Such monitoring may include analysis of on-time performance, vehicle headways, running times, passenger loads, and operator behavior. A key benefit of performance monitoring is the feedback loop that can lead to improvements to planning, scheduling, and operations (Furth (2000), Levinson (1991)).

TriMet service standards require that all transit services undergo periodic review. The Service Evaluation and Adjustment Process is an annual review of existing services where each route is analyzed to see if it meets the agency's standard for vehicle loading and OTP. For routes which violate vehicle loading and OTP standards, the first course of action is to fix any problems without adding additional vehicle hours of service. If additional hours are necessary, all route needs are compared and are fixed according to the budget allocated for that year. Service adjustments are informally undertaken at TriMet on a quarterly basis. Part of the service adjustment process includes schedule modifications to address OTP and overloading problems. Schedule changes of +/- three minutes are allowed between signup periods. The agency standards for on-time performance and schedule efficiency are both 75%. TriMet defines *on-time* as a bus departure no more than one minute early and five minutes late. *Schedule efficiency* is defined as the ratio of revenue hours to vehicle hours and measures the effectiveness of service provision.

Similar to most transit agencies, level of service at TriMet is determined by policy and demand. Consistent with observations by Furth and Wilson (1981), TriMet service standards represent a combination of policy and rules of thumb (Coffel (1993), TriMet (1989)). During peak periods of operation, level of service is driven by demand subject to vehicle capacity considerations. Loading standards at TriMet seek to balance passenger comfort and operating costs. Loading standards are based on the average number of passengers per vehicle passing the peak load point during the highest hour of passenger loadings on a per line basis. The agency calculates a *load factor* for each vehicle during the peak hour of service. The load factor is simply the passenger load divided by the seating capacity of the vehicle. The agency does not tolerate passups due to overcrowded buses although they do occur sometimes and are regularly monitored. Service frequencies are determined by calculating the *boarding rate* which is the average number of passengers per vehicle on a per minute basis crossing the peak load point.

TriMet produces hourly capacity reports which monitor passenger loads at the maximum load point for the peak hour of service. Fig. 6 shows an hourly capacity report by route and direction. *Begin time* and *end time* define the peak hour of service and are presented along with the number of *trips* operated during the peak hour. *Hourly load* is the average passenger load summed over all trips at the maximum load point. *Seating capacity* is the average number of scheduled seats available during the service period. *Achievable capacity* is a statistic measuring the average vehicle design capacity (seating capacity plus standees). This variable is calculated by multiplying the vehicle design capacity by 80% at the trip level, then averaging over all trips. TriMet sets this value at 80% of the design standard due to concerns about passenger comfort and the number of passups. The *load to seat* ratio is the amount of achievable capacity divided by the scheduled seating capacity. *Load to achievable capacity* is the ratio of the passenger load divided by achievable capacity averaged over all trips. A load to achievable capacity ratio of 95% indicates that

Peak Hour Service Standards Report   *Weekdays*   *Fall 2003 Quarter*

Hourly Capacity Report: Based on Achievable Capacity

Route and Direction	Begin Time - End Time	Trips	Hourly Load	Seating Capacity	Achievable Capacity	Load to Seat Ratio	Load to Achievable Capacity
64 -Marquam Hill / Tigard - Outbound	3:55 PM 4:54 PM	2	100	86	102	116.1%	97.5%
64 -Marquam Hill / Tigard - Inbound	6:37 AM 7:36 AM	3	148	129	153	114.3%	96.0%
179 -Canby-Clackamas TC - Inbound	3:36 PM 4:35 PM	2	51	56	58	91.4%	91.4%
5 -Interstate - Outbound	3:05 PM 4:04 PM	6	278	258	307	107.0%	89.8%
5 -Interstate - Outbound	3:15 PM 4:14 PM	6	275	258	307	106.7%	89.6%
31 -Estacada - Outbound	3:49 PM 4:48 PM	2	91	86	102	106.3%	89.3%
64 -Marquam Hill / Tigard - Outbound	3:19 PM 4:18 PM	2	91	82	102	110.6%	88.6%
1 -Greeley - Outbound	3:02 PM 4:01 PM	3	136	129	153	105.1%	88.3%
179 -Canby-Clackamas TC - Inbound	3:05 PM 4:04 PM	3	74	84	84	88.0%	88.0%
5 -Interstate - Outbound	2:54 PM 3:53 PM	6	270	258	307	104.8%	88.0%
108 -Jackson Park - Outbound	9:55 AM 10:54 AM	4	180	156	204	115.5%	87.9%
4 -Fessenden - Outbound	3:38 PM 4:37 PM	5	225	195	256	115.3%	87.8%
71 -60th-122nd Ave - Outbound	2:52 PM 3:51 PM	4	179	172	204	104.3%	87.6%
5 -Interstate - Outbound	3:25 PM 4:24 PM	6	268	258	307	103.8%	87.2%
179 -Canby-Clackamas TC - Inbound	2:06 PM 3:05 PM	3	73	84	84	86.5%	86.5%
108 -Jackson Park - Outbound	9:40 AM 10:39 AM	6	265	234	307	113.2%	86.2%
64 -Marquam Hill / Tigard - Inbound	7:05 AM 8:04 AM	3	132	125	153	105.6%	85.9%
5 -Interstate - Outbound	2:43 PM 3:42 PM	6	262	258	307	101.5%	85.3%
108 -Jackson Park - Outbound	9:10 AM 10:09 AM	8	348	312	409	111.5%	84.9%
108 -Jackson Park - Outbound	9:25 AM 10:24 AM	7	304	273	358	111.4%	84.9%
72 -Killingsworth-82nd Ave - Inbound	11:15 AM 12:14 PM	6	261	234	307	111.4%	84.8%
4 -Fessenden - Outbound	3:50 PM 4:49 PM	5	217	195	256	111.3%	84.8%
179 -Canby-Clackamas TC - Inbound	2:35 PM 3:34 PM	3	71	84	84	84.5%	84.5%
14 -Hawthorne - Inbound	8:34 AM 9:33 AM	8	344	312	409	110.4%	84.1%
5 -Interstate - Outbound	3:35 PM 4:34 PM	6	258	258	307	100.1%	84.1%
72 -Killingsworth-82nd Ave - Inbound	10:52 AM 11:51 AM	5	215	195	256	110.3%	84.0%
61 -Marquam Hill / Beaverton - Outbound	3:54 PM 4:53 PM	2	88	86	102	99.8%	83.8%
72 -Killingsworth-82nd Ave - Inbound	11:03 AM 12:02 PM	6	257	234	307	109.8%	83.7%

Fig. 6. Hourly Capacity Report by Route

additional service will soon be needed. This same information is presented in Fig. 7 except that the focus is on individual routes. Fig. 7 includes additional information for each trip operating during the peak hour. Data are presented for Route 5- Interstate in the outbound direction. In addition to trip *start time*, *start location*, and *train* (block), the scheduled departure time at the maximum load point (*TP time*) is shown. The report includes trip level variables related to passenger activity and vehicle utilization including *average boarding rides*, *average maximum load*, and the *maximum load factor*. The maximum load factor is similar to the load to seat ratio mentioned previously. *Percent overcapacity* refers to the percentage of bus trips that are operating at more than 130% of seated capacity. Reported *passups* are initiated by operators who communicate to dispatch that passengers are being passed up due to overload situations. This measure is used for informational purposes only as not all operators use this feature consistently. *Headway adherence* represents the percentage of trips that are within +/-50% of scheduled headway at the peak load point. The report also includes standard OTP measures including the percentage of trips that are *early*, *on-time*, and *late*. OTP is averaged over all time points. The data show that on-time performance problems associated with the 3:15 p.m. departure (14% early departures) is causing passenger overloading problems for the subsequent trip departing at 3:26 p.m. This trip experiences excessive delays (36% late departures) resulting in poor vehicle capacity utilization for the trip departing at 3:37 p.m.

Peak Hour Service Standards Report	Weekdays	Fall 2003 Quarter

Hourly Capacity Report: Based on Achievable Capacity

5 -Interstate - Outbound   3:05 PM - 4:04 PM   @ Rose Quarter Transit Center

Hourly Load 276	Seat Capacity 258	Achievable Capacity 307	Load to Seat Ratio 107%	Load to Achievable Capacity 90%

Start Time	TP Time	Train	Start Location	Avg Boarding Rides	Avg Max Load	Max Load Factor	Percent Over Capacit	# of Pass Ups	APC Obs	Headway Adherence	On Time	Early	Late
2:52 PM	3:09 PM	510	SW 6th & College	78	44	102%	11%		37	87%	79%	2%	19%
3:04 PM	3:21 PM	513	SW 6th & College	77	44	103%	6%		31	82%	70%	5%	25%
3:15 PM	3:31 PM	506	SW 6th & College	66	45	104%	17%	5	30	85%	68%	14%	17%
3:26 PM	3:42 PM	504	SW 6th & College	83	57	133%	52%	1	23	81%	59%	4%	36%
3:37 PM	3:53 PM	508	SW 6th & College	63	39	90%	9%		33	87%	72%	8%	19%
3:48 PM	4:04 PM	503	SW 6th & College	65	47	110%	19%		36	87%	83%	3%	15%

5 -Interstate - Outbound   3:15 PM - 4:14 PM   @ Rose Quarter Transit Center

Hourly Load 275	Seat Capacity 258	Achievable Capacity 307	Load to Seat Ratio 107%	Load to Achievable Capacity 90%

Start Time	TP Time	Train	Start Location	Avg Boarding Rides	Avg Max Load	Max Load Factor	Percent Over Capacit	# of Pass Ups	APC Obs	Headway Adherence	On Time	Early	Late
3:04 PM	3:21 PM	513	SW 6th & College	77	44	103%	6%		31	82%	70%	5%	25%
3:15 PM	3:31 PM	506	SW 6th & College	66	45	104%	17%	5	30	85%	68%	14%	17%
3:26 PM	3:42 PM	504	SW 6th & College	83	57	133%	52%	1	23	81%	59%	4%	36%
3:37 PM	3:53 PM	508	SW 6th & College	63	39	90%	9%		33	87%	72%	8%	19%
3:48 PM	4:04 PM	503	SW 6th & College	65	47	110%	19%		36	87%	83%	3%	15%
3:58 PM	4:14 PM	514	SW 8th & College	89	43	100%	12%	1	34	80%	79%	2%	19%

Fig. 7. Hourly Capacity Report by Route and Direction

A number of performance reports are readily accessible by agency staff through customized query interfaces connected to an enterprise-level database. Examples of such interfaces are shown in Figs. 8 and 9. Fig. 8 shows the TriMet BDS Data Query Engine interface which provides access to a number of performance reports. The service performance Productivity Improvement Process (PIP) reports interface is displayed in Fig. 9. The PIP process is an interdepartmental program aimed at improving overall OTP. To a large degree, these reports focus on the performance of individual operators and routes. Performance reports are typically generated on monthly, quarterly, or annual bases. Summary reports are also available for the current signup period, the previous week, and the previous service day as well. Performance reports generated at the route level help identify operational problems requiring closer scrutiny. Of particular relevance to schedulers are performance reports pertaining to trips, time points, and peak load points.

Fig. 10 is a service delivery report showing ridership and performance statistics at the route level. The *frequent bus* variable is a flag depicting one of the 15 frequent service routes operated by TriMet. The variable *trips* represents the number of daily scheduled trips. The report shows information on the number of *revenue hours*, *vehicle hours*, and *schedule efficiency*. Schedule efficiency is calculated as the amount of revenue hours divided by vehicle hours times 100. Also presented is the ratio of *recovery to service hours* which is a measure that determines the percent of recovery

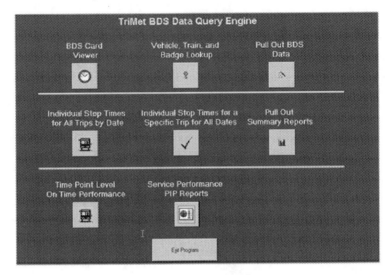

Fig. 8. BDS Data Query Engine

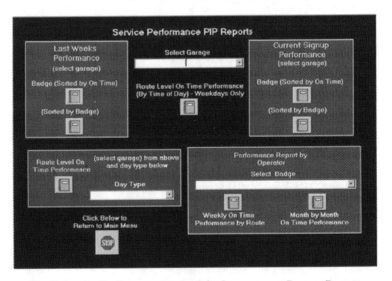

Fig. 9. Service Performance Productivity Improvement Process Reports

or layover time compared to the total number of vehicle hours. This is a comparative measure used for identifying routes with potential excess recovery time. The *recovery ratio* is defined as the median recovery/layover time divided by the scheduled recovery/layover time multiplied by 100. Recovery ratio values over 100 indicate that the typical operator is receiving more layover/recovery time than scheduled and may point to excess running time. *Actual speed* is a variable based on the median run time divided by scheduled distance. It is essentially a measure of revenue speed since

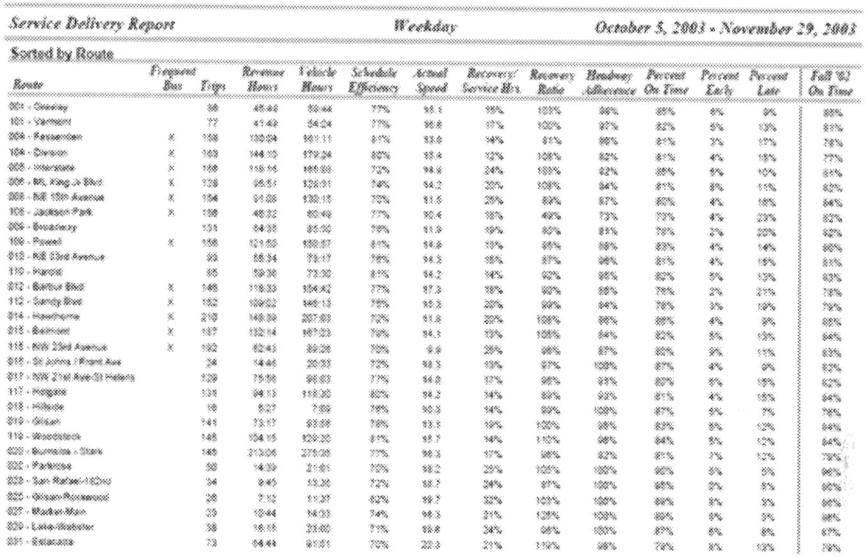

Service Delivery Report *Weekday* *October 5, 2003 - November 29, 2003*

Sorted by Route

Route	Frequent Bus	Trips	Revenue Hours	Vehicle Hours	Schedule Efficiency	Actual Speed	Recovery/ Service Hrs.	Recovery Ratio	Headway Adherence	Percent On Time	Percent Early	Percent Late	Fall '02 On Time

Fig. 10. Service Delivery by Route

it includes dwell time. The report also includes measures of *headway adherence* and *OTP* defined previously. Finally, *previous year on-time* percentage is presented for comparison purposes.

Fig. 11 displays much of same information as the previous figure except that it is broken out by time of day and direction for Route 104- Division. The route performance report includes measures of service efficiency and service quality consisting of the number of *boarding rides*, *rides per revenue hour*, average *maximum load*, average *load factor*, the percentage of trips that are *overcapacity* as well as the number of reported *passups*. TriMet calculates a variable that describes the percentage of trips that are *overcapacity due to headway* spacing problems. The purpose of this variable is to determine if large loads may be attributable to unscheduled gaps in service rather than true demand. This points towards possible schedule or operator issues rather than vehicle loading or demand problems. *Actual speed* is the same as the median speed variable presented previously. The route-level performance report includes information related to average *scheduled headway* as well and the *headway adherence* of all trips operating in the time period. The report also includes various performance measures such as *OTP* and measures related to passenger wait time experience (*excess wait time*, average *wait time per trip* and *total wait time* over all trips). Excess wait time is a measure adapted from Hounsell and McLeod (1998) and presented in more detail in Strathman *et al.* (2002b). The excess wait time measure places a heavy penalty on highly variable service and is used primarily as a basis for comparison among routes. The report contains important information about the performance of an individual route by time of day which can help the agency make more informed decisions about where to target resources. For example, overloading due to

Time of Day Route Performance Report   Weekdays   October 3, 2003 - November 29, 2003

Route: 104 - Division

	Trips	Boarding Rides	Rides/ Rev.Hr	Max Load	Load Factor	% Over Capacity	% Due to Headway	# of Passups	On Time	Early	Late	Sched. Headway	Headway Adhere.	Excess Wait Time	Wait Time Per Trip	Total Wait Time
Inbound																
Early AM	6	234	48.9	29	74%	0%	0%	0	97%	2%	2%	0:13	99%	00:07	0:04	0:29
AM Peak	15	772	60.1	35	89%	7%	71%	19	91%	2%	7%	0:08	71%	00:36	0:30	7:43
Midday	31	1,674	60.4	29	73%	2%	38%	9	87%	4%	9%	0:14	87%	00:25	0:22	11:23
PM Peak	10	526	57.4	23	56%	0%	0%	1	81%	3%	16%	0:13	73%	00:43	0:37	6:19
Night	21	549	33.5	14	36%	0%	0%	0	85%	3%	13%	0:19	91%	00:32	0:14	4:56
Total by Direction	83	3,796	53.0	25	65%	2%		29	87%	3%	10%				0:22	30:52
Outbound																
Early AM	5	130	34.4	17	43%	0%	0%	0	82%	0%	18%	0:21	100%	00:17	0:07	0:36
AM Peak	10	280	32.2	22	58%	1%	0%	0	77%	5%	18%	0:12	82%	00:58	0:27	4:29
Midday	30	1,619	55.2	30	78%	3%	45%	28	72%	6%	22%	0:14	82%	01:19	1:03	31:40
PM Peak	15	749	49.1	35	89%	5%	54%	13	68%	5%	27%	0:08	52%	01:40	1:22	20:44
Night	20	607	37.3	24	62%	0%	33%	2	80%	5%	14%	0:19	93%	00:50	0:25	8:21
Total by Direction	80	3,384	46.2	28	71%	2%		43	75%	5%	20%				0:49	65:52
Route Total	163	7,180	49.5	26	68%	2%		72	81%	4%	15%				0:35	96:45

Fig. 11. Route Performance Report by Direction and Time of Day

uneven headways is most pronounced during peak periods in the primary direction of travel. Passups appear to be an important issue in the midday time period in the outbound direction largely because of uneven headways. The percentage of trips that are late is much higher in the outbound direction in all time periods compared to the inbound direction. Highly variable service in the midday and p.m. peak time periods in the outbound direction forces passengers to arrive at bus stops early in order to compensate for unreliable service. The estimated total amount of passenger wait time for the partial signup is 96h:52m. Solutions to a number of these problems will either require scheduling adjustments or operations control actions.

Fig. 12 is a service delivery report showing information at the level of the individual trip for Route 104- Division. This report is primarily used by schedulers and service planners as a more specific diagnostic tool to investigate problems identified from higher level summary reports. Information such as *train* number, *start location*, and the number of valid *APC observations* are presented in addition to the passenger activity and OTP information. The main difference between this and the previous figure is the inclusion of trip time information. The amount of *scheduled run time* and *scheduled recovery time* as well as the *median run time* and *median recovery* are presented. The *run time ratio* is defined as the median actual run time divided by the scheduled run time multiplied by 100. A value greater than 100 indicates that the typical operator requires more time to complete the trip than what is scheduled. The *run time coefficient of variation* (CV) is the standard deviation of actual run time divided by the mean actual run time times 100. The run time CV is a unit free measure and is useful for making comparisons across trips with varying scheduled run times. The report provides useful information to schedulers including the *median speed, maximum load factor*, the *run time ratio*, the *recovery ratio*, and *OTP*. For example, three trips show a run time ratio greater than 100% indicating that too

Route: 104 - Division *Weekdays*

Direction Outbound to Division & 145th or Gresham TC *October 5, 2003 to November 28, 2003*

Start Time	Train	Start Location	Avg. Boarding Rides	Avg. Max Load	Max Load Factor	Percent Over Capacity	# of Pass Ups	APC Obs.	Sched. Run Time	Median Run Time	Run Time Ratio	Run Time CV	Median Speed	Headway Adherence	Sched. Recovery	Median Recovery	Recovery Ratio	On Time	Early	Late
									h:mm:ss	h:mm:ss			mph		mm:ss	mm:ss				
2:15 PM	401	SW 5th & Oak	63	35	91%	12%	2	35	1:01:00	0:59:24	97%	7%	13.6	63%	14:30	15:48	120%	59%	11%	29%
3:31 PM	403	SW 5th & Oak	87	42	108%	8%	4	38	1:03:00	1:09:30	110%	7%	11.5	53%	22:05	12:06	55%	20%	1%	79%
2:46 PM	407	SW 5th & Oak	66	34	86%	6%		38	1:04:00	1:02:05	97%	9%	12.9	59%	18:30	19:04	106%	71%	3%	27%
2:55 PM	437	SW 5th & Oak	77	36	97%	8%		34	1:04:00	1:03:34	100%	6%	12.6	77%	17:30	18:30	106%	68%	9%	23%
3:10 PM	404	SW 5th & Oak	85	42	108%	6%	13	32	1:04:00	1:06:32	104%	3%	12.2	63%	18:02	17:02	95%	48%	1%	51%
3:20 PM	438	SW 5th & Oak	61	38	98%	0%	1	37	1:04:00	1:01:45	97%	3%	13.0	68%	20:02	23:16	117%	64%	4%	32%
3:30 PM	482	SW 5th & Oak	65	37	94%	0%	1	28	1:04:00	1:02:48	97%	9%	12.8	74%				61%	6%	33%
3:45 PM	402	SW 5th & Oak	70	45	105%	12%		38	1:05:00	1:03:31	98%	6%	12.6	69%	14:30	15:44	120%	75%	12%	13%
3:50 PM	447	SW 5th & Oak	73	40	102%	13%	1	36	1:08:00	1:05:53	97%	3%	12.3	72%	16:02	17:43	111%	62%	12%	26%
4:01 PM	438	SW 5th & Oak	75	46	118%	14%	5	36	1:05:00	1:04:28	99%	3%	12.4	88%	20:05	21:44	109%	59%	7%	39%
4:11 PM	444	SW 5th & Oak	60	32	82%	3%	1	37	1:05:00	1:03:44	98%	5%	12.6	47%				69%	4%	27%
4:18 PM	445	SW 5th & Oak	36	27	69%	0%		34	0:59:00	0:55:07	93%	3%	14.3	53%	25:02	28:41	119%	75%	7%	18%
4:27 PM	443	SW 5th & Oak	58	37	95%	3%	1	37	1:04:00	1:03:42	100%	8%	12.6	47%				51%	2%	47%
4:36 PM	448	SW 5th & Oak	60	33	85%	2%		36	0:59:00	0:58:35	99%	5%	13.7	48%	20:30	22:34	113%	77%	5%	18%
4:43 PM	440	SW 5th & Oak	45	28	71%	0%		38	1:04:00	1:02:03	97%	4%	12.9	74%	25:02	24:16	103%	57%	25%	17%
4:50 PM	445	SW 5th & Oak	44	30	80%	0%		33	1:02:00	0:59:24	96%	3%	13.3	88%	16:02	17:36	109%	78%	6%	16%
4:57 PM	431	SW 5th & Oak	48	32	82%	3%		36	1:03:00	1:00:54	97%	4%	13.2	61%				85%	5%	10%
5:03 PM	442	SW 5th & Oak	53	41	104%	12%		37	1:02:00	0:59:24	99%	6%	13.5	39%				55%	1%	44%
5:08 PM	7252	SW 5th & Oak	42	28	71%	5%	1	37	1:01:00	0:59:34	98%	6%	13.7	24%				73%	13%	14%
5:10 PM	435	SW 5th & Oak	34	26	68%	0%		39	0:59:00	0:56:38	96%	4%	14.2	11%	16:02	17:28	109%	72%	1%	27%
5:25 PM	406	SW 5th & Oak	64	45	114%	12%	4	38	0:59:00	1:00:37	102%	7%	13.2	74%	21:30	18:38	86%	41%	0%	59%
5:30 PM	403	SW 5th & Oak	46	34	87%	0%		33	0:59:00	0:59:08	100%	4%	13.6	58%				91%	2%	7%
5:40 PM	450	SW 5th & Oak	92	42	107%	16%		31	0:58:00	0:56:32	96%	7%	14.2	88%	17:02	18:56	111%	65%	0%	35%
5:55 PM	434	SW 5th & Oak	43	37	95%	0%	1	36	0:56:00	0:53:17	95%	6%	14.5	53%	23:30	26:30	114%	71%	3%	26%
6:02 PM	439	SW 5th & Oak	60	34	88%	0%		36	0:55:00	0:55:34	99%	4%	14.4	74%	26:00	29:32	114%	88%	3%	9%
6:17 PM	441	SW 5th & Oak	82	40	104%	6%	2	39	0:58:00	0:55:13	96%	3%	14.3	88%				68%	1%	31%
6:32 PM	401	SW 5th & Oak	48	33	85%	0%		36	0:54:00	0:52:34	98%	7%	15.2	89%	13:00	15:42	105%	62%	5%	33%
6:47 PM	448	SW 5th & Oak	37	30	78%	0%		38	0:53:00	0:51:34	97%	5%	15.5	84%	17:30	18:30	102%	83%	1%	16%
7:02 PM	407	SW 5th & Oak	39	26	67%	0%		38	0:52:00	0:50:13	97%	5%	16.0	80%	18:30	18:30	103%	64%	3%	33%
7:17 PM	437	SW 5th & Oak	34	27	70%	0%		37	0:52:00	0:49:34	95%	7%	16.2	96%	17:05	18:48	110%	71%	5%	24%
7:32 PM	436	SW 5th & Oak	32	27	69%	0%		37	0:52:00	0:50:10	96%	9%	16.5	97%	16:00	17:58	114%	79%	7%	14%

Fig. 12. Service Delivery by Trip

little run time is scheduled for the typical operator on these trips. *Headway adherence* is particularly problematic for trips departing between 5:03 and 5:10 p.m. This may be due to interactions between regular service vehicles and trippers (shown as trips with no scheduled recovery times). Trippers are vehicles that are brought online to serve periods of heavy demand and are typically operated by part-time operators. With respect to OTP, 25% of trips associated with the 4:43 p.m. departure are leaving time points early. The trip departing at 3:31 p.m. has the highest percentage of trips operating late and is also the same trip with the lowest recovery ratio and the highest run time ratio.

Figs. 13–15 represent graphs developed exclusively for scheduling purposes. The unit of analysis is an individual trip operating between time points. Fig. 13 shows the relationship between actual and scheduled running time for Route 4- Fessenden from the time point at the intersection of Albina & Killingsworth to the time point at Lombard & Interstate in the outbound direction. The Y-axis represents run time in minutes and the X-axis shows the trip number along with the scheduled departure time. The *20th*, *50th*, and *80th percentile run times* are presented in relation to *scheduled run time*. Again, the 50th percentile run time is the time required for the typical operator to operate the schedule. The 20th and 80th percentile run time are somewhat arbitrary measures used to set bounds on variation in run time. The run time graphs are used to assess whether the amount of existing scheduled run time is adequate. If run times are set too low or too high, then scheduling adjustments are necessary. The data show that the amount of scheduled run time is set slightly higher than the median run time for most trips.

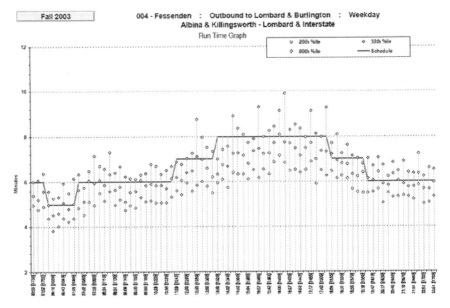

Fig. 13. Run Time Graph

Fig. 14 displays the median and the scheduled run time along with the average maximum passenger load and the average number of actual stops. The secondary Y-axis shows the number of stops and the mean passenger load for each trip over the segment. One can readily see the relationship between actual stops and passenger loads in relation to scheduled run time and the 50th percentile run time. Fig. 15 is an optimal run time graph showing the median and scheduled run times along with the suggested optimum scheduled run time. The optimal run time procedure uses cluster algorithms based on nonparametric density estimates applied to median run times on an interpolated per minute basis for grouping similar observations.

Fig. 16 is a time point level run time report showing data for two trips on a radial through-route, the 4 Fessenden/104 Division. The report shows cumulative *scheduled run time* from the origin location to the destination location. The two trips departing at 3:50 p.m. and 4:01 p.m. have scheduled run times of 1h:51m and 1h:53m, respectively. Cumulative statistics for the *median*, *20th*, and *80th percentile run times* are presented along with the *optimal* run time. Passenger activity information in the form of *average ons*, *average offs*, *average maximum load*, *average maximum speed* and *average actual stops* are presented. Also shown is *OTP* as well as *average minutes late* at the beginning time point. Overall, both the amount of scheduled run time and that predicted by the cluster algorithms are fairly close to the median. With respect to scheduled run time, the most problematic time point is associated with the departure at SW 5th and Oak on the first trip which is off by 4m:46s relative to the median. The optimal run time predicted by the cluster algorithms is much closer at 0m:46s. Note that this time point is also associated with the greatest amount of pas-

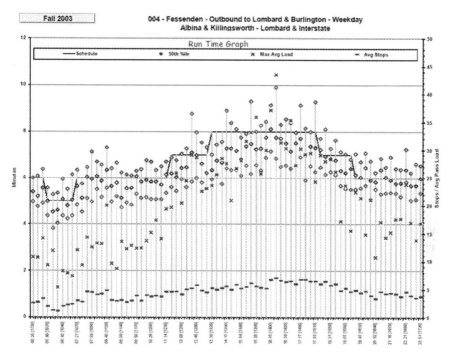

Fig. 14. Run Time Graph with Stops and Passenger Loads

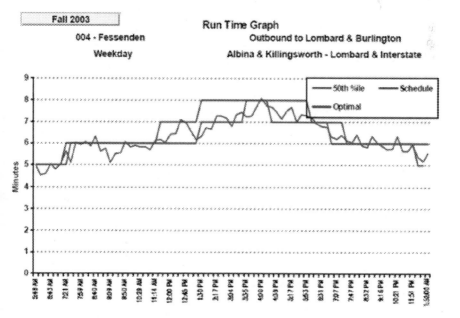

Fig. 15. Optimal Run Time Graph

senger activity. Consistent with theory, delays tend to propagate as buses proceed along a route, then begin to decline as passenger activity drops off. Overall, OTP is much better on the second trip compared to the first trip. As mentioned previously, the quality of a given performance measure is contingent on having accurate schedules. Regarding the first trip in Fig. 16, it is evident that the severity of late departures subsequent to the Rose Quarter Transit Center time point can be attributed to not enough scheduled run time rather than poor performance.

Weekday Time Point Level Run Time Report - Winter 2000/2001

Route 4 - Fessenden/Division - to 148th & Division or Gresham TC

Beginning Time Point	Scheduled Run Time	Median	20th %ile	80th %ile	Optimal	Avg Ons	Avg Offs	Max Load	Avg Speed	Avg Stops	On Time	Early	Late	Avg Min Late
Key Time 1: 3:50 PM		*Train Number:* 450												
Burlington & Leonard	06:00	06:12	05:38	06:50	06:00	18	4	13	29.8	5	87%	3%	0%	1.0 min.
Fessenden & Columbia Way	14:00	14:22	13:46	15:20	13:00	12	9	18	26.8	8	96%	3%		0.9 min.
Willis & Chautauqua	22:00	23:02	22:00	24:04	23:00	3	8	15	27.7	4	85%	0%	9%	1.4 min.
Lombard & Interstate	28:00	29:00	28:04	31:02	29:00	11	6	18	27.6	4	82%	5%	13%	2.5 min.
Albina & Killingsworth	42:00	43:42	42:34	46:54	43:00	14	27	22	28.2	9	76%	2%	20%	3.1 min.
Rose Quarter Transit Center	47:00	46:44	47:14	51:44	48:00	7	5	15	31.8	4	62%	0%	38%	4.7 min.
SW 5th & Oak	1:01:00	1:05:46	1:03:36	1:10:24	1:05:00	33	12	35	28.1	9	49%	3%	48%	6.1 min.
Division & 12th	1:10:00	1:13:42	1:11:01	1:19:54	1:11:00	6	10	35	25.5	7	28%	0%	72%	7.9 min.
Division & 39th	1:21:00	1:23:50	1:21:30	1:29:53	1:22:00	8	18	32	27.8	9	31%	3%	66%	7.5 min.
Division & 82nd	1:30:00	1:32:39	1:30:16	1:38:24	1:31:00	8	9	27	31.4	7	43%	0%	57%	8.9 min.
Division & 122nd	1:35:00	1:37:12	1:35:06	1:42:43	1:36:00	4	7	24	31.3	4	46%	2%	52%	8.4 min.
Division & 148th	1:43:00	1:45:32	1:43:08	1:51:54	1:43:00	7	14	21	31.1	8	49%	3%	48%	8.7 min.
Division & 182nd	1:51:00	1:52:49	1:50:12	1:58:12	1:51:00	1	11	12	33.6	3	44%	2%	54%	8.2 min.
Key Time 1: 4:01 PM		*Train Number:* 451												
Burlington & Leonard	06:00	06:20	05:46	06:52	06:00	14	3	12	30.4	4	68%	2%	0%	0.8 min.
Fessenden & Columbia Way	14:00	14:23	13:32	15:22	14:00	11	8	17	27.4	8	90%	10%	0%	0.3 min.
Willis & Chautauqua	22:00	23:00	21:58	24:38	23:00	4	6	16	27.9	5	65%	15%	0%	0.3 min.
Lombard & Interstate	28:00	29:06	28:08	30:22	29:00	9	6	19	28.9	4	67%	10%	3%	1.2 min.
Albina & Killingsworth	42:00	43:42	42:22	46:18	43:00	16	30	24	29.8	10	92%	3%	5%	1.4 min.
Rose Quarter Transit Center	48:00	48:56	47:30	51:34	48:00	8	6	18	33.1	3	82%		12%	1.7 min.
SW 5th & Oak	1:03:00	1:06:07	1:04:03	1:09:43	1:05:00	32	18	36	28.9	8	74%	3%	23%	3.3 min.
Division & 12th	1:12:00	1:14:03	1:11:52	1:19:34	1:13:00	5	11	37	27.2	7	80%	0%	20%	3.4 min.
Division & 39th	1:23:00	1:24:40	1:22:25	1:30:10	1:23:00	5	15	32	29.0	8	61%	13%	26%	3.0 min.
Division & 82nd	1:32:00	1:34:33	1:32:03	1:39:47	1:32:00	6	10	25	31.7	7	61%	16%	23%	2.4 min.
Division & 122nd	1:37:00	1:38:58	1:36:40	1:44:24	1:38:00	3	7	19	32.3	4	66%	7%	28%	3.3 min.
Division & 145th	1:45:00	1:46:30	1:44:06	1:51:13	1:44:00	3	7	13	33.6	6	66%	8%	26%	2.8 min.
Division & 182nd	1:53:00	1:53:28	1:50:58	1:59:30	1:51:00	1	7	9	34.3	2	66%	11%	23%	2.3 min.

Fig. 16. Time Point Level Run Time Report

TriMet has made considerable progress in the use of AVL and APC data for scheduling purposes. These efforts are largely made possible because 1) operations data archived at the level of the individual bus stop, 2) the extent of AVL and APC deployment on the fleet, and 3) the willingness of the agency to continuously improve the BDS system through provision of adequate resources in the areas of data validation, database design and management, and performance monitoring.

4 Operator Behavior and Service Reliability

TriMet is keenly interested in addressing service quality issues related to operator variability and inconsistent operator behavior. Operators are in a unique position to adequately gauge operating conditions encountered along a route. In the absence of communication from operators to schedulers about inadequate or excess run time in schedules, operators may instead choose to vary departure times and/or operating speeds. Operators may intentionally deviate from scheduled departure times at origins in order to maximize recovery/layover time or to receive the minimum break specified in the work contract given delays on the previous trip. Efforts to reduce the amount of variability in departure times through better management of bus operators would ultimately lead to more efficient schedules. It should be noted that labor agreements prohibit the ability of the agency to discipline operators based on BDS information; however, it is perfectly appropriate for the information to be passed on to supervisors so that remedial actions can take place. The agency is trying to identify operators who may possibly need additional training or admonishment from a supervisor. For each of the figures which follow, the name of each operator has been omitted and badge number has been truncated due to privacy considerations.

A graph depicting the relationship between late origin departures and OTP for TriMet's frequent service bus routes is presented in Fig. 17. The unit of analysis is the individual operator. The Y-axis shows the percentage of time point departures classified as late (averaged over all time points over all days). The X-axis shows the percentage of trips during the signup period where an operator left the beginning of the line late. For example, at 25% leaving late, an operator is identified as leaving the terminal more than three minutes late 25% of the time. A high value for leaving late with a low average percent late at time points is indicative of either fast operators or too much run time built into schedules. A low value for leaving late with a high average percent late characterizes either slow operators or not enough running time built into schedules.

Operator OTP at time points for the current signup is presented in Fig. 18. In addition to the percentage of trips *on-time*, *early*, and *late*, *peer on-time* is also presented along with the *on-time difference* in relation to the operator peer group. Peer on-time is the weighted average on-time percentage of all operators driving the same route and direction during similar time periods throughout the day. The first step in the calculation is to determine the overall percentage of trips that are on-time for each route, direction, and time of day component, irrespective of operator. The second step involves calculating a weighted average on-time percentage representing the peer group (weighted by the number of trips operated by the operator of interest). For example, assume that a given operator operates ten inbound trips in the a.m. peak on a given route and 15 outbound trips during the p.m. peak on a different route. If the overall on-time percentage for the first route in is 80% and 70% for the second route, then the estimate for peer group on-time percentage is 74%. The peer on-time measure is important because it provides a more accurate basis for comparison of OTP among operators. It is not appropriate to compare an operator who drives a radial trunk during the peak hour to someone who operates a low ridership suburban feeder during the midday time period. Since the peer on-time measure is a weighted average, it also works for extra-board operators who operate different runs from day to day. *Begin of line late* represents the percentage of departures more than three minutes late. The report also shows information related to *average lifts per 8 hours* of work since some lateness can be explained by lift operation activity. The data

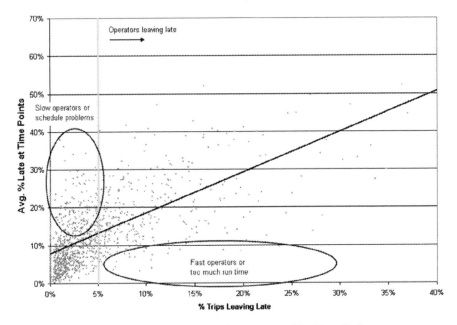

Fig. 17. Relationship between Late Departures and On-Time Performance

in Fig. 18 are sorted by on-time difference. Four operators have an OTP difference relative to their peer group of 25% or greater. Of these four, three are leaving the beginning of the line late more than 30% of the time. Fig. 19 shows operator OTP by month of service for an individual operator. The variables are similar to the ones presented in the previous figure except that the percentage of trips departing late from the beginning of the line is broken out into three and five minute intervals. The report depicts a consistent operator who is almost never late or early, is well above average with respect to peer OTP, and departs route origins promptly. The data presented in the operator OTP report are particularly relevant to operations management personnel. The importance of this report is to track the performance of an individual operator to see if the recently identified problem is a continuing trend, a new pattern, or an isolated incident.

Of interest to the agency are operators that have a pattern of leaving their respective garages consistently late. Information related to late operator pull outs is presented in Fig. 20 including *badge, name,* and type of *duty, percent pull out late, average minutes late,* and the *number of pull out* observations used in the analysis. The data are sorted by average minutes late so the table is showing the worst operators from Center Garage for a partial signup. Fig. 21 shows more detailed information about a single operator including *scheduled pull out time* and *actual pull out time, minutes late, pull out notes,* and *pull out to location.* Information related to *scheduled departure time, actual arrival time,* and *minutes late at the first time point* are also presented. The report also includes information related to *minutes late at the second time point.* Situations where operators pulled out late and were more

Operator On Time Performance

Center Garage — **1 Minute Early to 5 Minutes Late** — ***Current Signup***

Badge	Operator Name	Percentage On Time	Percentage Early	Percentage Late	Peer On Time	On Time Difference	Begin of Line Late	Avg. Lifts Per 8 hr.
690		25%	0%	75%	67%	-42%	47%	2.3
4		31%	0%	68%	72%	-41%	33%	7.1
343		31%	0%	68%	68%	-37%	32%	2.4
155		48%	0%	52%	75%	-27%	9%	
023		49%	39%	11%	75%	-25%	0%	1.5
227		46%	0%	54%	71%	-25%	9%	5.2
271		51%	7%	43%	75%	-25%	0%	2.3
7		53%	1%	46%	77%	-24%	5%	1.6
14		62%	34%	5%	85%	-23%	0%	1.2
043		58%	6%	36%	81%	-23%	12%	2.9
170		57%	0%	43%	80%	-23%	7%	0.6
535		58%	0%	42%	81%	-23%	16%	6.2
134		55%	0%	45%	77%	-22%	18%	3.2
963		60%	2%	38%	82%	-22%	32%	0.5
778		62%	0%	38%	84%	-22%	21%	0.6
010		61%	1%	39%	82%	-22%	19%	1.2
705		60%	0%	39%	82%	-22%	25%	1.0
279		60%	0%	40%	81%	-21%	0%	3.0
948		63%	6%	31%	83%	-20%	20%	3.1
416		60%	5%	35%	80%	-20%	26%	4.3
647		61%	4%	35%	81%	-20%	14%	2.7
39		63%	0%	37%	82%	-19%	21%	2.2
365		58%	3%	40%	77%	-19%	7%	1.0
800		60%	14%	27%	79%	-19%	7%	2.1

Peer On Time = Average On Time of all Operators driving the same route and direction during the same time period.

Page 1

Fig. 18. Operator OTP by Badge

Operator Monthly On Time Performance Report

Operator: 135

Time Point Level On Time Performance — **1 Minute Early to 5 Minutes Late** — *Leaving Begin of Line Late*

MONTH	Percentage On Time	Percentage Early	Percentage Late	Avg. Min. Late	Peer On Time	On Time Difference	> 3 Min. Late	> 5 Min. Late
November 2003	98%	2%	0%	0.2	85%	13%	0%	0%
October 2003	98%	1%	1%	0.3	82%	16%	0%	0%
September 2003	98%	1%	2%	0.5	82%	16%	1%	1%
August 2003	99%	1%	1%	0.4	83%	16%	2%	1%
July 2003	99%	1%	0%	0.2	85%	14%	1%	0%
June 2003	99%	1%	0%	0.3	84%	15%	0%	0%
May 2003	98%	1%	1%	0.3	83%	15%	0%	0%
April 2003	98%	2%	1%	0.3	83%	15%	1%	1%
March 2003	97%	1%	2%	0.6	84%	13%	3%	1%
February 2003	99%	1%	0%	0.4	84%	16%	0%	0%
January 2003	99%	1%	0%	0.3	81%	18%	0%	0%
December 2002	97%	2%	2%	0.6	83%	14%	3%	0%
November 2002	98%	1%	1%	0.4	86%	12%		
October 2002	99%	0%	1%	0.4	84%	15%		
September 2002	100%	0%	0%	0.4	86%	14%		
August 2002	99%	0%	0%	0.1	85%	14%		
July 2002	99%	0%	0%	0.1	87%	12%		
June 2002	100%	0%	0%	0.2	86%	14%		

On Time Peformance is calculated by comparing departure time from a Timepoint compared to the scheduled time.

Page 1

Fig. 19. Individual Operator OTP by Month

than three minutes late at the first two time points are denoted by a double asterisk. Highly variable pull out delays have important scheduling implications since buses do not operate independently from each other. Designing schedules to accommodate extreme variability in operator behavior is a second best solution compared to better management of operators at the beginning of the line.

Pull Out Late Report
November 30, 2003 - December 17, 2003

Center Garage

Badge	Operator Name	Duty	Percent Pull Out Late	Average Minutes Pull Out Late	Number of Pull Outs in Analysis
887		Full Time	81.8%	05:35	11
085		Full Time	80.0%	07:52	5
39		Extra Board	77.8%	09:01	9
96		Full Time	72.7%	08:37	11
043		Extra Board	71.4%	03:40	7
690		Mini Run	70.0%	05:23	10
705		Extra Board	60.0%	05:04	5
769		Full Time	53.8%	06:14	13
112		Full Time	50.0%	08:07	10
25		Full Time	50.0%	06:31	10
896		Full Time	50.0%	04:07	8
63		Full Time	46.2%	07:11	13
930		Full Time	46.2%	05:22	13
696		Full Time	45.5%	04:57	11
468		Extra Board	42.9%	06:14	14
168		Full Time	42.9%	04:31	7
90		Full Time	40.0%	05:04	10
554		Regular Relief	40.0%	05:04	10
343		Mini Run	38.1%	04:04	21
474		Regular Relief	37.5%	06:56	16
86		Full Time	37.5%	04:45	8
990		Extra Board	36.4%	05:19	11
58		Full Time	36.4%	03:18	11
67		Mini Run	35.7%	04:18	14
235		Mini Run	35.7%	03:05	14

Late = Pull out greater than 3 minutes late and more than 3 minutes late at the first and second time points.

Page 1

Fig. 20. Pull Out Late by Operator

The question arises as to whether the data analysis capabilities afforded by the TriMet BDS has made a difference with respect to schedule efficiency and service reliability. Long term trends for OTP and schedule efficiency are presented in Fig. 22. The data pertain to OTP for weekday fixed route bus service and schedule efficiency trends for TriMet's 15 frequent service routes. Although subject to monthly variation, the general trend is that OTP has been increasing over time. In 2004-Jan., OTP is at 82.4% which is considerably higher than the agency standard of 75%. With respect to TriMet's 15 frequent service bus routes, schedule efficiency is at 75.5% which is slightly above the agency standard of 75%. Figures were not available at the time of this publication regarding schedule efficiency for all fixed route bus service. These trends are encouraging given higher ridership levels and worsening traffic congestion in the region over the past couple of years.

Pull Out by Operator Report

Operator: 288

Date	Train	Sched. Pull Out Time	Actual Pull Out Time	Pull Out Late	Pull Out Note	Pull Out To:	1st TP Scheduled	1st TP Arrive	1st TP Leave Late	2nd TP Leave Late
15-Dec-03	7703	7:10 AM	7:18:44 AM	08:44		27th & Vaughn at Montgomery P	7:25 AM	7:31:44 AM	08:06	08:10
14-Dec-03	506	8:36 AM	8:43:20 AM	07:20		6th & College	8:46 AM	8:48:24 AM	07:26	05:20
13-Dec-03	7201	10:25 AM	10:28:02 AM	03:02		Anchor & Channel	10:45 AM	10:47:14 AM	03:50	05:46
12-Dec-03	7703	7:10 AM	7:15:54 AM	05:54		27th & Vaughn at Montgomery P	7:25 AM	7:27:42 AM	03:16	03:04
11-Dec-03	7703	7:10 AM	7:12:32 AM	02:32		27th & Vaughn at Montgomery P	7:25 AM	7:25:56 AM	02:36	03:02
09-Dec-03	7703	7:10 AM	7:13:12 AM	03:12		27th & Vaughn at Montgomery P	7:25 AM	7:25:24 AM	03:06	02:58
06-Dec-03	7201	10:25 AM	10:29:24 AM	04:24		Anchor & Channel	10:45 AM	10:39:14 AM	03:42	03:20
05-Dec-03	7703	7:10 AM	7:15:18 AM	05:18		27th & Vaughn at Montgomery P	7:25 AM	7:26:52 AM	04:18	06:52
04-Dec-03	7703	7:10 AM	7:14:06 AM	04:06		27th & Vaughn at Montgomery P	7:25 AM	7:27:32 AM	04:28	05:14
01-Dec-03	7703	7:10 AM	7:14:26 AM	04:26		27th & Vaughn at Montgomery P	7:25 AM	7:28:30 AM	05:22	06:26
26-Nov-03	1405	7:08 AM	7:10:58 AM	02:58		Foster & 94th (i-205 Overpass)	7:24 AM	7:21:34 AM	02:00	01:28
26-Nov-03	106	3:08 PM	3:13:28 PM	05:28		Vermont & 45th	3:25 PM	3:26:02 PM	04:10	04:34
26-Nov-03	1405	7:08 AM	7:11:04 AM	03:04		Foster & 94th (i-205 Overpass)	7:24 AM	7:22:42 AM	01:44	01:26
25-Nov-03	106	3:08 PM	3:11:16 PM	03:16		Vermont & 45th	3:25 PM	3:26:24 PM	03:00	04:54
24-Nov-03	106	3:08 PM	3:08:00 PM		Pulled Out Early	Vermont & 45th	3:25 PM	3:22:46 PM	01:34	04:46
24-Nov-03	1405	7:08 AM	7:08:00 AM		Pulled Out Early	Foster & 94th (i-205 Overpass)	7:24 AM	7:19:34 AM	02:06	02:56
23-Nov-03	7509	7:14 AM	7:16:38 AM	02:38		Milwaukee Transit Center	7:23 AM	7:21:58 AM	02:32	01:58
20-Nov-03	1202	6:14 AM	6:14:00 AM		Pulled Out Early	Parkrose/ Sumner Transit Center	6:39 AM	6:33:48 AM	-01:48	00:08
19-Nov-03	9904	4:35 PM	4:37:34 PM	02:34		5th & Oak	4:50 PM	4:48:34 PM	00:54	01:50
18-Nov-03	106	3:08 PM	3:10:56 PM	02:56		Vermont & 45th	3:25 PM	3:25:24 PM	01:22	06:24
17-Nov-03	106	3:08 PM	3:11:16 PM	03:18		Vermont & 45th	3:25 PM	3:26:20 PM	02:04	04:30
17-Nov-03	1405	7:08 AM			BOS Problem	Foster & 94th (i-205 Overpass)	7:24 AM	7:21:40 AM	01:36	01:20
16-Nov-03	7509	7:14 AM	7:18:54 AM	04:54		Milwaukee Transit Center	7:23 AM	7:23:50 AM	01:58	01:00
12-Nov-03	1405	7:08 AM	7:08:00 AM		Pulled Out Early	Foster & 94th (i-205 Overpass)	7:24 AM	7:18:54 AM	02:02	03:02
12-Nov-03	106	3:08 PM	3:16:36 PM	08:36		Vermont & 45th	3:25 PM	3:27:12 PM	02:34	04:12
11-Nov-03	1405	7:08 AM	7:11:08 AM	03:08		Foster & 94th (i-205 Overpass)	7:24 AM	7:22:46 AM	02:36	02:52
11-Nov-03	1902	4:56 PM	4:58:14 PM	02:14		6th & Main	5:10 PM	5:10:14 PM	01:54	02:00
04-Nov-03	1405	7:08 AM	7:08:32 AM	00:32		Foster & 94th (i-205 Overpass)	7:24 AM	7:18:56 AM	01:14	02:14
04-Nov-03	106	3:08 PM	3:09:36 PM	01:36		Vermont & 45th	3:25 PM	3:25:14 PM	01:20	04:26
03-Nov-03	106	3:08 PM	3:12:12 PM	04:12		Vermont & 45th	3:25 PM	3:26:40 PM	03:18	08:42
02-Nov-03	7509	7:14 AM	7:20:04 AM	06:04		Milwaukee Transit Center	7:23 AM	7:24:42 AM	02:30	02:02

Fig. 21. Pull Out Late by Operator by Stop and Date

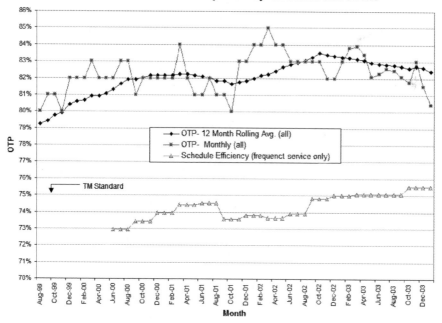

OTP and Schedule Efficiency- Weekday Fixed Route Bus Service

— OTP- 12 Month Rolling Avg. (all)
— OTP- Monthly (all)
— Schedule Efficiency (frequent service only)

TM Standard

Fig. 22. OTP and Schedule Efficiency Trends

5 Conclusions

Scheduling and operations management have benefited from analysis of data collected by the TriMet BDS. These benefits stem from a number of factors. First, the AVL and APC system was designed to collect data at the level of the individual bus stop. This provides a level of detail that is ideally suited to analysis of bus operations. Second, the widespread deployment of AVL and APC technologies on the vehicle fleet generates large amounts of high quality data providing the agency with a complete snapshot of the system at any point in time and space. Third, the agency has key personnel in place that continuously strive to improve the efficacy of the system and to refine data collection and analysis practices and procedures. Over time, the number of agency personnel making direct use of information derived from BDS data has increased substantially. While this report is limited to uses of BDS information in the areas of scheduling and operations management, other areas of the agency are making use of BDS data as well including service planning, finance, marketing, and maintenance. The BDS has also played an important role in identifying corridors and intersection locations suitable for transit signal priority.

Consistent with observations by Koffman (1992), the true value of integrated AVL and APC technologies lies in report generation capabilities. Of note is the gradual increase in performance monitoring capabilities within the agency, both in the number of different reports being generated as well as the types of performance measures being calculated. Measures such as peer on-time performance, late pullouts and late departures from trip origins, % of overloaded trips due to headway spacing problems, recovery ratio, achievable capacity, and excess wait time are providing a better picture of service quality and operational efficiency compared to the standard performance measures reported by most agencies. Currently, the agency is attempting to incorporate work rules related to operator breaks and to identify operators that habitually speed or speed at excessive levels. It should be noted that the agency is not specifically targeting operators in order to realize efficiency gains. The primary aim is to provide better service to passengers in a more cost effective manner. At the same time, the agency realizes that certain bus trips do not have sufficient recovery/layover time and that additional steps are needed to ensure that operator needs are addressed.

TriMet is in a unique position with respect to the data analysis opportunities provided by the BDS. It is not simply a question of having an AVL system in place or having a small percentage of the vehicle fleet equipped with APCs. While poll-based AVL systems are useful, they do not provide the level of detail necessary to undertake comprehensive analyses of transit operations. Likewise, limited APC deployment on vehicle fleets forces agencies to spend considerable effort assigning vehicles to specific trips for data collection purposes. Furthermore, by relying on limited APC deployment for collection of passenger activity information, estimates of boardings, alightings, and loads are not nearly as robust as those based on larger quantities of data.

The present study has highlighted many of the efforts at TriMet to improve schedule efficiency and service reliability related to operator behavior through analysis of empirical data collected by the BDS. Efforts to improve schedule efficiency center on reducing excess run and layover times in schedules as well as through use of supervisory actions to reduce operator variability. Much of what has been presented in this report would not have been possible without the data collection capabilities of the BDS and an ongoing commitment by the agency to make the most use of the information.

References

Abkowitz, M. D. (1978). Transit service reliability. Technical report, Cambridge, MA: USDOT Transportation Systems Center and Multisystems, Inc.

Benn and Barton-Aschman Associates (1995). Bus route evaluation standards. *TCRP Synthesis 10*. Transportation Research Board, Washington, D.C.

Boyle, D. K. (1998). Passenger counting technologies and procedures. *TCRP Synthesis 29*. Transportation Research Board, Washington, D.C.

Casey, R. (1999). Advanced public transportation systems deployment in the United States. Technical report, Volpe National Transportation Systems Center, Cambridge, MA.

Coffel, B. (1993). TriMet scheduling practices. Technical report, Tri-County Metropolitan Transportation District of Oregon: Portland, OR.

Furth, P. G. (2000). Data analysis for bus planning and monitoring. *TCRP Synthesis 34*. Transportation Research Board, Washington, D.C.

Furth, P. G., Hemily, B. J., Muller, T. H. J., and Strathman, J. G. (2003). Uses of archived AVL-APC data to improve transit performance and management: Review and potential. *TCRP Web Document H-28*. Transportation Research Board, Washington, D.C.

Furth, P. G. and Wilson, N. H. M. (1981). Setting frequencies on bus routes: Theory and practice. *Transportation Research Record*, **818**, 1–7.

Guenthner, R. P. and Hamat, K. (1983). Distribution of bus transit on-time performance. *Transportation Research Record*, **1202**, 7–13.

Hounsell, N. and McLeod, F. (1998). AVL implementation application and benefits in the UK. Technical report, Paper presented the 77th annual meeting of the Transportation Research Board, Washington, D.C.

Kimpel, T. J., Strathman, J. G., Griffin, D., Callas, S., and Gerhart, R. L. (2004). Automatic passenger counter evaluation: Implications for National Transit Database reporting. *Transportation Research Record*, **1835**, 93–100.

Koffman, J. (1992). Automatic passenger counting data: Better schedules improve on-time performance. Technical report, Paper presented at the Fifth Workshop on Computer-Aided Scheduling of Public Transport, Montreal.

Levinson, H. S. (1980). Bus route and schedule planning guidelines. *NCHRP Synthesis of Highway Practice 69*. Transportation Research Board, Washington, D.C.

Levinson, H. S. (1991). Supervision strategies for improved reliability of bus routes. *TCRP Synthesis of Transit Practice 15*. Technical report, Transportation Research Board, Washington, D.C.

Strathman, J. G., Dueker, K. J., Kimpel, T. J., Gerhart, R. L., Turner, K., Callas, S., and Griffin, D. (2001). Bus transit operations control: Review and an experiment involving Tri-Met's automated bus dispatch system. *Journal of Public Transportation*, **41**(1), 1–26.

Strathman, J. G., Kimpel, T. J., and Callas, S. (2002a). Headway deviation effects on bus passenger loads: Analysis of Tri-Met's archived AVL-APC data. Report PR126. Center for Urban Studies, Portland, OR.

Strathman, J. G., Kimpel, T. J., Dueker, K. J., Gerhart, R., and Callas, S. (2002b). Evaluation of transit operations: Data applications of Tri-Met's automated bus dispatch system. *Transportation*, **29**, 321–345.

TriMet (1989). TriMet sevice standards. Technical report, Tri-County Metropolitan Transportation District of Oregon, Portland, OR.

Wilson, N. H. M., Bauer, A., Gonzalez, S., and Shriver, J. (1984). Short range transit planning: Current practice and a proposed framework. Technical report, Urban Mass Transit Administration, Washington, D.C.

Woodhull, J. (1987). Issues in on-time performance of bus systems. Unpublished paper. Southern California Rapid Transit District, Los Angeles, CA.

Parallel Auction Algorithm for Bus Rescheduling

Jing-Quan Li[1], Pitu B. Mirchandani[1], and Denis Borenstein[2]

[1] Department of Systems and Industrial Engineering, University of Arizona, Tucson AZ
85719, USA jingquan@email.arizona.edu; pitu@sie.arizona.edu
[2] Business School, Federal University of Rio Grande do Sul, Porto Alegre, RS, Brazil
denisb@ea.ufrgs.br

Summary. When a bus on a scheduled trip breaks down, one or more buses need to be
rescheduled to serve the customers on that trip with minimum operating and delay costs. The
problem of reassigning buses in real-time to this *cut trip*, as well as to other scheduled trips
with given starting and ending times, is referred to as the *bus rescheduling problem* (BRP).
This paper considers modeling, algorithmic, and computational aspects of the single-depot
BRP. The paper develops the sequential and parallel auction algorithm to solve the BRP. Computational results show that our approach solves the problem quickly.

1 Introduction

The bus rescheduling problem arises when a trip is disrupted. Severe weather conditions, an accident, a traffic jam, and the breakdown of a bus are examples of possible
disruptions that demand the rescheduling of bus trips. The BRP can be approached
as a dynamic version of the classical vehicle scheduling problem (VSP) where assignments are generated dynamically.

Although the literature describes several different approaches to solve the VSP
(Daduna and Paixão (1995)), the BRP has not been sufficiently addressed by researchers. However, when the fleet size is limited and disruptions are frequent, good
automated rescheduling tools to assist decision makers become important. As a consequence of this gap in research, very few companies use automated rescheduling
policies. The objective of this research is to address this gap. In particular, the single-
depot BRP is modeled, and algorithms that solve this problem in a reasonable amount
of time are proposed.

The most pertinent decision for the BRP is on which vehicle should backup the
disrupted trip. The existence of several alternatives generates, in comparison to the
VSP, several possible *feasible networks* for the problem, each one corresponding to
a possible choice of backup vehicle. The selection of the backup vehicle involves
several factors, such as the time when the trip was disrupted, the position of the remaining vehicles, the available capacity of the potential backup vehicles, and the

itinerary compatibility among trips. The existence of several possible feasible networks makes the BRP a very interesting but difficult problem to solve.

This paper has the following major objectives: (i) to model the single depot BRP; and, (ii) based on previous algorithms developed for the VSP, to develop a parallel auction algorithm specifically implemented to solve the BRP. The major contributions of this paper to the literature are: (i) definition of the BRP, dealing with issues such as common itineraries, available capacities and time constraints, and backup trip candidates; and, (ii) implementation of a fast parallel auction algorithm for solving the BRP, using message passing to speed up communication among several processors.

2 Literature Review

Because automatic recovery from disruptions is a relatively new operational strategy, the literature related to the topic is scarce. Most transit companies typically avoid reassigning trips during operational disruptions because reassignment could complicate crew assignment and passenger service. Nevertheless, there is a vast literature on the VSP. Since the BRP is strongly related to the VSP, we start our literature review discussing the state-of-the art on modeling and solving the VSP.

Overviews of algorithms and applications for the single-depot VSP (SDVSP) and some of its extensions can be found in Bodin and Golden (1981), Ceder (2002), Daduna and Paixão (1995). The SDVSP has been formulated as a linear assignment problem, a transportation problem, a minimum-cost flow problem, a quasi-assignment problem, and a matching problem in the literature.

Bokinge and Hasselstrom (1980) propose a minimum-cost flow approach that uses a significant reduction of the size of the model in terms of the number of variables, at the price of an increased number of constraints. Dell'Amico et al. (1993), Jonker and Volgenant (1986) and Song and Zhou (1990) propose an $O(n^3)$ successive shortest-path algorithm and variations for the SDVSP.

Paixão and Branco (1987) propose an $O(n^3)$ quasi-assignment algorithm that is especially designed for the SDVSP. Haase and Friberg (1999) propose an exact algorithm for the vehicle and crew scheduling problem (VCSP). Both the vehicle and crew scheduling aspects are modeled by using set-partitioning type of constraints. A branch-and-cut-and-price algorithm is proposed, i.e., column generation and cut generation are combined in a branch-and-bound algorithm. The column generation master problem corresponds to an LP relaxation, while the pricing problem corresponds to a shortest path problem for generating crew duties. Freling et al. (2001) use a quasi-assignment model and employ a forward/reverse auction algorithm for the solution. Computational results show that the approach relating to quasi-assignment significantly outperforms approaches based on the minimum-cost flow and linear-assignment models.

Currently, one of the best models and algorithms for the SDVSP is the quasi-assignment with auction algorithm (Freling et al. (2001)). Bertsekas and Eckstein (1988) also show that if ϵ-scaling is used, i.e., applying the auction algorithm starting with a large value of ϵ and gradually reducing it to a final value that is less than $1/n$, the complexity is $O(nm \log nC)$, where n is the number of elements to assign, m is the number of possible assignments between pairs of elements, and C is the maximum absolute benefit.

To the best of our knowledge, the only contribution towards solving the dynamic VSP is due to Huisman *et al.* (2004) who proposed an approach to the problem by solving a sequence of optimization problems. Their work is motivated to design robust vehicle schedules that avoid trips starting late in environments characterized by significant traffic jams.

Whereas the above cited articles address a related research topic in considerable depth, they do not deal with the issue of this paper – the modeling and solving of the single-depot bus rescheduling problem (SDBRP).

3 Problem Description

We first introduce some definitions and notation to describe the bus rescheduling problem. To relate to a cut or a broken cycle in a graph, we refer to a disrupted trip due to a disabled bus, or a bus that is effectively inoperable, as a *cut trip*. *Breakdown point* is the point on the cut trip where the trip is disrupted. *Current trip* is the trip on which a vehicle is running. It includes both regular and *deadheading* (a movement of vehicles without serving passengers) trips. *Backup trip* is the trip which the backup vehicle is serving. Trips i and j are a *compatible pair of trips* if the same bus can reach the starting point of Trip j after it finishes the Trip i. A *route* is a sequence of trips in which each consecutive pair of trips in the sequence is compatible. Trip i is an *itinerary compatible trip* with cut Trip j if Trip i shares the same itinerary of Trip j from the breakdown point until its ending point.

The SDBRP can be defined as follows. Given a depot and a series of trips with fixed starting and ending times, given the travel times between all pairs of locations, and given a cut trip, find a feasible minimum-cost reschedule in which (1) each bus performs a feasible sequence of trips, and (2) all passengers (if there are any) on the cut trip are served. Unlike the SDVSP in which the fixed capital cost is dominant, the SDBRP problem focuses on the operating and delay costs. Furthermore, in order that transit crew can be reassigned on a new schedule, the computation of the SDBRP needs to be completed as fast as possible.

There are two possible situations in the SDBRP. The first is when the cut trip is a regular one. Unless the disruption is of a nature that it is impossible to reach the breakdown point, the passengers of the cut trip have to be served. The solution comprises of sending a backup bus to the breakdown point, and from the breakdown point completing the cut trip, and serving its passengers. However, since it is very likely some trips have common itineraries, the passengers can also be served incidentally by the buses that cover compatible itineraries after the breakdown point. Consider the following situation: a backup bus changes its schedule and travels towards the breakdown point, but all the passengers from the disabled vehicle have been incidentally picked up by vehicles that cover compatible itineraries with the cut trip. This situation needs to be avoided. If the cut trip is a deadheading trip, the solution is to assign a backup bus for the starting location of the next trip of the deadheading bus. In both cases, it is very likely that the SDBRP provides new routes for a subset of the pre-assigned buses. Also, we can expect some delays in the cut trip, mainly in the first situation.

In the VSP, there is no need to consider assigning a specific vehicle to the trips, since all vehicles are identical, and we can assign them arbitrarily after the schedule is determined. However, unlike the VSP, the BRP has to take into account this issue, since many buses are not at the depot and they are at different locations when a bus becomes disabled. The corresponding operating costs are also different. Furthermore, this situation creates different possible feasible networks depending on the selected backup trip, making the BRP a collection of several VSPs.

In the VSP, a vehicle can be generally assigned from the depot to any trip before its starting time. Nevertheless, assigning a vehicle from the depot to some future trips in the rescheduling problem may fail if the arrival time of a rescheduled vehicle from the depot to the starting point of a trip is later than the starting time of this trip. We may treat the depot as a special trip (or node) and define its starting time to be the breakdown time. This time is used to determine if a backup vehicle from the depot will be on time to serve a future trip.

From the viewpoint of the cut trip, the remaining trips can be divided into two categories: (1) unfinished trips that have compatible itineraries with the cut trip from the breakdown point, and (2) the remaining unfinished trips. Fig. 1 illustrates these two categories. The breakdown point is point X on Trip 1. The set of compatible trips with Trip 1 from point X is $\{3\}$.

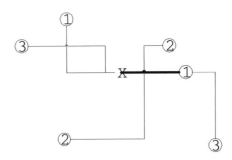

Fig. 1. Example of Itinerary Compatible Trips

Define set A to be the set of unfinished compatible itineraries with the cut trip from the point X, ordered by the travel time from their current position to point X. Define set B to be the remaining unfinished trips (including the trip directly from the depot). If the backup trip alternatives are from set A, the backup vehicles can pick up the passengers incidentally. Although a reschedule may not be necessary, it may be necessary to assign a bus from set B to cover unfinished trips originally assigned to the disabled bus. If the backup trip alternatives are from set B, backup vehicles need to travel toward the breakdown point for picking up the passengers on the disabled bus.

Whereas there is a unique feasible network in the VSP, the BRP may have several feasible networks (sharing the same nodes, but with different arcs connecting them). Suppose that a regular trip becomes disrupted, and a backup vehicle needs to go there to pick up the passengers. The starting time of this backup trip is dependent on the backup vehicle. The cost and compatible trips are different for alternative backup vehicles, since the serving vehicles are in different positions of the network, rather than at the depot, as usually assumed in the VSP. However, although there may exist many feasible networks, the differences among them are the arcs associated with the cut trip and the backup trip candidates.

In this paper, we make the following assumptions: (i) a bus can only change its route after finishing its current trip; (ii) only the cut trip will suffer delays; and (iii) there are no restrictions on the number of rescheduled buses. The next section describes our model and solution approach for the SDBRP.

4 Modeling the Bus Rescheduling Problem

The objective of the SDBRP is to minimize operating and delay costs over all possible feasible networks. As a consequence, any solution approach needs a procedure to explicitly or implicitly generate the set of feasible networks.

4.1 Generating Feasible Networks

The most important aspect of the SDBRP is that the solution is dependent on the existing situation and alternatives to serving the cut trip. Each possible configuration of a recovery can be translated as a possible feasible network. These feasible networks share the nodes (the trips), but have different arcs connecting them. The definition of the set of all possible feasible networks is dependent on the pre-assigned configuration of the trips, the available capacity of the involved vehicles, and times to carry out deadheading and regular trips. As commented in Section 3, it is possible to have a different feasible network for each possible backup trip. This subsection describes a procedure to generate feasible networks based on the available capacity of the involved vehicles, the times to complete the trips in the network, and the compatibility of itineraries and trips.

A capacity problem appears if the backup trip is from set A. It is quite possible that some passengers are in the disabled bus. If the number of passengers remaining in the cut trip is greater than the vacant capacity of the bus serving the backup trip, this vehicle is not enough for picking up all of the passengers. So, it is possible that more than one bus needs to be sent to the breakdown point of the cut trip. The first vehicle to arrive at the breakdown point picks up some passengers, the next vehicle picks up some more passengers, and so forth until all passengers from the cut trip are served. If the vehicle is from set B, it is an empty vehicle. In that case, we assume that one bus is enough for picking up all passengers.

In addition to the capacity problem, we need to consider time constraints related to the travel time of vehicles in current trips. It is not possible to select a vehicle serving a trip in set A if it has already passed the breakdown point when the disruption has occurred. Also, it is important to note that if a vehicle serving a trip from set B reaches the breakdown point later than a vehicle serving on a backup trip from set A, which has enough vacant capacity, then the bus from set B cannot backup the cut trip.

In order to generate the set of feasible networks, we first need to determine how many backup trips from set A are sufficient to serve all the passengers from the disabled vehicle. Let $C(i)$ be the empty seats of the backup vehicle from Trip i when it reaches the breakdown point. And let $T(i)$ be its arrival time at the breakdown point. Actually, $C(i)$ and $T(i)$ are random variables, but in this deterministic model, we use average values. Let $A(n)$ be the subset of A that includes the first n elements of A. Let P be the number of passengers in the disabled vehicle. Let T_d be the disruption time. We can get n^*, the number of backup trips in A that are sufficient for picking up all passengers from the cut trip, by solving the following system of inequalities,

$$
\begin{aligned}
\sum_{i \in A(n^*)} C(i) &\geq P \\
\sum_{i \in A(n^*-1)} C(i) &< P \\
T(i) &\geq T_d, i \in A(n^*).
\end{aligned}
\tag{1}
$$

If these inequalities have a feasible solution n^*, and an associated time $T(a_{n^*})$ by which the n^* buses serve the passengers on the disabled bus, then, we can determine B^*, the set of candidate backup trips from set B, by

$$
B^* = \{m | T_d \leq T(m) < T(a_{n^*}), \forall m \in B, a_i \text{ is the } i\text{-th element in set } A(n^*)\}.
$$

If B^* is empty, all backup trips are from set $A(n^*)$. In this situation, there is only one feasible network, resulting from eliminating the cut trip from the original network; the problem can then be treated as a VSP. If at least one backup trip candidate is from set B^*, we can connect an arc from this backup trip to the breakdown point in the corresponding feasible network. In this situation, we may have several feasible networks since several backup candidates may exist.

If the Inequalities (1) do not have a feasible solution, we can set $T(n^*) \leftarrow \infty$, and set B^* as B. In this case, a vehicle from set B has to backup the cut trip although it is possible that vehicles from set A may pick up some passengers from the disabled bus.

A feasible network is defined formally as follows. Each regular trip is a "node" of the feasible network, which is graphically represented as a short line segment to indicate starting and ending points of the trip (see, e.g., Fig. 2). Let b denote the cut trip and K be the set of possible backup trips. "Arcs" in the network correspond to vehicle assignment to trips. For example, an arc from node 2 to node 4 implies the same vehicle may be assigned to Trip 4 after it has served Trip 2 (e.g., see Fig. 2(a)). Let s and t denote the same depot in the network, where s simply means the depot as a vehicle's starting point, and t as its terminating point. Let $N' = N - \{b\}$ be the set of total remaining trips excluding the cut trip, numbered according to non-decreasing starting times. Let $P \in N$ denote the trips that existing vehicles are currently serving. If Trip $i \in P$ is a deadheading trip, its starting time and ending time are set as the current time, since the vehicle on this deadheading trip can be rescheduled right away. Define arc-set $E(k) = E \cup \{(k, b)\}$, where $E = \{(i, j) \in \{N \cup s\} \times N' | [i < j] \wedge [i$ and j are compatible trips]\}$ is the set of arcs that correspond to the deadheading trips. A feasible network for backup Trip k can be defined as $G(k) = \{V, X(k)\}$ with nodes $V = N \cup \{s, t\}$ and arcs $X(k) = E(k) \cup (s \times P) \cup (N \times t)$, for $k \in K$, where k is the backup trip. Since the trip in P is currently being served by an existing vehicle, there is no need to allocate another vehicle to cover it. The arcs, $(s \times P)$, are included only for modeling convenience. We define $G = \{G(k) | k \in K\}$ as the set of all feasible networks.

We illustrate feasible networks and our procedure with an example. Suppose we have to complete four trips with the travel times indicated in Table 1. Suppose the travel time from the ending point of each trip (or depot) to the starting point of a trip is a constant (4 time units).

Table 1. Travel Times

Trip	Starting Time	Ending Time	Duration
1	8	14	6
2	1	16	15
3	18	25	7
4	20	28	8

Suppose a vehicle breaks down on Trip 1 at the point X at time 11. Thus, the travel time from point X to the ending point of Trip 1 is 3 units. Assume that: (a) the cut vehicle is carrying 11 passengers at point X, (b) on the average, all vehicles have more than 16 available seats, (c) Trip 2 is an itinerary compatible trip with Trip 1 from the breakdown point X, and the vehicle serving Trip 2 has not passed the point X, (d) the required time for any vehicle serving a trip from the ending point of the regular trip to the breakdown point is a constant, 3 time units, and (e) the time of a vehicle from the depot to the breakdown point is 12 time units. Thus, set $A = \{2\}$; and set $B = \{0, 3, 4\}$, where the element 0 denotes an assignment of a bus from the depot. Since the expected vacant capacity of the vehicle on Trip 2 is 16, this vehicle can pick up all passengers. If the vehicles serving trips from set B reach point X later than the deadline (time when the vehicle on Trip 2 arrives at point X), they cannot be used as the backup vehicle candidates. Times of vehicles to reach X from set B are as follows: for Trip 3, 25 + 3 = 28, and for Trip 4, 28+ 3 = 31.

The following cases are described in Fig. 2 to illustrate the generation of the possible feasible networks, where Fig. 2(a) shows the initial schedule.

Case 1: Suppose the vehicle on Trip 2 reaches X at time unit 11. In this case, the only backup trip candidate is Trip 2. Although we do not need any backup vehi-

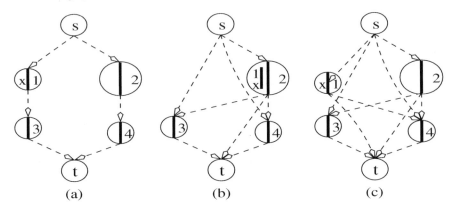

Fig. 2. Example of Feasible Networks

cle to go to the breakdown point, it is possible to require an additional vehicle to cover the remaining trips assigned to the disabled vehicle. In this case, there is only one feasible network (see Fig. 2(b)). Trip 2 is finished on time. The feasible network can be constructed by removing Trip 1 and associated arcs.

Case 2: Suppose the vehicle on Trip 2 reaches X at time unit 13. In this case, the backup vehicle candidates are: (i) the vehicle assigned to Trip 2, and (ii) an extra vehicle from the depot. If the backup vehicle is the vehicle on Trip 2, the feasible network is given in Fig. 2(b). Fig. 2(c) presents the feasible network if the backup vehicle is the extra vehicle from the depot. The time for this vehicle to finish Trip 1 would be 12 + 3 = 15. Then the time to the starting point of Trip 3 is 15 + 4 =19, if this vehicle was assigned to Trip 3, which is later than the starting time of Trip 3. Therefore, Trips 1 and 3 become incompatible (19 > 18), and this new vehicle cannot be assigned to Trip 3 (therefore, there is no arc from Trip 1 to Trip 3 in Fig. 2(c)).

Based on these feasible networks, we can model the SDBRP as a VSP in each feasible network, and the SDBRP optimal schedule is the one with the minimum total cost over all possible feasible networks. It is quite likely that the remaining vehicles have their routes changed to accommodate the disturbances caused by the disrupted trip. If there are a large number of feasible networks, then in order to decrease the number of feasible networks, it is possible to define a time limit by which a bus has to arrive at the breakdown point. If there are large number of elements in B^*, some candidates that exceed this time limit can be deleted using this constraint.

4.2 Mathematical Formulation

The SDBRP can be modeled as a minimization problem over several SDVSPs, each corresponding to a possible feasible network. Let y_{ij} be a binary decision variable, with $y_{ij} = 1$ if a vehicle is assigned to Trip j directly after Trip i, $y_{ij} = 0$ otherwise. Let c_{ij} be the vehicle cost of arc $(i, j) \in X(k)$, which is a function of travel and idle time. Let D_k be the delay cost related to the solution of Trip k as the backup trip. The quasi-assignment based formulation for the SDBRP is as follows:

$$\min_{G}\{\min \sum_{(i,j)\in X(k)} c_{ij}y_{ij} + D_k\}$$

subject to

$$\sum_{j:(i,j)\in X(k)} y_{ij} = 1 \qquad \forall i \in N$$

$$\sum_{i:(i,j)\in X(k)} y_{ij} = 1 \qquad \forall j \in N$$

$$y_{ij} \in \{0,1\} \qquad \forall(i,j) \in X(k)$$

where G is the set of all feasible networks.

The objective of our formulation is to find a schedule with the minimal operating and delay cost. The constraints in the formulation assure that each trip is assigned to exactly one predecessor and one successor.

Freling *et al.* (2001) compared the efficiency of several algorithms for the VSP, including the Hungarian algorithm (Paixão and Branco (1987)), successive shortest path algorithm (Dell'Amico (1989)), and the minimum cost flow approach (Bokinge and Hasselstrom (1980)) and showed that auction based algorithms are the fastest and most stable on average. Since solving the single-depot vehicle rescheduling problem is equivalent to solving $|G|$ vehicle scheduling problems, the auction algorithm was selected as our approach due to its excellent results for the VSP (Freling *et al.* (2001)). The auction method is also well suited for implementation on parallel machines (Bertsekas and Castañon (1991)), improving overall computational performance. This property is important to the vehicle rescheduling problem since it needs to be solved very quickly. The next section presents these algorithms.

5 Auction-Based Algorithms for Solving the SDBRP

Before describing the developed algorithms, we will introduce the basic concepts related to auction algorithms.

5.1 Auction Algorithms: An Introduction

An auction algorithm was originally proposed by Bertsekas (1992) for the classical symmetric assignment problem. Given its outstanding performance, it was further developed for the shortest path problem, the asymmetric assignment problem, and the transportation problem (Bertsekas (1992)). In the classical symmetric assignment problem, we need to match n persons and n objects on an one-to-one basis. Let a_{ij} be the benefit of matching person i and object j. The objective function is to maximize the total benefit. In the auction algorithm, each object j has a price p_j, and this price is updated upwards as persons bid for their best object, that is, the object for which the corresponding benefit minus the price is maximal. The auction algorithm is composed of two phases: the bidding phase and the assignment phase.

In the bidding phase, every unassigned person looks for its "best" object; in the assignment phase, the object determines the highest bid, since it may receive more than one bid. Meanwhile, if some objects that have already been assigned to some persons in a preceding iteration are now assigned to new persons, the persons who lose their objects are inserted into an unassigned set. After all the persons and objects are matched, the auction algorithm is terminated.

The combined forward and backward auction algorithm consists of forward and backward auction iterations, where, in a forward auction iteration the persons bid for the objects, while in a backward auction iteration objects bid for the persons. The combined auction algorithm has also been used for quasi-assignment problems (Freling et $al.$ (2001)). The combined auction algorithm for these problems is similar to the combined algorithm for the classical assignment problem, except that the person and object which represent the depot do not participate in the bidding. In the combined auction algorithm for the VSP, the person can be seen as the trip that is forward assigned, and the object can be seen as the trip that is backward assigned. The algorithms developed in the paper to solve the SDBRP are based on the combined auction algorithm by Freling et $al.$ (2001).

The performance of the auction algorithm is often improved by using ϵ scaling in Bertsekas (1992), where an integer ϵ is added to the prices, with ϵ gradually decreasing in subsequent iterations. As suggested by Bertsekas and Castañon (1991), a possible implementation of ϵ scaling is as follows: the integer benefits of a_{ij} are first multiplied by $n + 1$ and the auction algorithm is applied with progressively lower values of ϵ, up to the point where ϵ becomes 1 or smaller. Using ϵ-scaling, the complexity of the algorithm is $O(nm \log nC)$, where n is the number of elements to assign, m is the number of possible assignments between pairs of elements, and C is the maximum absolute benefit.

Freling et $al.$ (2001) describes the auction algorithm as follows. The value of a bid of Trip i (or person i) for another Trip j (or object j), which is candidate for forward assignment, is denoted by $f_{ij} = a_{ij} - p_j$. The value of a bid of Trip i for the depot is denoted by $f_{it} = a_{it}$. Let N be all trips and A be all arcs in the feasible network, respectively. Introduce π_j to denote the price of object j, when the backward auction is conducted.

Step 1: Perform the forward auction algorithm for each Trip $i \in N$ (or person i) which is currently not assigned to a Trip j (or object j) or depot.

Step 2: Determine the trip or depot j_i with the maximum bid value $\beta_i = max\{f_{ij}|j : (i,j) \in A\}$. Determine also the second highest value $\gamma_i = max\{f_{ij}|j : (i,j) \in A, j \neq j_i\}$. If Trip i (or person i) has only one arc $(i,j) \in A$, set $\gamma_i = -\infty$; If $j_i = t$ go to Step 4.

Step 3: Update the prices: $p_{j_i} = p_{j_i} + \beta_i - \gamma_i + \epsilon = a_{ij_i} + \gamma_i + \epsilon$, and $\pi_i = a_{ij_i} - p_{j_i}$. Update the assignments. If Trip j_i was already backward assigned, then remove the previous assignment. Return to Step 1.

Step 4: Update the price: $\pi_i = a_{it}$, update the assignment, and return to Step 1.

The reverse auction procedure is similar, with bids for candidates for forward assignments replaced by bids for candidates for backward assignments (Freling *et al.* (2001)).

5.2 Sequential Auction Algorithm for the BRP

The sequential auction algorithm is based on the combined forward-backward auction algorithm developed by Freling *et al.* (2001), considering the existence of several possible feasible networks to be solved. The algorithm is described as follows:

Step 1: Based on the starting and ending times of trips and travel time between trips, apply the procedure described in Section 3 to build the set of all possible feasible networks. Calculate the costs for the compatible trip pairs and the total delay cost of each feasible network.

Step 2: For each feasible network, apply the forward-backward combined auction algorithm (Freling *et al.* (2001)) to find the minimum cost scheduling of each feasible network as follows:

Step 2.1: Set the initial prices to 0. Set the initial $\epsilon = (n + 1) * C$, where C is the maximum absolute benefit.

Step 2.2: Using current ϵ and prices from the last iteration, conduct the bidding and assignment until all trips are both forward and backward assigned (see Freling *et al.* (2001) for details).

Step 2.3: If $\epsilon \leq 1$, the auction algorithm for current feasible network terminates. Otherwise, set $\epsilon = 0.5 * \epsilon$ and clear the assignment, go to Step 2.2.

Step 3: Select the minimal operating and delay cost scheduling as the solution.

As pointed out by Bertsekas and Castañon (1991), the auction method is well suited for implementation on parallel machines, improving its computational performance. The next section discusses our parallel implementation of the auction-based algorithm for the SDBRP.

5.3 Parallel Auction Algorithm

A parallel synchronous model is used to implement the algorithm. The system is composed of an assignment processor and several bidding processors, where the assignment processor is in charge of determining the prices and making the assignment, and a bidding processor is in charge of conducting the bidding. We employ the Jacobi method to implement the parallel auction algorithm since this method needs less synchronization than the Gauss-Seidel method (Bertsekas and Castañon (1991)). Suppose there are T bidding processors that conduct bidding, and in the forward (backward) auction, the unassigned persons (objects) are partitioned into T subsets. Every bidding processor simultaneously conducts the bidding for a different subset. After bidding in each processor is completed, the results, including the partial assignment and prices of persons and objects for the specific subset, are sent to the assignment processor. When the assignment processor receives all results from the T

bidding processors, it combines them to determine the new assignment and prices for all the unassigned persons and objects. If some objects (persons) that have already been assigned to some persons (objects) in a preceding iteration are now assigned to new persons (objects), the persons (objects) who lose the objects (persons) will be put into the unassigned person (object) set.

Then, the new assignment information is sent back to T bidding processors and the auction continues. After all the persons and objects are assigned, the auction algorithm is terminated. A method which partitions the unassigned trips will be presented later. Fig. 3 illustrates the parallel synchronous implementation of the Jacobi method.

Since the forward-backward combined auction algorithm is used to solve the SDBRP, we have to determine if the auction is forward or backward at each new iteration. The first iteration always uses a forward auction operation. We employ the method from Bertsekas (1992) to refrain from switching between forward and backward auctions until at least one more person-object pair has been added to the assignment.

In order to partition the unassigned trips and simultaneously conduct the bidding, a simple partitioning method is used to allocate each unassigned person (object)

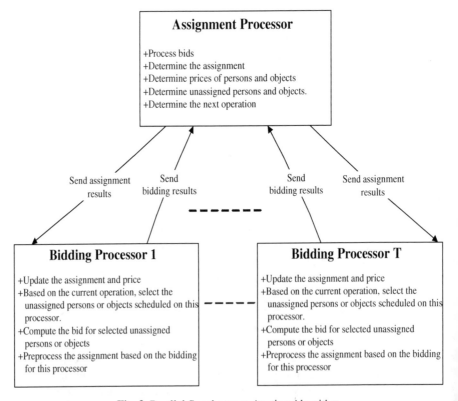

Fig. 3. Parallel Synchronous Auction Algorithm

on the bidding processors. Every bidding processor is assigned an ID, in the range $0, 1, \ldots, T - 1$. Considering that there are M unassigned persons (objects) stored in a list L, the unassigned persons (objects) for the bidding processor are defined by $Q[a_i] = i \bmod T, 0 \le i \le M - 1$, where a_i is the i-th unassigned person (object) in list L, and $Q[a_i]$ is the designed bidding processor of person (object) a_i.

A preprocessing technique is also employed for accelerating the computing and reducing the data-handling traffic. Consider the following situation: If there are an excessive number of unassigned persons for each bidding processor (this typically happens in the early stage of auction algorithms), it is quite likely that several persons bid for the same object in the same bidding processor. It is possible to make partial assignments in each bidding processor rather than in the assignment processor, considering the most dominant person requesting an object in the bidding processor. After the partial assignment is carried out in each processor, only one person bids for the same object in this bidding processor. This partial assignment can reduce the amount of data sent to the assignment processor. Computational experiments show that this method significantly reduces the running time of the parallel implementation.

The algorithm is described as follows. Steps 1 and 3 are the same as the corresponding steps in the sequential auction algorithm. Step 2 is as follows:

Step 2: For each feasible network, apply the forward-backward combined parallel auction algorithm to find the minimum cost scheduling of each feasible network as follows:

Step 2.1: Set the initial prices to 0. Set the initial $\epsilon = (n + 1) * C$. Send the information to bidding processors.

Step 2.2a: Upon receiving the current ϵ, assignment and prices from the assignment processor, conduct the bidding for the persons or objects allocated on each processor. Then, carry out the partial assignment and send the results to the assignment processor.

Step 2.2b: Based on the information received from the bidding processors, determine the assignment and prices. If all persons and objects are assigned, go to Step 2.3. Otherwise, send the assignment results to bidding processors.

Step 2.3: If $\epsilon \le 1$, then the current feasible network terminates. Otherwise, set $\epsilon = 0.5 * \epsilon$, clear the assignment and send the information to the bidding processors, and go to Step 2.2.

Table 2. Computational Results

Remaining Trips	Initial # of Buses	Backup Trips	New # of Buses	Objective Value	CPLEX	SBRP	PBRP2	PBRP4
100	28.6	2	29.0	135407	0.09	0.01	0.05	0.04
		3	28.8	134516	0.13	0.02	0.08	0.05
		5	28.6	133601	0.22	0.03	0.13	0.08
		10	28.6	133511	0.45	0.07	0.25	0.18
300	76.0	2	76.6	356042	0.88	0.21	0.18	0.17
		3	76.4	355100	1.32	0.31	0.27	0.25
		5	76.2	354189	2.20	0.51	0.48	0.41
		10	76.0	353284	4.42	1.01	0.91	0.75
		15	76.0	353256	6.65	1.49	1.32	1.13
		20	76.0	353233	8.87	1.98	1.75	1.52
500	121.6	2	122.1	564555	2.96	0.72	0.50	0.35
		3	122.0	564071	4.41	1.07	0.75	0.53
		5	121.9	563603	7.31	1.78	1.16	0.91
		10	121.6	562277	14.55	3.50	2.29	1.82
		15	121.6	562267	21.80	5.16	3.30	2.69
		20	121.6	562232	29.05	6.78	4.32	3.56
		25	121.6	562219	36.33	8.43	5.29	4.41
		30	121.6	562214	43.58	10.08	6.34	5.24
		35	121.6	562193	50.82	11.73	7.35	6.05
		40	121.6	562179	58.09	13.38	8.35	6.92
700	165.2	2	165.8	767241	6.71	1.73	0.82	0.70
		3	165.5	766037	10.07	2.60	1.26	1.07
		5	165.3	765181	16.82	4.30	2.07	1.78
		10	165.1	764322	33.56	8.33	4.06	3.53
		15	165.1	764213	50.41	12.47	6.10	5.22
		20	165.1	764154	67.31	16.43	8.11	6.84
		25	165.1	764151	84.16	20.33	10.03	8.39
		30	165.1	764141	101.03	24.21	11.94	9.96
		35	165.1	764135	117.96	28.05	13.97	11.51
		40	165.1	764128	134.85	31.90	15.87	13.12
900	211	2	211.0	1029542	13.15	4.16	1.45	1.18
		3	210.9	1028997	19.81	6.23	2.17	1.74
		5	210.9	1028957	33.16	10.31	3.55	2.87
		10	210.7	1028106	66.46	20.62	7.10	5.73
		15	210.7	1028091	99.85	30.86	10.46	8.52
		20	210.7	1028066	133.31	41.07	13.84	11.35
		25	210.7	1028059	166.81	51.19	17.16	14.07
		30	210.7	921768	200.22	61.16	20.45	16.79
		35	210.7	818583	233.61	71.19	23.69	19.56
		40	210.7	818576	267.11	81.17	26.99	22.32
1100	253.6	2	253.5	1218215	21.54	10.99	2.84	2.03
		3	253.5	1218200	32.32	16.47	4.30	3.15
		5	253.5	1218162	54.12	27.23	7.02	5.35
		10	253.2	1216918	108.87	54.03	13.48	10.62
		15	253.2	1216906	163.53	80.40	19.82	15.73
		20	253.2	1216885	217.79	106.77	26.38	20.97
		25	253.1	1216458	271.89	133.17	32.80	26.11
		30	253.1	1216446	325.98	159.13	39.20	31.24
		35	253.1	1216446	380.22	185.34	45.63	36.43
		40	253.1	1216444	434.49	211.48	51.95	41.60
1300	302.8	2	302.8	1440810	33.04	18.91	4.34	3.53
		3	302.6	1439972	49.38	28.31	6.51	5.31
		5	302.6	1439964	82.52	46.66	10.97	8.88
		10	302.5	1439139	165.47	91.86	21.64	17.50
		15	302.4	1438699	248.13	136.87	32.19	25.74
		20	302.4	1438686	331.27	182.19	42.68	34.25
		25	302.4	1438686	414.24	226.62	52.70	42.25
		30	302.4	1438674	496.81	271.51	62.88	50.52
		35	302.4	1438662	580.11	315.62	73.05	58.45
		40	302.4	1438662	662.99	359.45	82.80	66.21

6 Computational Experiments

The main objective of the computational experiments is to compare the performance of the developed algorithms, in terms of the required CPU time, to obtain the optimum solution. Therefore, for convenience, we only included the cost of reallocating buses (including allocation cost of the bus for the backup trip) and not the cost of delay to passengers on the disabled bus in the objective function for the SDBRP.

Since the constraint matrix is totally unimodular, the solution of the linear relaxation for the SDBRP provides an optimal solution. Nevertheless, solving the linear relaxation may require longer times than the auction algorithm, since the latter was specially designed to solve the VSP. We used CPLEX 7.0 Network Optimizer to solve the linear relaxation of the SDBRP. CPU times of the linear relaxation and of the auction algorithms were compared for verification purposes.

The algorithms were implemented in C++ on 900Mhz Sun Workstations. The communication protocol used for the parallel implementation was developed based on the Socket/Stream protocol. The following nomenclature is used to define the implemented algorithms:

(a) *CPLEX*: The use of CPLEX7.0 to solve the linear relaxation of the SDBRP;
(b) *SBRP*: The sequential auction algorithm;
(c) *PBRP2*: The parallel auction algorithm using 2 processors;
(d) *PBRP4*: The parallel auction algorithm using 4 processors.

The experiments were designed using the random data generation method for the VSP of Carpaneto *et al.* (1989). Let $\rho_1, \rho_2, \ldots, \rho_v$ be *relief points* (i.e., points where trips can start or finish) of a transportation network. We generate them as uniformly distributed random points on a (60×60) square and compute the corresponding travel times θ_{ρ_a, ρ_b} as Euclidean distances between relief points ρ_a and ρ_b. To simulate the trips, we generate for each Trip T_j ($j = 1, \ldots, n$) the starting and ending relief points, ρ'_j and ρ''_j, randomly selected from $\rho_1, \rho_2, \ldots, \rho_v$. The time between Trips T_i and T_j is defined as $\theta_{\rho''_i \rho'_j}, \forall i, j$. The starting and ending times, s_j and e_j, of Trip T_j are generated by considering first two classes of trips: short trips and long trips. For short trips, s_j is a uniformly distributed random integer in the interval (420,480) time units, say, minutes, with probability 15%; in (480,1020) with probability 70%; and in (1020,1080) with probability 15%. Since ending time e_j for Trip T_j must include a travel time between ρ'_j and ρ''_j, and dwell time at bus stops, we generate e_j as a uniformly distributed random integer in $(s_j + \theta_{\rho'_j, \rho''_j} + 5, s_j + \theta_{\rho'_j, \rho''_j} + 40)$. For long trips, we assume they start and end at the same point, and the travel time depends on the length of the resultant cycle and associated stops. Then we generate s_j as a uniformly distributed random integer in (300,1200) time units and e_j as a uniformly distributed random integer in $(s_j + 180, s_j + 300)$. Costs c_{ij}, c_{si} and c_{jt} are defined to include travel time and waiting time; we used

1. $c_{ij} = 10\theta_{i,j} + 2(s_j - e_i - \theta_{i,j})$, for all compatible pairs (T_i, T_j);
2. $c_{si} = 2000$, for trips from the depot to route T_i; and
3. $c_{jt} = \lfloor 10(\text{Euclidean distance between depot and Trip } T_j) \rfloor + 2000$, for trips from T_j to the depot.

In order to compare the computational efficiency of the sequential and parallel implementations of the auction algorithm, we consider a situation in which the total number of trips is composed of a 40:60 combination of short and long trips.

To evaluate the performance of the algorithms, we first generated a VSP and solved it. Then, a disruption was introduced so that an early trip is chosen as a cut trip (Trip T_b). We assumed that vehicles break down in the middle of the cut trip in time and distance. The arrival time to the breakdown point is calculated as follows: (i) for a backup vehicle on a regular trip, the arrival time is the ending time of the current trip plus the travel time from the ending point to the breakdown point; and, (ii) for a vehicle on a deadheading trip, the arrival time is the current time plus the travel time from its current location to the breakdown point. Euclidean distance is used in the calculation of travel distances.

Since in real-life situations, determination of backup trips requires knowledge of bus capacity and common itineraries, whereas in the simulation trips are generated only by distance and travel times, we simply assumed the possible number of backup trips to be among (5,10,15,20,25,30,35,40). For each value of G, ten instances were generated and solved.

Table 2 (p. 294) compares the performance of algorithms SBRP, PBRP2, and PBRP4. The first five columns give the number of the remaining trips, the original number of buses, the number of backup trips considered, the number of new buses required to finish the remaining trips, and the optimal cost, respectively (fractional buses are because each entry is an average of ten instances). The remaining columns show the average CPU seconds, excluding input and output time, for the four algorithms.

The table shows that an increase in the possible number of backup trips decreases the optimal cost, characterizing a trade-off between CPU time and optimal cost, defined by the number of possible backup trip alternatives being considered. Taking into consideration that (i) the small differences in the average optimum cost between 5 and 40 backup trip alternatives for large problems, and (ii) the considerable increase in the average CPU time for these problems, it may be worthwhile to develop heuristics to prune the number of possible backup trip alternatives, especially for a large number of remaining trips. The idea is to select and solve the problem only for a representative subset, in a way that we include, with a high probability, the feasible network that leads to the optimum solution.

An extra vehicle is needed when the number of buses in the rescheduling problem is equal to the number of buses in the original scheduling. Table 2 shows that at most one extra vehicle is needed to serve all the remaining trips. The average CPU time for all algorithms is highly dependent on the problem size. The table shows that for small problems (100 remaining trips) all algorithms are extremely fast, solving the problem, even for the high value of G, in less than 1s CPU time. It seems that parallel processing does not improve the CPU times for small problems. This fact can be explained by the required communication cost between the different processors. The communication time becomes relevant for small problems, but it is compensated by the fast processing time to solve the auction algorithm for large problems. Therefore, the parallel implementation is more efficient for large problems, in terms of the number of remaining trips and possible backup trips.

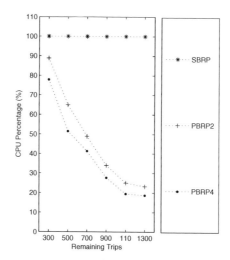

Fig. 4. Average CPU Time Percentage, Considering SBRP as the Comparison Basis

The two parallel algorithms (PBRP2 and PBRP4) become more efficient as the problem size increases (more remaining trips and more backup trips). Fig. 4 presents a pairwise comparison, considering SBRP as the comparison basis, on the average CPU time percentage for the problems for 300 remaining trips onwards. The CPU percentage is computed as follows

$$Percentage = 100 \times \frac{CPU_p}{CPU_{SBRP}}$$

where CPU_{SBRP} is the CPU seconds required by the algorithm SBRP, and CPU_p is the CPU seconds required by the parallel algorithms. For problems with more than 300 remaining trips, the use of parallelism results in significant reductions on the average CPU seconds required to solve the problems. Reductions in the CPU seconds are more significant for larger problems.

7 Conclusions

This paper models the single depot bus rescheduling problem and presents several algorithms to solve this problem. The solution approach is based on (i) the generation of all possible feasible networks obtained when a trip is disrupted, and (ii) the application of auction algorithms for solving the resultant vehicle scheduling problem. In addition, parallel processing was used as a possible approach to improve the efficiency of the auction algorithms.

From the extensive computational experiments performed using randomly generated data, the following important observations summarize the results:

- For small problems (less than 300 remaining trips), both sequential and parallel implementations are fast. The sequential algorithm, without using parallel processing, provided the solutions with the smallest CPU time, due to the added communication time between processors for parallel algorithms.
- For large problems (more than 300 remaining trips), the parallel algorithms outperform the sequential implementation, in terms of CPU time.

In summary, we can conclude that the developed solution approaches are computationally efficient to be used in automatic schedule recovery tools. As a follow up of this research we plan to develop a method to speed up the computational performance of the auction algorithm. Since the difference among feasible networks is small, one promising approach is to perform the algorithm in two stages. In the first stage, the algorithm is carried out for a reduced network that does not include the cut trip. Then an assignment close to the original one can be obtained with prices of persons and objects, since this reduced network is very similar to the original feasible network. In the second stage, we include the cut trip and backup trips to reconstruct feasible networks and apply the auction algorithms again, taking the initial prices obtained in the first stage. The second stage is performed for all possible backup trips. Preliminary experiments are promising.

We included two major assumptions in this study: (i) only the cut trip can suffer delays; and (ii) there is no restriction on the number of rescheduled trips. These assumptions may not be true for some applications. In some cases, a vehicle breakdown may also delay other trips (e.g., when the starting point of the next trip that the breakdown vehicle is scheduled to cover is too far from the depot and other vehicles). As a next step, a trip cancellation strategy is being introduced to handle such cases. The research team is also planning to include a strategy to limit the number of trips that can be rescheduled.

References

Bertsekas, D. (1992). Auction algorithms for network flow problems: a tutorial introduction. *Computational Optimization and Applications*, **1**, 7–66.

Bertsekas, D. and Castañon, D. (1991). Parallel synchronous and asynchronous implementations of the auction algorithm. *Parallel Computing*, **17**, 707–732.

Bertsekas, D. and Eckstein, J. (1988). Dual coordinate step methods for linear network flow problems. *Mathematical Programming*, **42**, 203–243.

Bodin, L. and Golden, B. (1981). Classification in vehicle routing and scheduling. *Networks*, **11**, 97–108.

Bokinge, U. and Hasselstrom, D. (1980). Improved vehicle scheduling in public transport through systematic changes in the time-table. *European Journal of Operational Research*, **5**, 388–395.

Carpaneto, G., Dell'Amico, M., Fischetti, M., and Toth, P. (1989). A branch and bound algorithm for the multiple depot vehicle scheduling problem. *Networks*, **19**, 531–548.

Ceder, A. (2002). Urban transit scheduling: framework, review and examples. *Journal of Urban Planning and Development*, **128**, 225–244.

Daduna, J. R. and Paixão, J. M. P. (1995). Vehicle scheduling for public mass transit – an overview. In J. R. Daduna, I. Branco and J.M.P. Paixão, editors, *Computer-Aided Transit Scheduling*, Lecture Notes in Economics and Mathematical Systems 430, pages 76–90. Springer, Berlin.

Dell'Amico, M. (1989). Una nuova procedura di assegnamento per il vehicle scheduling problem. *Ricerca Operativa*, **5**, 13–21.

Dell'Amico, M., Fischetti, M., and Toth, P. (1993). Heuristic algorithms for the multiple depot vehicle scheduling problem. *Management Science*, **39**, 115–125.

Freling, R., Wagelmans, A., and Paixão, J. M. (2001). Models and algorithms for single-depot vehicle scheduling. *Transportation Science*, **35**(165–180).

Haase, K. and Friberg, C. (1999). An exact branch and cut algorithm for the vehicle and crew scheduling problem. In N. H. M. Wilson, editor, *Computer-Aided Transit Scheduling*, pages 63–80. Springer, Berlin.

Huisman, D., Freling, R., and Wagelmans, A. (2004). A robust solution approach to the dynamic vehicle scheduling problem. *Transportation Science*, **38**, 447–458.

Jonker, R. and Volgenant, T. (1986). Improving the Hungarian assignment algorithm. *Operations Research Letters*, **5**, 171–176.

Paixão, J. M. and Branco, I. (1987). A quasi-assignment algorithm for bus scheduling. *Networks*, **17**, 249–269.

Song, T. and Zhou, L. (1990). A new algorithm for the quasi-assignment problem. *Annals of Operations Research*, **24**, 205–223.

Schedule-Based and Autoregressive Bus Running Time Modeling in the Presence of Driver-Bus Heterogeneity

Rabi G. Mishalani[1], Mark R. McCord[1], and Stacey Forman[2]

[1] The Ohio State University, Department of Civil and Environmental Engineering and Geodetic Science, 2070 Neil Avenue, Room 470, Columbus, OH 43210, USA
mishalani@osu.edu
[2] TranSystems, 5747 Perimeter Drive, Suite 240, Dublin, OH 43017, USA

Summary. Bus route running time represents a key element of transit performance. An understanding of running time behavior and the factors that influence it is essential for off-line planning and operations design purposes including fleet size planning, schedule design, and passenger travel time performance assessment. Such an understanding is also critical for real-time applications including bus operations control and passenger information systems. This paper focuses on developing models of running time and estimating them using field data. Two model structures are considered. The schedule-based model specifies the upcoming running time as a function of the most recent deviation from the schedule the bus has exhibited at the terminus. This model characterizes the situation where a late running bus attempts to catch up with the schedule and, hence, reflects an upcoming running time shorter than the target running time, and vice versa. The autoregressive model specifies the upcoming running time as a function of the most recent running time. This model characterizes one of two situations depending on the sign of the parameter estimate. On the one hand, when the most recent running time is longer than the mean, the upcoming running time would also be longer than the mean if the operation is dominated by exogenous factors that cause delays such as other traffic or weather. On the other hand, the upcoming running time would be shorter than the mean if the driver is capable of speeding up to reduce the delay in the operation. Irrespective of the model structure, the characteristics of the driver-bus pair may also influence the extent to which the upcoming running time will deviate from the target or the mean. To capture this potential heterogeneous phenomenon, the fixed effects formulation is adopted whereby driver-bus pair dummy variables are included in the model. Field data are utilized in estimating the two types of models in the presence of driver-bus heterogeneity. In general, the schedule-based model is superior to the autoregressive model in describing running time behavior. Moreover, driver-bus heterogeneity is found to be a significant contributor to this behavior.

1 Introduction and Motivation

Running time is defined as the amount of time it takes a bus to complete one cycle of its assigned route and represents a key element of transit performance. An understanding of running time behavior and the factors that influence it is essential for off-line planning and operations design purposes including fleet size planning, schedule design, and passenger travel time performance assessment. Moreover, such an understanding is critical for real-time applications including bus operations control and passenger information systems. A bus's running time may not be equal to the target or expected running time on a single run due to exogenous variables such as vehicle and pedestrian traffic, passenger demand, weather, bus characteristics, or driver characteristics. Nevertheless, it might be possible to more accurately predict a future bus running time when incorporating knowledge of some of these factors and recent information on the bus's location.

In light of the developments in automatic vehicle location (AVL) systems, their application to public transit, and their use for control and passenger information purposes, numerous researchers including Wall and Dailey (1999), Lin and Zeng (1999), Hickman (2001), Dueker et al. (2001), Bertini and El-Geneidy (2004), and Shalaby and Farhan (2004) have been studying bus travel times. While various variables influencing bus travel times are considered, none of the mentioned studies examine the effects of different driver-bus pairs – i.e., driver-bus heterogeneity – on travel times. Moreover, none take into account the possible value of considering the most recent running time of a particular bus in predicting the next running time of that same bus. Furthermore, only Lin and Zeng (1999) from the above mentioned studies explicitly take into account the effect of deviations from the schedule in modeling travel time. The study presented in this paper focuses on the possible presence of driver-bus heterogeneity, considers the effect of the most recent running time of a bus in predicting its future running time, and captures the effect of schedule deviations.

The effects of driver-bus characteristics are of particular interest. Confirming the presence of such effects of driver and bus heterogeneity and understanding them are valuable in various ways. In a planning context, the transit agency can take such considerations into account in vehicle and crew scheduling or in after-the-fact evaluation. In a real-time operations control and traveler information context, such understanding has the potential to improve running time forecasts, an essential input to real-time functions. This paper focuses on bus running time modeling in the presence of driver-bus pair heterogeneity.

2 Running Time Models

Two running time model specifications are developed for three Campus Area Bus Service (CABS) routes operated by the Ohio State University's Transportation and Parking Services. The developed schedule-based and the autoregressive models take advantage of the panel nature (Greene (2003)) of the CABS data set, whereby bus numbers represent different cross-sections each observed over several consecutive time periods. In the case of the CABS operation, these bus numbers can be good proxies for driver-bus pair characteristics including bus age, bus size or type, driver experience, driver age, and driver gender.

The bus number is expected to be a good proxy in this case because, in general, the operator assigns the same small subset of drivers to buses (Basinger (2003)). Thus, more often than not, a specific bus is paired with only a few possible drivers over the course of an academic year (the data set used in this study spans a period of time falling within a single academic year; see Section 3). Hence, bus numbers are used as proxies for driver-bus pairs in capturing the possible heterogeneity reflected by such a pair. Ideally, driver information would be explicitly used. However, such information is not available in the CABS data set, and its use is, therefore, reserved for future research. It is worth emphasizing that in the context of the available data set, any identified heterogeneity based only on bus number information would strengthen the motivation for capturing driver information explicitly in future studies.

As discussed in Greene (2003), several specifications capturing driver-bus pair heterogeneity are possible including the fixed and random effects formulations. While both have been investigated, the focus of this paper is on the former. In the developed models, the time step reflecting the time dimension of the panel data set is an index indicating a particular running time by a particular bus. That is, the time step does not capture a specific point or period of time, but rather is a variable index that increases by an increment of one as soon as a bus run (across the entire route) is complete and a new run commences.

In addition to capturing driver-bus heterogeneity, in general it is possible to attempt to model the effect of time-of-day using the CABS data set. However, given the university campus context of the service and the consequent prominent effect of class schedules and distribution across campus, a typical peak and off-peak pattern is not apparent. Therefore, such treatment is reserved for future research. Further discussion regarding various influencing factors is presented in the final section of this paper.

2.1 Schedule-based Fixed Effects Specification

The schedule-based model in the absence of driver-bus heterogeneity, referred to as the homogeneous schedule-based model henceforth, takes the following form:

$$r_{t+1} - \overline{r} = \beta + \alpha d_t + \epsilon_{t+1} \tag{1}$$

where t = time index specific to a set of consecutive running time observations (referred to as a stream), r_{t+1} = bus running time at time step $t+1$, $\overline{r}$ = mean bus running time, d_t = actual bus arrival time at a pre-specified location minus the scheduled arrival time at time step t, ϵ_{t+1} = random term representing unobserved explanatory variables and measurement errors with a mean of zero, and β and α = parameters. The dependent variable $(r_{t+1} - \overline{r})$ is the difference between the running time at time step $t + 1$ and the mean running time. The intercept parameter β is hypothesized to represent the difference between the target running time and the mean running time. Ideally, the target would be the scheduled running time. A positive value of β would imply that buses are running faster on average than the target running time, and a negative value would imply that buses are running slower on average than the target running time. To illustrate, consider a bus arriving precisely on schedule at time step t (i.e., $d_t = 0$). In this case, the next running time r_{t+1} is modeled to be the target running time $(\bar{r} + \beta)$ plus the random variable ϵ_{t+1}.

The parameter α of the explanatory variable d_t models the upcoming running time in relation to schedule deviations. If α is negative, a bus arriving ahead of schedule at time step t (i.e., $d_t < 0$) would lead to an expected running time at time step $t + 1$ greater than the target running time, and vice versa. In this way, the value of α reflects the ability of a driver to adjust his or her running time to maintain the schedule. A larger absolute value indicates a greater ability of a driver to maintain the schedule. Staying on schedule is important even when the schedule is not known to passengers, since otherwise bus bunching might occur and increase the expected waiting time for passengers (Larson and Odoni (1981)). If α is positive, a bus arriving behind schedule at time step t (i.e., $d_t > 0$) would lead to an expected running time at time step $t + 1$ greater than the target running time, and vice versa. This might occur due to the persistence of exogenous factors, such as the route characteristics mentioned above. In this case, the value of α reflects the magnitude of this persistence.

Introducing driver-bus pair heterogeneity, the schedule-based fixed effects specification reflects the addition of dummy variables to the homogeneous schedule-based model as follows:

$$r_{t+1,i} - \bar{r} = \alpha d_{ti} + \gamma_1 W_1 + \ldots + \gamma_i W_i + \ldots + \gamma_N W_N + \epsilon_{t+1,i} \qquad (2)$$

where i = index identifying buses (thus representing driver-bus pairs), $r_{t+1,i}$ = running time of bus i at time step $t + 1$, d_{ti} = actual bus arrival time of bus i at a pre-specified location minus the scheduled arrival time of bus i at time step t, $W_i = 1$ for bus i and 0 otherwise, N = total number of buses, $\epsilon_{t+1,i}$ = random term representing unobserved explanatory variables and measurement errors with a mean of zero, and β, α, and γ_i = parameters. Notice that $\bar{r}$ reflects the mean running time over all individual runs and is estimated by:

$$\bar{r} = \frac{\sum_{i=1}^{N} \sum_{j=1}^{M_i} \sum_{t=1}^{T_{ji}} r_{tji}}{\sum_{i=1}^{N} \sum_{j=1}^{M_i} T_{ji}} \qquad (3)$$

where r_{tji} = tth running time of bus i on its jth stream of consecutive runs, T_{ji} = total number of runs for bus i on its jth stream, and M_i = total number of streams for bus i. Note that since a dummy variable W_i for each bus cross-section is included, the intercept β of the homogeneous schedule-based model of (1) is dropped to avoid collinearity (Pindyck and Rubinfeld (1998)).

Unlike the homogeneous model where the same mean running time, intercept β, and parameter α would be used to predict a future running time for all buses over all times, the fixed effects model allows different driver-bus pairs to have different target running times due to differences in bus performance or driver behavior. More specifically, this model reflects different target running times $(\bar{r} + \gamma_i W_i)$ for different buses through the introduction of bus specific dummy variables W_i. Thus, each parameter γ_i in (2) represents the deviation from the mean running time $\bar{r}$ for bus i.

2.2 Autoregressive Fixed Effects Specification

The autoregressive model in the absence of driver-bus heterogeneity, referred to as the homogeneous autoregressive model henceforth, takes the following form:

$$r_{t+1} - \bar{r} = \rho(r_t - \bar{r}) + \epsilon_{t+1} \tag{4}$$

In this model, the explanatory variable $(r_t - \bar{r})$ is the difference between the running time at time step t and the mean running time. Since the expectations of both the explanatory and dependent variables $(r_t - \bar{r})$ are zero, the intercept is zero. The parameter ρ in this model represents the correlation between the running times at time steps t and $t + 1$ (Wei (1990)). If ρ is negative, then a running time at time step $t + 1$ would be expected to be less than the mean if the running time at time step t were greater than the mean, and vice versa. In this way, the drivers' attempts to maintain the mean running time dominate, and a larger absolute value of ρ indicates a greater ability of the drivers to correct a running time to maintain the mean. If ρ is positive, then a running time at time step $t + 1$ would be expected to be greater than the mean if the running time at time step t were greater than the mean. This might happen if exogenous influences, such as vehicular and pedestrian traffic or passenger demand, were high and caused a bus to continuously have a greater than the mean running time despite any attempts on the part of the drivers to correct for such effects. However, when the running time drops below the mean, the drivers attempt to sustain lower running times to compensate for the previously longer running times.

Introducing driver-bus pair heterogeneity, the autoregressive model specification is easily extended to the fixed effects formulation by again adding dummy variables:

$$r_{t+1} - \bar{r} = \rho(r_{ti} - \bar{r}) + \gamma_1 W_1 + \ldots + \gamma_i W_i + \ldots + \gamma_N W_N + \epsilon_{t+1,i} \tag{5}$$

Again notice that $\bar{r}$ reflects the mean running time over all individual runs and is estimated by (3). Just as in the schedule-based specification, without the dummy variables, the same mean running time and parameter ρ would be used to predict a future running time for all buses over all times. The fixed effects model, however, allows different driver-bus pairs to have different expected future running times for the same present deviation from the mean $(r_{ti} - \bar{r})$ due to differences in bus performance or driver behavior. More specifically, this model reflects different systematic deviations from the mean running time $\bar{r}$ for different buses on their respective next runs due to the introduction of bus specific dummy variables W_i. Thus, each parameter γ_i in (5) represents that deviation for bus i.

3 Data

The Ohio State University's Campus Area Bus Service (CABS) provides students, staff, and guests of the university with a transit bus service whereby 15 to 20 40-foot buses run simultaneously on several routes on and in the areas surrounding campus. Buses on these routes follow a schedule determined by the operators of CABS.

Because the routes vary in length and characteristics, they have expected running times that vary from route to route. CABS uses the Bus Location Information System (BLIS) (Bus Location Information System (2003)) to obtain AVL data for buses on several of its routes for both real-time and planning applications. BLIS includes Global Positioning System (GPS) receivers on each bus and wireless communications devices that send position and time data to a central computer server. The data sent from the GPS receivers were used to determine streams of consecutive running times experienced by each bus operating on a specific route. The bus schedule was provided by the Transportation and Parking Services (T&P) (Transportation and Parking Services (2001–2002)).

The data used in this study cover three separate bus routes from September 17, 2001 through March 29, 2002 (i.e., two academic quarters, Fall 2001 and Winter 2002) on weekdays between 6:45 a.m. and 7:00 p.m. when the schedule remains unchanged. Specifically, the data set includes a series of bus running times, the difference between the actual bus arrival time at a pre-specified stop and the scheduled arrival time, the route, and the bus number. The three routes are Campus Loop North, East Residential, and Core Circulator. These routes represent a range of characteristics that are discussed in detail in the next section. The basic features of these routes are given in Table 1.

Table 1. Bus Route Characteristics

Route	Length [km]	Scheduled Run. Time [min]	Scheduled Headway [min]	No. of Stops	Average Stop Spacing [m]
Core Circulator	2.20	12.0	6.0	11	200
Campus Loop North	8.29	30.0	10.0	20	414
East Residential	8.08	30.0	10.0	20	404

4 Results and Discussion

4.1 Estimation

The Ordinary Least Squares (OLS) estimation results of the homogeneous models represented by (1) and (4) – the specifications assuming the absence of driver-bus pair heterogeneity – are first presented to serve as a reference in the subsequent discussion. These results are shown in Tables 2 and 3 for the schedule-based and autoregressive models, respectively.

In order to capture driver-bus pair heterogeneity, the schedule-based and autoregressive models represented by (2) and (5), respectively, are estimated using OLS in LIMDEP (Greene (2003)). The dummy variable parameters of these preliminary estimations were examined. If the t-statistic of a dummy variable was large enough (approximately greater than 1.3), suggesting that the parameter was significantly different from zero at the 10% level, the data for that bus were kept as an individual cross-section group. Otherwise, that bus's data were grouped with data from other buses with dummy variable parameters not significantly different from zero into a larger cross-section group. Each of the models was then re-estimated using the new cross-sectional groups. The results are presented for each route separately.

Table 2. Estimated Homogeneous Scheduled-Based Models

Variable	Est. parameter	Standard error	t-statistic
Core Circulator route:			
Intercept	0.13447	0.047695	2.8194
d_t	- 0.37880	0.026545	- 14.270
No. of observations = 850, $\overline{R^2}$ = 0.19268			
Campus Loop North route:			
Intercept	0.14992	0.025943	5.7508
d_t	- 0.25939	0.0099484	- 26.074
No. of observations = 4828, $\overline{R^2}$ = 0.12329			
East Residential route:			
Intercept	0.52370	0.047900	10.933
d_t	- 0.45844	0.018768	- 24.426
No. of observations = 2142, $\overline{R^2}$ = 0.21765			

Table 3. Estimated Homogeneous Autoregressive Models

Variable	Est. parameter	Standard error	t-statistic
Core Circulator route:			
Intercept	0.077082	0.052259	1.4750
$r_t - \overline{r}$	- 0.15551	0.033806	- 4.6000
No. of observations = 850, $\overline{R^2}$ = 0.02320			
Campus Loop North route:			
Intercept	0.019630	0.026263	0.74745
$r_t - \overline{r}$	- 0.24339	0.013470	- 18.070
No. of observations = 4828, $\overline{R^2}$ = 0.06318			
East Residential route:			
Intercept	- 0.030171	0.044769	- 0.67392
$r_t - \overline{r}$	- 0.35829	0.020560	- 17.427
No. of observations = 2142, $\overline{R^2}$ = 0.12386			

Core Circulator

In the preliminary estimation, five of the dummy variable parameters for each of the seven cross-sections were significantly different from zero at the 10% level. Therefore, the data corresponding to the two insignificant bus dummy variable parameters were grouped into one cross-section, and the model was re-estimated using six dummy variables. The results are given in Table 4. As expected, the value of α (the parameter of d_{ti}) is negative, implying that buses with an early arrival time on one run (i.e., $d_{ti} < 0$) are on average expected to have running times greater than the target running time on the next runs reflecting the drivers' attempts to maintain the schedule. This effect is similar to that of the homogeneous schedule-based model of Table 2. However, the fixed effects model accounts for heterogeneity by allowing an extra term for each cross-section to adjust the next running time by a fixed amount, depending on the value of i. If the dummy variable parameter is positive, the value of $r_{t+1,i}$ would be increased for all data points in cross-section i, and vice versa.

Table 4. Fixed Effects Schedule-Based Model for the Core Circulator Route

Variable	Est. parameter	Standard error	t-statistic
d_{ti}	- 0.41991	0.027304	- 15.379
W_1 (bus # 220)	- 0.30961	0.09526	- 3.2500
W_2 (bus # 322)	0.41391	0.11624	3.5607
W_3 (bus # 321)	0.23205	0.09101	2.5498
W_4 (bus # 302)	0.28165	0.20170	1.3964
W_5 (bus # 299)	0.40835	0.18859	2.1653
W_6 (all other buses)	0.05675	0.10512	0.53988
No. of observations = 850, $\overline{R^2}$ = 0.21566			

For example, on the Core Circulator route, for bus 220 the parameter corresponding to the dummy variable W_1 is estimated to be -0.30961. The parameter α is estimated to be -0.41991. Therefore, for bus 220, the schedule-based model is as follows:

$$r_{t+1,1} - \overline{r} = -0.41991d_{t1} - 0.30961 + \epsilon_{t+1,1} \qquad (6)$$

The negative dummy variable parameter implies that a reduction of 0.30961 minutes from the mean $\overline{r}$ in the next running time is specifically attributable to bus 220 regardless of the value of d_{t1}. This might take into account behavioral differences due to the driver or mechanical differences due to the bus, such as acceleration and deceleration capabilities. Given the statistical significance of the parameter estimates corresponding to the various dummy variables (except for γ_6), such differences clearly exist. Also, notice that the goodness-of-fit (corrected for the additional dummy variables) $\overline{R^2}$ improves over the case where heterogeneity is not captured (see Table 2).

For the autoregressive model for the Core Circulator route, none of the dummy variable parameters is significantly different from zero at the 10% level. Therefore, heterogeneity is not detected in this case, and the fixed effects autoregressive model would be identical to the homogeneous model of Table 3.

Campus Loop North

The data for the Campus Loop North route encompassed 16 cross-sections (buses), six of which exhibited dummy variable parameters significantly different from zero at the 10% level in the preliminary estimation. The estimation results for the subsequent model including seven dummy variables (one for each of the six buses with dummy variables significantly different from zero at the 10% level, and one for all the other buses) are given in Table 5.

Table 5. Fixed Effects Schedule-Based Model for the Campus Loop North Route

Variable	Est. parameter	Standard error	t-statistic
d_{ti}	- 0.26872	0.010110	- 26.579
W_1 (bus # 526)	0.20046	0.04379	4.5774
W_2 (bus # 527)	1.7414	0.05554	3.1356
W_3 (bus # 250)	0.32967	0.15246	2.1624
W_4 (bus # 322)	0.76693	0.28232	2.7165
W_5 (bus # 574)	0.65221	0.17407	3.7467
W_6 (bus # 314)	- 0.29485	0.16726	- 1.7629
W_7 (all other buses)	0.04460	0.04219	1.0570
No. of observations = 4828, R^2 = 0.12740			

The value of α is again negative. Also notice, for example, that bus 527 is associated with a specific increase from the mean $\bar{r}$ of 1.7414 minutes in the next running time. Again, the parameter estimates corresponding to the dummy variables are statistically significant (except for γ_7), and $\overline{R^2}$ reflects a slight increase over the case where heterogeneity is not captured (see Table 2).

For the autoregressive model for the Campus Loop North route, only three of the dummy variable parameters were significantly different from zero at the 10% level in the preliminary estimation. The estimation results for the subsequent model including four dummy variables are given in Table 6. First notice that the value of ρ (the parameter of $(r_{ti} - \bar{r})$) is negative, implying that if a bus has a running time greater than the mean running time on one run, the following running time is expected to be lower than mean, and vice versa. This result indicates that corrections for a long or short running time with respect to the mean (presumably resulting from the drivers' attempts to maintain the schedule) are dominant. Also, in the autoregressive case, positive dummy variable parameters would increase the expected value of $r_{t+1,i}$ for the corresponding bus, and vice versa.

Table 6. Fixed Effects Autoregressive Model for the Campus Loop North Route

Variable	Est. parameter	Standard error	t-statistic
$r_{ti} - \bar{r}$	- 0.24666	0.013461	- 18.324
W_1 (bus # 527)	0.09572	0.05731	1.67013
W_2 (bus # 299)	3.25117	0.81521	3.9881
W_3 (bus # 322)	0.49351	0.29164	1.6922
W_4 (all other buses)	- 0.02399	0.02964	- 0.8092
No. of observations = 4828, R^2 = 0.06687			

For example, on the Campus Loop North route, for bus 322 the dummy variable parameter for W_3 is estimated to be 0.49351. The parameter ρ is estimated to be -0.24666. Therefore, for bus 322 the autoregressive model is as follows:

$$r_{t+1,3} - \bar{r} = -0.2467(r_{t3} - \bar{r}) + 0.49351 + \epsilon_{t+1,3} \qquad (7)$$

The positive dummy variable parameter implies that an increase of 0.49351 minutes from the mean $\bar{r}$ in the next running time is specifically attributable to bus 322 regardless of the value of $(r_{t3} - \bar{r})$. Given the statistical significance of the various parameters (except for γ_4), heterogeneity clearly exists. Furthermore, the value of $\overline{R^2}$ reflects an improvement over the case where heterogeneity is not captured (see Table 3).

Comparing the schedule-based model with the autoregressive model in the presence of heterogeneity, notice that the corrected goodness-of-fit of the schedule-based model is superior. This result is revisited in more detail subsequently when the various models are compared across the routes.

East Residential

The data for the East Residential route encompassed twelve different cross-sections (buses), nine of which exhibited dummy variable parameters significantly different from zero at the 10% level in the preliminary estimation. The estimation results for the subsequent model including ten dummy variables are given in Table 7. Notice that the value of α is again negative, the parameters corresponding to the dummy variables W_i are statistically significant (except for γ_{10}), and $\overline{R^2}$ reflects a slight improvement over the case where heterogeneity is not captured (see Table 2).

For the autoregressive model only two of the dummy variable parameters were significantly different from zero at the 10% level in the preliminary estimation. The estimation results for the subsequent model including three dummy variables are given in Table 8. Notice that the value for ρ is negative, the parameters corresponding to dummy variables W_i are statistically significant (except for γ_3), and $\overline{R^2}$ reflects a slight improvement compared to the value in Table 3 where driver-bus pair heterogeneity is assumed absent. Also, as in the case of the Campus Loop North route, the autoregressive fixed effects model does not fit the data as well as the schedule-based fixed effects model based on the corrected goodness-of-fit measure. Again, this result is revisited subsequently.

4.2 Prediction

In addition, the two model formulations for both the schedule-based and autoregressive models were used to predict future running times for subsets of data randomly removed from the three original route data sets. The predicted running times using models estimated on the complements of these data subsets were then compared to the actual running times. The first step in this prediction exercise was to randomly select subsets of data constituting 10% of each of the three different bus routes' data

Table 7. Fixed Effects Schedule-Based Model for the East Residential Route

Variable	Est. parameter	Standard error	t-statistic
d_{ti}	- 0.50146	0.019102	- 26.252
W_1 (bus # 239)	0.90307	0.10078	8.9609
W_2 (bus # 574)	1.3614	0.61008	2.2316
W_3 (bus # 250)	4.5481	0.69181	6.5743
W_4 (bus # 322)	0.76305	0.08906	8.5681
W_5 (bus # 218)	0.73392	0.48187	1.5231
W_6 (bus # 319)	1.7311	0.42385	4.0842
W_7 (bus # 573)	0.15424	0.10416	1.4808
W_8 (bus # 321)	0.36737	0.11932	3.0790
W_9 (bus # 302)	0.58556	0.09826	5.9596
W_{10} (all other buses)	0.17251	0.16249	1.0616

No. of observations = 2142, R^2 = 0.24282

Table 8. Fixed Effects Autoregressive Model for the East Residential Route

Variable	Est. parameter	Standard error	t-statistic
$r_{ti} - \overline{r}$	- 0.36096	0.020558	- 17.558
W_1 (bus # 574)	0.91679	0.65485	1.4000
W_2 (bus # 250)	1.7488	0.73192	2.3894
W_3 (all other buses)	- 0.01255	0.04490	- 0.27943

No. of observations = 2142, R^2 = 0.12621

sets. The remaining records in the original data set (90% of the entire data set) were used to estimate the models, while the selected records were used to predict running times and to conduct comparisons between predicted and actual running times.

The homogeneous and fixed effects models were estimated for both the schedule-based and the autoregressive specifications. This step was completed using each route's 90% data subset, producing a total of four different estimated models for each route. Note that the estimated parameters of these models would be slightly different from those presented earlier, since the latter are based on the entire data sets. In addition, a naive model was considered whereby the mean running time is used as the predictor. Thus, five distinct models were examined in the prediction analysis.

Next, the five models were applied to each datum in the 10% subsets for each route. This produced five different predictions of upcoming running times for each datum. The actual running time for each datum is also available in the data set. The differences between the predicted running times and the actual running times were therefore computed, thus reflecting the prediction error for each of the five models. Finally, summary prediction error statistics for each of the routes were computed on the basis of the absolute values and the squares of the errors for each of the five models. These statistics are examined and compared to gain an understanding of which model performed best in terms of predicting upcoming running times for each route. The summary statistics employed in the comparisons – standard deviation of the prediction error, mean of the absolute value of the error, and mean of the squared error – are estimated as follows:

$$\sigma_e = \sqrt{Var[e]} \qquad (8)$$

$$E[|e|] = \frac{1}{S}\sum_{i=1}^{S}|e_i| \qquad (9)$$

$$E[e^2] = \frac{1}{S}\sum_{i=1}^{S}e_i^2 \qquad (10)$$

where i = index representing each element of the randomly selected subset representing 10% of the entire set, e_i = predicted minus the actual running time, and S = number of observations in the 10% subset.

As will be seen in the subsequent tables, each model discussed in this paper outperformed the naive model for all routes. This result, though not surprising, indicates that incorporating either the most recent deviation from the schedule or the most recent deviation from the mean running time improves predictions of the next running time, as compared with simply using the mean running time as the predictor. The remainder of the results are discussed for each route separately.

Core Circulator

For the Core Circulator route, from the original 850 data records in the set, 765 were used to estimate the models, while 85 were used for prediction. Table 9 summarizes the prediction error statistics for each of the five models. In this and the subsequent two tables, Tables 10 and 11, the naive model summary statistics are noted for both models to make comparisons easier. Naturally, they are the same for both models. For the schedule-based model, as highlighted in Table 9, the fixed effects model has the lowest values for all three error statistics. For the autoregressive model, while the fixed effects model again has the lowest mean absolute value of the errors, the homogeneous model is superior on the basis of the standard deviation of the error and mean of squared error. Nevertheless, the lowest error statistics for the schedule-based model are consistently superior to those of the autoregressive model. Therefore, the fixed effects schedule-based model was better at predicting future running times for the Core Circulator data.

Campus Loop North

For the Campus Loop North route, from the original 4828 data records in the set, 4345 were used to estimate the models, while 483 were used for prediction. Table 10 summarizes the prediction error statistics for each of the five models. Again, as highlighted in the table, the fixed effects model has the lowest error statistics for the schedule-based model. In the case of the autoregressive model, the homogeneous model has the lowest values for all three error statistics, although the values of the statistics for the homogeneous and fixed effects formulations are very close to one

Table 9. Prediction Error Statistics for the Core Circulator Route

		Model		
	Statistic	Homogeneous	Fixed Effects	Naive
Schedule-based:	σ_e	1.4998	**1.4906**	1.6527
	$E\left[\|e\|\right]$	1.1525	**1.1440**	1.2768
	$E\left[e^2\right]$	2.2281	**2.2144**	2.7003
Autoregressive:	σ_e	**1.6231**	1.6357	1.6527
	$E\left[\|e\|\right]$	**1.2555**	1.2531	1.2768
	$E\left[e^2\right]$	**2.6110**	2.6444	2.7003

another. Moreover, when comparing the schedule-based model with the autoregressive model, the statistics are slightly superior for the homogeneous autoregressive model, indicating that this model was slightly better at predicting future running times for the Campus Loop North route.

Table 10. Prediction Error Statistics for the Campus Loop North Route

		Model		
	Statistic	Homogeneous	Fixed Effects	Naive
Schedule-based:	σ_e	1.8805	**1.8697**	1.9028
	$E\left[\|e\|\right]$	1.3988	**1.3951**	1.4242
	$E\left[e^2\right]$	3.5496	**3.5080**	3.6437
Autoregressive:	σ_e	**1.8404**	1.8441	1.9028
	$E\left[\|e\|\right]$	**1.3825**	1.3837	1.4242
	$E\left[e^2\right]$	**3.4106**	3.4240	3.6437

East Residential

For the East Residential route, from the original 2142 data records in the set, 1928 were used to estimate the models, while 214 were used for prediction. Table 11 summarizes the prediction error statistics for each of the five models. Once again, the fixed effects formulation has the lowest error statistics for the schedule-based model, while the homogeneous formulation has the lowest values for the autoregressive model. When comparing the two models, the corresponding statistics are seen to be lower for the fixed effects schedule-based model, indicating that this model was better at predicting future running times for the East Residential route.

4.3 Route Comparison

While the Core Circulator, Campus Loop North, and East Residential bus routes are all operated by CABS, each has its own characteristics as partly indicated in Table 1. The East Residential route has a scheduled running time of 30 minutes, a headway of 10 minutes, and operates mostly off-campus in residential areas. Moreover, the

Table 11. Prediction Error Statistics for the East Residential Route

	Statistic	Model Homogeneous	Fixed Effects	Naive
Schedule-based:	σ_e	1.9165	**1.8899**	2.2430
	$E[|e|]$	1.4448	**1.4330**	1.7182
	$E[e^2]$	3.6566	**3.5562**	5.0076
Autoregressive:	σ_e	**2.0992**	2.1035	2.2430
	$E[|e|]$	**1.5742**	1.5800	1.7182
	$E[e^2]$	**4.3873**	4.4056	5.0076

distances between the bus stops are large compared to the other two routes. The Campus Loop North route also has a scheduled running time of 30 minutes and a headway of 10 minutes. However, it operates mostly on campus where pedestrian traffic is relatively high. Finally, the Core Circulator route runs entirely on campus and has the shortest scheduled running time, 12 minutes, and a headway of 6 minutes. The distances between stops are also comparatively short.

Due to the short running time for the Core Circulator route, the 6-minute headway, short stop spacing, and high pedestrian traffic, drivers are expected to be less likely to pay as much attention to the schedule. If this is indeed the case, the autoregressive model might capture the operating behavior more accurately than the schedule-based model. In contrast, drivers on the East Residential route are expected to be more effective at maintaining the schedule given its favorable characteristics. Therefore, in this case it is expected that the schedule-based model would best capture its behavior. As for the Campus Loop North route, the longer running time and 10-minute headway might suggest that the schedule-based model would be the better of the two. However, the effects of uncontrollable exogenous factors such as the high pedestrian traffic might interfere with the drivers' attempts to adhere to the schedule. In this case, it is unclear on an *a priori* basis which of the two models would capture the behavior more accurately. In what follows, the various models are compared across the routes with a focus on the fixed effects formulation. The fixed effects formulation is chosen because of its statistical superiority over the homogeneous formation for the schedule-based model on all three routes and for the autoregressive model on two of the three routes.

First, the fixed effects schedule-based model results are considered. The East Residential route exhibits the highest $\overline{R^2}$ value of 0.24282, and the Campus Loop North the lowest value of 0.12740. This result indicates that the schedule-based fixed effects model fits the East Residential data set best, which is consistent with the *a priori* expectations discussed above. As for the estimated values of the parameter α, all three are negative and significantly different from zero at the 5% level, which would be consistent with attempts by drivers to meet the schedule in all three cases. In addition, the magnitude of the estimated parameters is also important. The further the value of α is from zero, the more indicative it is that a driver has a greater ability to correct for recent deviations from the schedule. Notice that the estimated value of α is the furthest from zero for the East Residential route. This result is again consistent with the *a priori* expectations based on the characteristics of the route. As for the other two routes, the value of α for the Core Circulator route is higher in absolute value than that of the Campus Loop North route. This result is not quite consistent with the *a priori* expectations. One possible explanation might relate to differences in the level of experience of the drivers assigned to each route. This difference could be fairly high for CABS due to the mix of professional and student drivers. However, this information is not available for separate routes at this time and, therefore, such considerations are reserved for future research.

Considering the autoregressive fixed effects models, the East Residential data set again exhibits the highest $\overline{R}^2$ value of 0.12621, and the Core Circulator data the lowest value of 0.024346. Although this result is not consistent with the *a priori* expectations, it is not surprising in light of the favorable performance of the schedule-based model in the case of the Core Circulator route. As for the estimated value of ρ, the parameters are negative and statistically significant at the 5% level, which would be consistent with drivers attempting to achieve regular running times from run to run. Again, the magnitude of the estimated values is also important. A relatively higher value in absolute terms would indicate that a driver has a greater ability to adjust the running time. The estimated value of ρ is largest in absolute value for the East Residential route, which is consistent with the *a priori* expectation that this route's characteristics are more amenable to adjustments in its operation. The value of ρ is the lowest in absolute terms for the Core Circulator route. This result is not inconsistent with *a priori* expectations that this route is subject to a high degree of exogenous influences with limited opportunity to adjust operations due to the relatively short stop spacing and route length.

When the corrected goodness-of-fit values across the schedule-based and autoregressive models are compared, these values are higher for the schedule-based model for all three routes. This result indicates that the schedule-based model is better at describing running time behavior on all three routes. While this conclusion is not entirely consistent with the *a priori* expectations, it does add credibility to the above discussed results associated with each model as it applies to each of the three routes. The prediction results are consistent with these findings for the Core Circulator and East Residential routes. However, on the basis of prediction performance, the autoregressive model is slightly better than the schedule-based model in predicting the behavior of the Campus Loop North route. This result supports the *a priori* expectation regarding the ambiguity associated with this route. Finally, it is interesting to note that the differences between the naive model and the other three models with regard to prediction errors is greatest for the East Residential route, further supporting the belief that favorable conditions on this route allow drivers to better adjust the operation to meet the schedule.

5 Summary and Future Research

An understanding of running time behavior and the factors that influence it is essential for off-line planning and operations design purposes. Moreover, such an understanding is critical for real-time applications including bus operations control and passenger information systems. In this paper, two different running time models – schedule-based and autoregressive – were presented, estimated, and evaluated both in the absence of (homogeneous formulation) and in the presence of (fixed effects formulation) driver-bus heterogeneity. For all three bus routes considered and both the homogeneous and fixed effects formulations, the schedule-based model fits the observed manifestations of the operating behavior better than the autoregressive model. All the schedule-based models have better corrected goodness-of-fit. While this was expected for East Residential route with a 30-minute scheduled running time, it was not expected for the Campus Loop North route with a 12-minute schedule running time due to the presence of unobserved influencing factors such as high vehicular and pedestrian traffic. Nevertheless, the results indicate that, even under such conditions, the schedule remains an important explanatory factor. Regarding the question of heterogeneity of driver-bus pairs, the incorporation of dummy variables capturing this effect in the schedule-based model consistently produced statistically significant results, indicating the presence of heterogeneity. The results of the prediction exercise further confirm this conclusion.

Much additional research could be conducted with rich AVL-based data sets, especially as more of the issues discussed by Furth *et al.* (2003) are addressed. The most obvious extension to the study presented in this paper is to examine more bus routes on CABS and other transit systems. The analysis of additional routes might allow for further comparisons between routes and might lead to the determination of additional explanatory variables relating to the characteristics of bus routes. The length of the route, distance between stops, passenger demand, pedestrian and other vehicular traffic, and time-of-day might all be factors that affect running times. For example, Hickman (2001), Bertini and El-Geneidy (2004), and Shalaby and Farhan (2004) considered the effect of passenger demand. In addition, Dueker *et al.* (2001) considered the effect of drawbridge interruptions. Moreover, Shalaby and Farhan (2004) considered the effect of time-of-day. Nevertheless, studies investigating such influencing variables in the presence of driver-bus heterogeneity would be worthwhile, especially when using empirical observations of actual bus transit operations.

Another possibility for future research might include a closer examination of driver-bus pairs. In this research, the bus number was considered a proxy for a driver-bus pair. However, CABS operates in three different shifts, and it is possible for a bus to be driven by different drivers during more than one shift. This information was not readily available from CABS, but it might be possible to obtain in future. If all driver-bus combinations could be considered as separate cross-sections, achieving better models might be possible. Furthermore, if specific data regarding drivers and buses are available, it might be possible to determine how different driver socioeconomic characteristics and bus characteristics affect running times. For example, a driver's gender, experience, or age might be the cause of heterogeneity across cross-sections. In a similar fashion, a bus's age, size, or model type affecting, for example, acceleration and deceleration capabilities, might cause heterogeneity. Both of these types of characteristics could be evaluated, and developing better models might be possible.

Acknowledgement: The support of Sarah Blouch and Steven Basinger of the Transportation and Parking Office at The Ohio State University in providing partial funding, access to the AVL data, and information on CABS operating policies is greatly appreciated.

References

Basinger, S. (2003). Personal communication. Transportation and parking services. The Ohio State University, Columbus, OH.

Bertini, R. L. and El-Geneidy, A. M. (2004). Modeling transit trip time using archived bus dispatch system data. *Journal of Transportation Engineering*, **130**(1), 56–67.

Bus Location Information System (2003). http://blis.units.ohio-state.edu.

Dueker, K. J., Kimpel, T. J., Strathman, J. G., Gerhart, R. L., Turner, K., and Callas, S. (2001). Development of a statistical algorithm for the real-time prediction of transit vehicle arrival times under adverse conditions. (Final report TransNow project PSU-92210). Portland State University.

Furth, P. G., Hemily, B. J., Muller, T. H. J., and Strathman, J. G. (2003). Uses of archived AVL APC data to improve transit performance and management: review and potential. Transit cooperative research program, transportation research board. Technical report, The National Academies. Washington, D.C.

Greene, W. H. (2003). *Econometrics*. Prentice Hall, Englewood Cliffs.

Hickman, M. D. (2001). An analytic stochastic model for the transit vehicle holding problem. *Transportation Science*, **35**(3), 215–237.

Larson, R. C. and Odoni, A. R. (1981). *Urban Operations Research*. Prentice Hall, Englewood Cliffs.

Lin, W. H. and Zeng, J. (1999). Experimental study of real-time bus arrival time prediction with GPS data. *Transportation Research Record 1666*, pages 101–109.

Pindyck, R. S. and Rubinfeld, D. L. (1998). *Econometric Models and Economic Forecasts*. McGraw Hill, New York.

Shalaby, A. and Farhan, A. (2004). Prediction model of bus arrival and departure times using AVL and APC data. *Journal of Public Transportation*, **7**(1), 41–61.

Transportation and Parking Services (2001–2002). Campus area bus service schedule. Technical report, The Ohio State University, Columbus, OH.

Wall, Z. and Dailey, D. J. (1999). An algorithm for predicting the arrival time of mass transit vehicles using automatic vehicle location data. Transportation Research Board 78th Annual Meeting Paper No. 990870.

Wei, W. W. S. (1990). *Time Series Analysis: Univariate and Multivariate Methods*. Addison Wesley, Boston.

A Train Holding Model for Urban Rail Transit Systems

André Puong[1] and Nigel H.M. Wilson[2]

[1] Harvard Business School. Soldiers Field. Boston, MA 02163.
apuong@mba2007.hbs.edu
[2] Department of Civil and Environmental Engineering, Massachusetts Institute of
Technology. 77 Mass. Ave. Cambridge, MA 02139. nhmw@mit.edu

Summary. Urban rail transit lines are subject to disruptions that can adversely affect passenger level of service and routine operations. This paper focuses upon the development of a real-time disruption response model with an emphasis on the train holding strategy. The paper also discusses the short-turning control strategy which is often used in conjunction with holding for longer disruptions. The holding problem is modeled as a non-linear mixed-integer program and a two-step solution procedure is designed to solve it quickly, yielding solution times of less than 10 seconds. The model is applied to a disruption scenario on a simplified representation of the MBTA Red Line. The sensitivity of the optimal holding strategy to the assumptions of finite train capacity and the value of in-vehicle time are also investigated. The results show a high level of regularity in the headway distribution for the control strategy when in-vehicle time is not considered. When accounting for in-vehicle delay, the optimal holding strategy consists of only a few trains being held at a few stations. Overall, the results suggest the present formulation yields control strategies that are simple enough to be implemented by transit practitioners and that the solution times are feasible for real-time implementation.

1 Introduction

Urban rail transit lines are subject to occasional disruptions or delays that can severely impact passenger level of service and routine transit operations. The goal of transit operators is to limit those negative impacts by using effective operations control strategies, given the infrastructure characteristics and operating plans of the system.

State of the art train regulation systems strive to keep regular headways between trains along the line: this minimizes total passenger in-station waiting time, assuming a *Poisson* passenger arrival process and non-binding train capacities. However, these systems do not address longer disruption durations in which train capacities can become critical. Nor do they evaluate the exact costs and benefits of any control action in determining the "optimal" strategy.

This gap has been addressed by researchers in recent years with the development of mixed integer program formulations for the train regulation problem (O'Dell and Wilson (1999) and Shen and Wilson (2001)). The objective of the problem is to minimize the *weighted sum* of:

- the total passenger in-station waiting time, and
- the extra passenger riding time due to train holding,

subject to the system's infrastructure and other operational constraints.

Although insightful in their findings and interpretation of the optimal response strategies, the prior models have not been suitable for implementation within transit agencies for several reasons. First, the formulations adopted in O'Dell and Wilson (1999) and Shen and Wilson (2001) are based on train arrival and departure times at stations. As dispatchers are interested in holding times–which are derived from the difference of those two times, these formulations artificially increase the number of variables and thus the size of the problem as well as solution times. As a result these models cannot be counted on to produce effective strategies in a real-time computational context. Second, the aforementioned objective function is linearized from its exact quadratic form to obtain a linear programming formulation of the problem. While this approximation significantly decreases solution times, no investigation has been made into its effects on the structure of the optimal control strategies. Indeed, the resulting strategies are usually too complex to be implemented by dispatchers in practice no matter how efficient they may be in theory at reducing the total passenger waiting time.

The work presented in this paper is motivated by the above shortcomings and also by recent advances in non-linear optimization software performance, allowing optimization problems with non-linear objective functions to be solved more quickly.

The focus in this paper is the train holding strategy, which is the core strategy for dealing with service interruptions of less than 20 minutes. For longer disruptions trains are often short-turned in conjunction with holding, and this paper also briefly discusses this more general problem. The core holding problem is modelled as a deterministic 0-1 integer program, using a different problem formulation but a similar objective function as in Shen and Wilson (2001). This formulation is presented here along with a solution procedure that minimizes the *exact* cost function with solution times comparable to those obtained in Shen and Wilson (2001). The model is applied to a disruption scenario on a simplified transit system based on the MBTA Red Line. The structure of the optimal control strategies is then analyzed. Finally, a general discussion of the short-turning strategy is provided, and it is shown how the developed holding model can be used to assess some forms of short-turning.

2 Model Description

2.1 Assumptions and Model Features

The following assumptions and limitations are made for the problem:

- *The duration of the delay is a known fixed parameter.* As discussed in the prior literature this assumption is not realistic, but the resulting model may become a module in the more efficient stochastic formulation of this problem which awaits future research.
- *Passenger arrival rates and alighting fractions are constant and station-specific.*
- *Train dwell-times are constant and station-specific.* Dwell-times are generally a function of boardings and alightings (see Lin and Wilson (1992)), and thus depend *a priori* on the adopted holding strategy. Nonetheless, dwell-time standard deviations at a station are in general under half a minute, which is a small fraction of the mean passenger waiting time. Thus, simplifying the dwell-time component may not be critical in developing holding strategies that seek to minimize passenger waiting time.
- *Inter-station running times are deterministic.* This assumption is made since train movements include variations that are difficult to model: they are a function of many factors such as weather, track conditions and the signal system.
- *The safe separation between trains is ensured* by imposing a minimum safe headway h_s between successive trains.
- *Trains are considered for holding for the remainder of the current trip, plus the next trip for trains located close to the disruption.* This limits the time window for the evaluation of any holding strategy and thus limits the capacity of the developed model to devise holding strategies whose benefits extend far into the future. On the other hand, extending the model to include stations visited on subsequent trips increases the size of the problem and affects its real-time tractability.

2.2 Data Requirements

The following set of data is required as input to the holding model:

- Passenger arrival rates and alighting fractions at each station for the time period of interest.
- Train capacity.
- Disruption location and estimated duration.
- Last station departed and headways for all trains in the system. This information is readily available from automatic vehicle location (AVL) systems.
- Maximum acceptable delay for all trains dispatched from the terminal.

2.3 Notation

The following notation is used:

λ_m is the passenger arrival rate at station m
α_m is the alighting fraction at station m
d_0 is the delay duration
h_s is the minimum safe headway between trains
Ξ is the minimum turnaround time at the terminal station
h_i is the uncontrolled departure headway of train i

C_i is the capacity of train i

m_i is the first station visited by train i after the disruption starts

Ω_i is the scheduled layover time of train i at the terminal after the disruption location

Ψ_i is train i's maximum dispatching time deviation from schedule at the terminal after the disruption location

M is the number of stations in the disruption direction, with station $M - 1$ being the queuing location[3] before the terminal.

M_0 is the index of the station immediately ahead of the blockage

S_i is the set of stations visited by train i and included in the model (i.e., all stations $m : m_i \leq m \leq 2M - 3$)

B, A, T, R denote the sets of trains behind and ahead of the blockage in the disruption direction, at the terminal and in the reverse direction, respectively

The following variables are used in the problem formulation:[4]

$r_{i,m}$ denotes the holding time of train i at station m

$R_{i,m}$ $= \sum_{p=m_i}^{m} r_{i,p}$, i.e., the cumulative holding time of train i up to station m. Thus, $r_{i,m} = R_{i,m} - R_{i,m-1}$, $\forall m \geq m_i$, $\forall i$

$L_{i,m}$ denotes train i's passenger load arriving at station m

$P_{i,m}$ denotes the number of passengers left behind by train i at station m

3 Problem Formulation

3.1 The Objective Function

The cost function to be minimized is the *total passenger time*, i.e., the total in-station waiting plus the extra riding-time due to train holding. This cost function can be written as the weighted sum of three costs, $F(\mathbf{R}, \mathbf{L}, \mathbf{P}) = F_1(\mathbf{R}) + \mu F_2(\mathbf{R}, \mathbf{L}) + F_3(\mathbf{R}, \mathbf{P})$, where we note $\mathbf{R} = \{R_{i,m}\}$, $\mathbf{L} = \{L_{i,m}\}$ and $\mathbf{P} = \{P_{i,m}\}$.

In the above sum, F_1 represents the total in-station waiting time for passengers boarding the first train arriving at each station, F_2 represents the total extra riding-time for on-board passengers due to train holding, F_3 accounts for the extra in-station waiting time incurred by passengers who are denied boarding fully-loaded trains, and μ is a positive coefficient that weights the negative effects of extra ride-time against in-station waiting time.

---

[3] In a standard stub-end terminal configuration, when both terminal platforms are occupied and another train is about to arrive at the terminal, this train must wait until a platform is cleared. In case the corresponding queuing location is not a station, we would then model it using a virtual station $M - 1$ with no associated passenger arrivals ($\lambda_{M-1} = 0$) or alightings ($\alpha_{M-1} = 0$). Hence, $2M - 3$ stations are represented in the model.

[4] Note that train $i + 1$ *precedes* train i in our model and that the disabled train has index 0. Also, stations are ordered consecutively starting with the disruption location. Also, we have the initial conditions $R_{i,m} = 0$, $\forall m < m_i$ since train i is not considered for holding before station m_i.

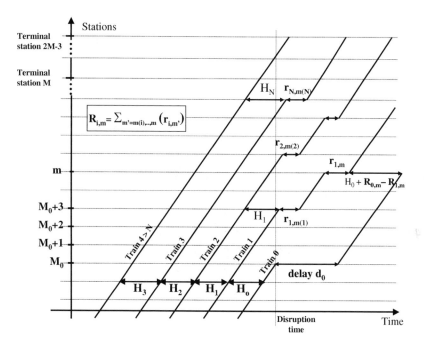

Fig. 1. Time-space Diagram

The expressions for F_1, F_2 and F_3 are derived from inspection of the headways from the time-space diagram shown in Fig. 1. The diagram shows that train i's departing headway from station m, $H_{i,m}$ is $(h_i + R_{i,m})$ for $m < m_{i+1}$ and $(h_i + R_{i,m} - R_{i+1,m})$ for $m \geq m_{i+1}$. Hence, the general form of the functions F_i can be written as follows:[5]

$$F_1(\mathbf{R}) = \sum_{i \in B \cup A \cup T \cup R} \sum_{m \in S_i} \frac{\lambda_m}{2} H_{i,m}^2 \tag{1}$$

$$F_2(\mathbf{R}, \mathbf{L}) = \sum_{i \in B \cup A \cup T \cup R} \sum_{m \in S_i} L_{i,m}\left(1 - \alpha_m\right)\left(R_{i,m} - R_{i,m-1}\right), \text{ and} \tag{2}$$

$$F_3(\mathbf{R}, \mathbf{P}) = \sum_{i \in B} \sum_{m \in S_i} P_{i,m} H_{i-1,m} \tag{3}$$

Since trains $i \in A \cup T \cup R$ are located ahead of the blockage, the disruption has no effect on these trains unless they are held. Thus, the capacity constraint is dealt with by restricting holding actions for these trains such that no passenger can be left

---

[5] Equations (1) - (3) are not suitable for implementation as is. Specifically, they do not consider the possible presence of a second train at the terminal station (which has a second platform). This also applies to the model constraints. This implementation issue is not addressed here for the sake of clarity.

behind. In contrast, trains behind the blockage might become overloaded and leave passengers behind as passengers trying to board these trains are accumulating during the disruption both ahead of and behind the blockage. Therefore, the cost component F_3 (and constraint (5) below) only applies to trains in B.

3.2 Constraints

The above objective function $F(\mathbf{R}, \mathbf{L}, \mathbf{P})$ is minimized, subject to the system operational constraints:

Load/capacity constraints for trains ahead of the blockage

$$L_{i,m+1} = (1 - \alpha_m)L_{i,m} + \lambda_m H_{i,m}, \, \forall m \in S_i, \, \forall i \in A \cup T \cup R \qquad (4a)$$

$$(1 - \alpha_m)L_{i,m} + \lambda_m H_{i,m} \leq C_i, \, \forall m \in S_i, \, \forall i \in A \cup T \cup R \qquad (4b)$$

Load/capacity constraints for trains behind the blockage

$$L_{i,m+1} = \min\left((1 - \alpha_m)L_{i,m} + \lambda_m H_{i,m} + P_{i+1,m}, C_i\right), \, \forall m \in S_i, \, \forall i \in B \quad (5)$$

Left-behind-passenger constraints for trains behind the blockage

$$P_{i,m} = (1 - \alpha_m)L_{i,m} + \lambda_m H_{i,m} - L_{i,m+1}, \, \forall m \in S_i, \, \forall i \in B \qquad (6)$$

Minimum safe headway constraints for non-terminal stations

$$H_{i,m} \geq h_s, \, \forall m \in S_i : m \neq M, \, \forall i \in B \cup A \cup T \cup R \qquad (7)$$

Terminal capacity/queuing constraints

$$R_{i+2,M} - R_{i+2,M-1} \leq h_{i+1} + h_i + R_{i,M-1}, \, \forall i \in B \cup A \qquad (8)$$

Queuing constraints behind the blockage

$$R_{i,M_0+i-1} = 0, \, \forall i < 0 \, (\text{i.e., } i \in B - \{0\}) \qquad (9)$$

Layover constraints at terminal

$$R_{i,M} \geq \Omega_i, \, \forall i \in B \cup A \cup T \qquad (10)$$

Turnaround constraints at terminal

$$R_{i,M} - R_{i,M-1} \geq \varXi, \, \forall i \in B \cup A \cup T \qquad (11)$$

Maximal deviation from schedule constraints

$$R_{i,M} - \Omega_i \leq \Psi_i, \, \forall i \in B \cup A \cup T \qquad (12)$$

Disruption duration constraint

$$R_{0,M_0} \geq d_0 \qquad (13)$$

Non-negativity constraints

$$R_{i,m} - R_{i,m-1} \geq 0 \text{ and } R_{i,m}, L_{i,m}, P_{i,m} \geq 0, \forall i, m \qquad (14)$$

Headway definition

$$
\begin{aligned}
H_{i,m} &= h_i + R_{i,m}, \forall m \in S_i : m < m_{i+1} \\
H_{i,m} &= h_i + R_{i,m} - R_{i+1,m}, \forall \in S_i : m \geq m_{i+1}
\end{aligned}
\qquad (15)
$$

Most of the above constraints are self-explanatory but some deserve further explanation. Terminal capacity constraints (8) require the second preceding train to have left the terminal to allow a train to enter. Equation (9) constrains trains behind the blockage not to be held until they reach the closest station to the blockage where they can queue (station $M_0 + i$, $i < 0$). In this case, the queuing time is incorporated into the holding variable R_{i,M_0+i} as queuing or holding has the same impact on headway. In the same fashion, layover times and the delay d_0 are incorporated into the cumulative holding times in Equations (10) and (13), respectively. Finally, Equations (10) - (12) ensure that operational constraints are respected at the terminal.

4 Model Analysis

4.1 A Mixed Integer Program

We first note that the min function in Equation (5) is modeled through the use of binary variables $\nu_{i,m}$ as follows:

$$
\begin{aligned}
L_{i,m+1} &\leq P_{i+1,m} + (1 - \alpha_m)L_{i,m} + \lambda_m H_{i,m}, &\forall m \in S_i, \forall i \in B \qquad (16a) \\
L_{i,m+1} &\leq C_i, &\forall m \in S_i, \forall i \in B \qquad (16b) \\
L_{i,m+1} &\geq P_{i+1,m} + (1 - \alpha_m)L_{i,m} + \lambda_m H_{i,m} - K\nu_{i,m}, \\
& &\forall m \in S_i, \forall i \in B \qquad (16c) \\
L_{i,m+1} &\geq C_i - K(1 - \nu_{i,m}), &\forall m \in S_i, \forall i \in B \qquad (16d)
\end{aligned}
$$

where K is a large constant.

Consequently, our holding problem is a 0-1 mixed integer program where train i is at capacity at station m iff $\nu_{i,m} = 1$. Although the problem is quite small, the number of binary variables (several thousand) makes it difficult to solve in real-time.

Clearly, a better understanding of the problem can potentially reduce the number of binary variables and feasible solutions to search, thus dramatically reducing the solution times of the problem.

4.2 A Two-Step Solution Procedure

To further reduce the number of binary variables, we use the following two-step solution procedure:

Step 1. Solve the train control problem for $(\mathbf{R}, \mathbf{L}, \mathbf{P}, \nu)$ by constraining holding times at stations to be zero. Find a feasible solution $(\mathbf{R}^0, \mathbf{L}^0, \mathbf{P}^0, \nu^0)$ to the resulting linearly constrained problem.

Step 2. Solve for $(\mathbf{R}, \mathbf{L}, \mathbf{P}, \nu)$ with variables $\nu_{i,m}$ for train i and station m such that $\nu^0_{i,m} \neq 0$. Constrain the other $\nu_{i,m}$ to be zero.

The rationale for this procedure is simple. We first locate in *Step 1* the trains and stations for which the train capacity constraint is active ($\nu^0_{i,m} = 1$ iff train i is at capacity at station m) when no train control will be applied. Given the information from this worst-case scenario, a better solution is sought in *Step 2*. In particular, the train capacity constraint should not be binding at stations where trains were not fully loaded in the no-hold case. As a consequence, this procedure removes a significant number of binary variables and thus dramatically reduces the number of feasible solutions.

4.3 Execution Time

We used version 12.0 of XPRESS-MP with a branch-and-cut strategy on an 800 MHz Pentium processor to solve the disruption scenario described above with the execution times shown in Table 1. We also present in this table the effectiveness of the two-step solution procedure described above. For each value of μ, we show the number of binary variables left after *Step 1*[6] of the solution procedure along with the solution time of each step. These times do not include the time needed to generate the model, which is independent of the model formulation.

We note that in all cases the number of binary variables, which is the bottleneck of the solution procedure, is considerably reduced so that less than 15 binary variables remain at *Step 2* of the procedure. The resulting solution times are significantly smaller: less than 6 seconds is needed to achieve optimality with the two-step solution procedure, while 56 seconds are necessary to solve the case $\mu = 0.1$ without the two-step solution procedure. For the other values of μ the decrease is less pronounced but still significant (it is reduced at least by a factor of 2).

Table 1. Execution Times

μ	# of $\nu_{i,m}$	# of $\nu_{i,m}$ Non-Fixed after Step 1	Solution Time without two-step procedure (sec)	of Step 1 (sec)	of Step 2 (sec)
0.0	203	13	14	2	4
0.1	203	13	56	1	3
0.5	203	13	14	2	3

---

[6] The solver was used here to solve the linear system of constraints. This is done by specifying no objective value and recording the first (and unique) feasible solution found.

5 Model Application

The model developed was applied to several disruptions on a simplified version of the MBTA Red Line, which is modeled as a single loop line with two terminal stations (Alewife and JFK) as shown in Fig. 2.[7] One disruption scenario is a 20-minute blockage at Harvard Square station (northbound) during the morning peak period. Train location (see Table 2) and passenger loads are derived from the scheduled running times as well as historical passenger counts. All initial train headways are assumed to be four minutes, and sensitivity analysis is performed by resolving this disruption for different values of the model parameter μ.

5.1 Results

Minimizing In-Station Waiting Time

The train holding model is first solved with infinite train capacities and without considering the costs to on-board passengers of holding trains ($\mu = 0$). The resulting optimal holding times and headways are shown in Tables 3 and 4,[8] respectively.

Under these conditions, the optimal holding pattern results in nearly perfectly even headways (at each station, across all trains). The regularity of the optimal headway distribution in this case is consistent with the result derived by Welding (1957), which states that passenger waiting time at a given station is minimized when the variance of headways between trains is minimized:

$$\overline{WT} = \frac{\bar{h}}{2} \left(1 + \frac{Var(h)}{\bar{h}^2} \right) \tag{17}$$

where:
$\overline{WT}$ = mean passenger waiting time
$\bar{h}$ = mean train headway
$Var(h)$ = variance of train headway

By inspecting the locations and the holding times in Table 3, along with the headway sequences across stations, we find that the optimal holding strategy generally has the following properties:

- No train is held at any station between stations m_i and m_{i+1}.
- The value of the constant headway decreases, as we move down the line.
- At any given station, a train's holding time is smaller than its preceding train's holding time.
- For any given train traveling in a given direction, its holding time (at holding stations) is monotonically decreasing.

---

[7] Details of this modeling procedure are omitted here for the sake of clarity.
[8] No holding action is taken for trains/stations that are not shown in the tables. Blocked train 0 and trains queued behind the blockage are not held at stations after the blockage is cleared, except at the terminal where they are held for the minimum turn-around time.

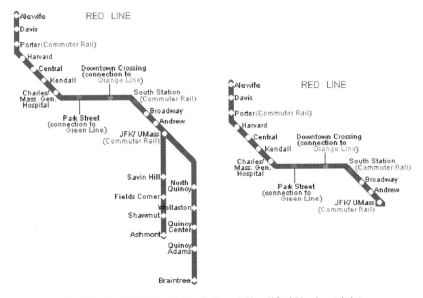

Fig. 2. The MBTA Red Line (left) and Simplified Version (right)

Table 2. Initial Train Locations: Harvard Northbound Disruption Case

Station	JFK	AND	BRW	STA	DTX	PKS	MGH	KEN	CEN	HAR	POR	DAV
Train -6	*											
Train -5		*										
Train -4			*									
Train -3				*								
Train -2							*					
Train -1									*			
Train 0										Blockage		
Train 1											*	
Train 2												*

Station	ALW	DAV	POR	HAR	CEN	KEN	MGH	PKS	DTX	STA	BRW	AND
Terminal Train T_1	*											
Terminal Train T_2	*											
Reverse Train 1_R			*									
Reverse Train 2_R					*							
Reverse Train 3_R							*					
Reverse Train 4_R									*			
Reverse Train 5_R											*	

Table 3. Holding Times (min): Harvard Northbound Disruption; $\mu = 0$, Infinite Capacity

Train	HAR	POR	DAV	QUE	ALW	DAV	POR	HAR	CEN	KEN	MGH	PKS	DTX	STA	BRW	AND
1	10.0	3.3	0.4	2.3	2.8	0.5	0.0	0.4	0.0	0.3	0.0	0.2	0.0	0.2	0.0	
2		6.7	2.3	4.5	2.1	1.1	0.0	0.8	0.0	0.6	0.0	0.4	0.0	0.3	0.0	
T_1			11.0	1.4	1.6	0.0	1.1	0.0	0.8	0.0	0.7	0.0	0.5	0.0		
T_2			4.5	0.7	2.1	0.0	1.5	0.0	1.1	0.0	0.9	0.0	0.7	0.0		
1_R					2.7	0.0	1.9	0.0	1.4	0.0	1.1	0.0	0.9	0.0		
2_R							2.3	0.0	1.7	0.0	1.3	0.0	1.0	0.0		
3_R									2.0	0.0	1.6	0.0	1.2	0.0		
4_R											1.8	0.0	1.4	0.0		
5_R													1.6	0.0		

Table 4. Preceding Departing Headway: Harvard Northbound Disruption; $\mu = 0$, Infinite Capacity

Train	HAR	POR	DAV	QUE	ALW	DAV	POR	HAR	CEN	KEN	MGH	PKS	DTX	STA	BRW	AND
0	24.0	14.0	10.7	10.3	10.0	7.2	6.7	6.7	6.3	6.3	6.0	6.0	5.8	5.8	5.6	5.6
1		14.0	10.7	8.7	6.5	7.2	6.7	6.7	6.3	6.3	6.0	6.0	5.8	5.8	5.6	5.6
2			10.7	13.0	6.5	7.2	6.7	6.7	6.3	6.3	6.0	6.0	5.8	5.8	5.6	5.6
T_1					6.5	7.2	6.7	6.7	6.3	6.3	6.0	6.0	5.8	5.8	5.6	5.6
T_2					6.5	7.2	6.7	6.7	6.3	6.3	6.0	6.0	5.8	5.8	5.6	5.6
1_R							6.7	6.7	6.3	6.3	6.0	6.0	5.8	5.8	5.6	5.6
2_R									6.3	6.3	6.0	6.0	5.8	5.8	5.6	5.6
3_R											6.0	6.0	5.8	5.8	5.6	5.6
4_R													5.8	5.8	5.6	5.6
5_R															5.6	5.6

Nevertheless, we note from Table 3 that the above mentioned properties do not always hold. In particular, trains are held at Davis Square Inbound (which is not a control station m_i) and the corresponding holding times are not decreasing. Also, headways are not even at either Alewife or the queuing location. Uneven headways are acceptable at the queuing location as there is no associated in-station waiting time: the objective function value is not a function of the headway distribution at this "virtual" station.

The two other points are explained by observing from Table 3 that the cumulative holding time of train 1 at Alewife is 16 minutes.[9] Since train 1's layover time at the beginning of the disruption is six minutes and the maximal deviation from schedule is ten minutes, this means that the constraint on the maximal deviation from schedule is binding, which forces it to depart from Alewife after being held for only 2.3 minutes. Limiting the hold at Alewife results in an uneven departure headway sequence at Alewife: train 0's headway is ten minutes while preceding trains left this station with six-minute headways. As the headway sequence "entering" Davis is uneven, trains are held at this station to achieve even departure headways and smaller waiting time even though this is not a station in the set $\{m_i\}$.

Solving the same problem with *finite* train capacities yields quite different optimal holding patterns. One reason might be that the train capacity constraint at stations with high travel demand limits the possibility of achieving perfectly even headways. However, a relatively high level of regularity in the headway distribution

---

[9] Train 1 is held 10 minutes at Porter Square, 3.3 minutes at Davis Square, 0.4 minutes at the queuing location and 2.3 at Alewife.

still exists. This supports the view that *the headway distribution must still be quite regular to be optimal.*

Minimizing Total Waiting Time

The same disruption scenario is solved for two non-zero values of μ (0.1 and 0.5), thus accounting for extra riding-time in our objective function. The results for $\mu = 0.1$ are shown in Tables 5 and 6.

Table 5. Holding Times (min): Harvard Northbound Disruption; $\mu = 0.1$, Capacity = 960 Passengers/Train

Train	HAR	POR	DAV	QUE	ALW	DAV	POR	HAR	CEN	KEN	MGH	PKS	DTX	STA	BRW	AND
0	20.0	0.0	0.0	0.0	2.0	0.0	0.0	0.0	0.0	0.0	0.0	0.0	0.0	0.0	0.0	0.0
1		10.2	0.8	0.0	4.9	1.5	0.0	0.0	0.0	0.0	0.0	0.5	0.0	0.0	0.0	0.0
2			4.9	0.0	9.4	0.8	0.0	0.0	0.0	0.0	0.0	0.0	0.0	0.0	0.0	0.0
T_1				12.1	0.5	0.3	0.0	0.0	0.0	0.0	0.0	0.0	0.0	0.0	0.0	0.0
T_2					5.4	0.2	1.6	0.0	0.0	0.0	0.0	0.0	0.0	0.0	0.0	0.0
1_R							3.9	0.0	0.0	0.0	0.0	0.0	0.0	0.0	0.0	0.0
2_R									1.8	0.0	0.0	0.0	0.0	0.0	0.0	0.0
3_R											0.0	0.5	0.0	0.0	0.0	0.0
4_R													0.0	0.0	0.0	0.0
5_R															0.0	0.0

Table 6. Preceding Departing Headway: Harvard Northbound Disruption; $\mu = 0.1$, Capacity = 960 Passengers/Train

Train	HAR	POR	DAV	QUE	ALW	DAV	POR	HAR	CEN	KEN	MGH	PKS	DTX	STA	BRW	AND
0	24.0	13.8	12.9	12.9	10.0	8.5	8.5	8.5	8.5	8.5	8.5	8.1	8.1	8.1	8.1	8.1
1		14.2	10.1	10.1	5.7	6.4	6.4	6.4	6.4	6.4	6.4	6.8	6.8	6.8	6.8	6.8
2			8.9	8.9	6.2	6.5	6.2	6.2	6.2	6.2	6.2	6.2	6.2	6.2	6.2	6.2
T_1				6.8	7.0	5.7	5.7	5.7	5.7	5.7	5.7	5.7	5.7	5.7	5.7	5.7
T_2					7.4	7.6	5.3	5.3	5.3	5.3	5.3	5.3	5.3	5.3	5.3	5.3
1_R							7.9	7.9	6.2	6.2	6.2	6.2	6.2	6.2	6.2	6.2
2_R									5.8	5.8	5.8	5.3	5.3	5.3	5.3	5.3
3_R											4.0	4.5	4.5	4.5	4.5	4.5
4_R													4.0	4.0	4.0	4.0
5_R															4.0	4.0

The main result obtained here is the striking simplicity of the optimal holding solutions: less than twenty train/station combinations are generally considered for holding. This suggests that even for small values of μ (e.g., 0.1), the costs of holding imposed on on-board passengers are large.

Moreover, trains are held only at station m_i and at a few subsequent stations, implying that early control actions yield significant benefits further down the line, since holding a train at a station not only modifies its departure headway at this station but also at later stations.[10] Hence, *holding a train at one of the earliest stations arrived at can yield significant benefits down the line and avoid the cost of holds at later stations.* As expected, holding actions are preferably applied at stations without high passenger through volumes to minimize in-vehicle passenger delay.

---

[10] The preceding train's hold also modifies it.

The model also shows that delay recovery is preferably performed at terminal stations to minimize the negative impacts of the disruption in the reverse direction: trains arriving at the terminal are held beyond the scheduled layover time but incur no (or few) later holds. Indeed, terminal holding and use of the scheduled layover time to "buffer" against the delay are preferred as no extra ride-time cost is associated with terminal holding.[11] For instance, terminal train T_1 is held 6.1 minutes more than its scheduled layover time of six minutes, and is held for only 0.5 and 0.3 minutes at Davis and Porter, respectively.

These observations are in line with operational practice, and can be contrasted with the more complex holding strategies obtained in O'Dell and Wilson (1999) and Shen and Wilson (2001). Furthermore, the resulting holding pattern is *no less efficient* than the more "complicated" holding strategies obtained for $\mu = 0$ (31.7% decrease in the objective value for $\mu = 0.1$ against 35.1% decrease for $\mu = 0$). This observation is comparable with the findings of Barnett (1978), who also highlights the simplicity of the optimal strategies derived analytically (Barnett (1978) assumed an *infinite* train capacity).

6 Comparison with a Heuristic Approach

The above solution structures may suggest that heuristics rather than a mathematical programming (MP) formulation could yield control strategies with comparable - albeit sub-optimal - total passenger waiting time. Heuristics also typically require significantly lower solution times. Such a solution technique was not investigated here, but a MP formulation is better suited to our problem for several reasons.

First, although a heuristic can strive to achieve even headways when minimizing in-station waiting time, we showed that accounting for in-vehicle waiting time presented no identifiable headway patterns. In this case, it is not clear that the knowledge of a limited number of holds at earlier stations suffices to formulate a heuristic. Second, our MP formulation provides greater flexibility in dealing with various disruption scenarios. For instance, in the case of two disruptions occurring simultaneously, only another disruption duration constraint (13) needs to be added to our formulation. Such a simple model modification is less evident in a heuristic-based approach. Third, given the small execution times presented in Section 4.3 and the simple holding strategies, it is unclear that any gain in solution times is worth achieving through heuristics, especially at the expense of control strategy optimality.

7 Model Limitations

Clearly, the model used here is *limited by the number of stations included in the model (2M − 3)*. Including only stations in the disruption direction and the reverse

---

[11] Holding has no associated costs other than the incurred additional waiting time for departure, since there are no through-standees at terminal stations ($\alpha_M = 1$ and thus, $(1 - \alpha_M)L_{i,M} = 0$).

direction may be unsatisfactory for long disruptions. In such cases, the number of trips needed to recover from the delay might be greater than the one trip considered in this model.

One could attempt to correct this limitation by "unfolding" the line more than once and setting the boundary of the system to a station with an index greater than $2M - 3$. This number could depend on the delay duration. Nevertheless, this approach obviously expands the size of the model and will increase the solution time as the delay duration increases. This would be a major impediment to the real-time tractability of this model. Additional difficulties arise from the implied longer analysis period as the system parameters (passenger arrival rates and alighting fractions) could probably no longer be assumed fixed.

8 The Short-Turning Control Strategy

Short-turning is essentially a complex control operation whereby, according to Wilson et al. (1992), "[...] a train [is turned] before it reaches its terminus with the aim of reducing headway variance in the reverse direction by filling in a large headway gap."

Indeed, in the case of longer disruptions, train capacity limits the possibility of holding trains ahead of the blockage to achieve even headways. Also, spreading a longer delay over the trains ahead results in longer headways and waiting times, which results in possible congestion concerns at stations ahead of the blockage. In this case, short-turning provides an effective (complementary) alternative to the holding strategy, by compensating for the loss of service in the peak demand direction.

The complexity of the short-turning strategy stems from selecting the set of trains to be short-turned and the sequence of trains in the after-short-turn direction that maximizes passenger time savings. The choice of the short-turned trains and their sequence varies greatly depending on the disruption location and duration, track configuration and train locations. Given this information, two types of short-turns are usually considered: short-turning ahead of, or behind, the blockage.

In all interesting cases, short-turning must provide additional train capacity to serve the Central Business District (CBD) and reduce the headway means and variances resulting from the service gap in front of the blockage. In practice, a short-turning action generally impacts four groups of passengers as identified by Wilson et al. (1992):

- Skipped segment boarders – passengers who, if the train had not been short-turned, would have boarded at stations outside the short-turn loop, in both directions.
- Skipped segment alighters – those passengers who are dumped by a short-turned train and must await a following train in order to reach their destination.
- Short-turn point boarders – those passengers who are waiting at the station before the crossover track and would have boarded a short-turned train had it continued.
- Reverse direction passengers – those traveling to the CBD who board a short-turned train.

The last group benefits from a short-turn decision while the first three groups are negatively affected. Depending on the type of short-turn, the benefits and levels of inconvenience experienced by each of these groups are different.

In this section we present the main characteristics of the two short-turning strategies and show how the previous holding model might be modified to evaluate the benefits of each type of short-turning action.

8.1 Short-Turning Ahead of the Blockage

Short-turning ahead of the blockage is considered in the AM peak period when the blockage is located before the CBD[12] as depicted in Fig. 3. In this case, trains in the reverse direction have already served the CBD and generally have low passenger loads. Hence, provided a crossover track is available between the CBD and the disruption, trains can be short-turned into the gap that is developing in front of the blockage.

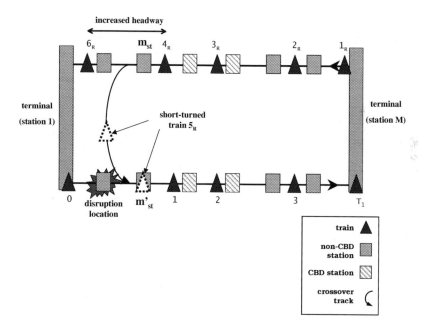

Fig. 3. Short-turning Ahead of the Blockage

In the reverse direction, at stations outside the short-turn loop, train service is reduced, resulting in headway gaps and uneven headway sequences if no further

---

[12] Most urban rail transit lines serve a CBD to which heavy passenger flows are focused during the peak periods, and the CBD generally consists of only a few stations located in the middle of the line.

control action is taken. Yet, there are a small number of these stations with low passenger flows (since the short-turn occurs near the terminal and passenger flows are focused on the CBD during the AM peak period). Hence, there are few benefits from holding trains at these stations: the uneven headway sequence would lead to a waiting time increase for the skipped segment boarders that is likely negligible in comparison to the time savings achieved in the peak direction.[13]

A similar argument – low passenger flows – holds for the negative impacts incurred by the skipped segment alighters and the short-turn point boarders. For each short-turn train, few passengers travel beyond the short-turn point and are forced to wait for another train. Moreover, due to the duration of a crossover operation, only a limited number of trains can be short-turned, so that the overall negative impacts of a short-turn option incurred by skipped segment alighters and short-turn point boarders are small, in comparison to the waiting time savings achieved in the disruption direction.

The above analysis suggests that trains in the reverse direction need not to be held to respond to the train service reduction.

In the disruption direction, trains are short-turned into the gap, behind the trains located immediately ahead of the blockage (see Fig. 3). This additional train service reduces the gap developing in front of the blockage, and thus the average headway at stations downline from the disruption. Moreover, complementary holds might further increase the benefits of the additional train service by evening out the headway sequences downline from the disruption.

Therefore, given a short-turn option – i.e., the set and sequence of trains to be short-turned – finding the complementary holds for the new train sequence simply amounts to solving a new holding problem with new train location/headway/load information. Since only a very few short-turn options are available –usually less than ten– and the corresponding holding problem can be quickly solved using our previously developed model, the best short-turning strategy can be determined in real-time.

8.2 Short-Turning Behind the Blockage

The short-turn behind-the-blockage strategy generally arises when the blockage occurs far enough beyond the CBD in the AM peak period (see Fig. 4). Trains behind the blockage then have low passenger loads and can be short-turned to service the reverse peak direction flow.

In the case of a short-turn behind strategy, we note that skipped segment alighters and short-turn point boarders incur the same detrimental effects of the short-turn decision, i.e., increased in-station waiting time. Nevertheless, removing trains from behind the blockage has specific consequences as described below.

First, the skipped segment boarders are affected by the train service reduction only if they would have boarded a short-turn train at a station located between the crossover track and the blockage. At stations located downline from the blockage,

---

[13] The results presented by Shen and Wilson (2001) provide support for such a statement.

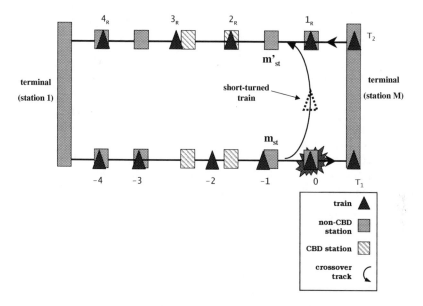

Fig. 4. Short-turning Behind the Blockage

passengers would board the first blocked train (train 0), assuming there is no train capacity issue at these stations (passenger arrival rates are low at these stations since the blockage is located near the terminal).

Second, and more importantly, train service removal can free platforms behind the blockage and limit the propagation of the queue of trains developing behind the blockage. If the disruption is long, this queue could propagate to the CBD area and hinder travel to the CBD. Thus, depending on the delay duration, removing trains from behind the blockage can yield benefits (decreased in-vehicle delay time) in the disruption direction. This beneficial consequence of short-turning in the before-short-turn direction was not relevant in the case of short-turning ahead since the end of the line was located between the short-turn location and the blockage: the terminal provided an additional platform for trains to queue behind the blockage and trains might be pulled out of service to a yard at the terminal.

In the after-short-turn direction, the new train sequence must achieve overall benefits from the additional train service. Nevertheless, this task is made more difficult in this short-turning case because there is no natural gap into which trains can be short-turned (see Fig. 4). Trains in the reverse direction are operating with a normal service headway of four minutes, which means that either train 1_R or train T_2 might have to be held to create a gap into which train -1 could be short-turned.

Hence, it appears that the choice of the train sequence must balance the following elements:

- The cost of holding trains travelling in the reverse direction to the CBD.
- The waiting time benefits from the additional trains in the peak direction.
- The negative effects of holding short-turned trains behind the blockage to achieve the desired train sequence, as trains can queue up behind the blockage.

We note that these tradeoffs are difficult to assess in general. Moreover, we recognize that, even for a given set of short-turn trains and a predetermined train sequence, the more complicated train sequence generally achieved in the after-short-turn direction does not lend itself to a simple use of the holding model to determine the optimal complementary holds. One reason is that holding trains ahead of the blockage (in both directions) might affect the train sequence that can be achieved, as timing is a critical factor for more complicated train sequences. Another reason is that several trains might now be preceded by a short-turn train, which makes short-turned trains difficult to represent in the holding model.

9 Conclusion

In this paper, we have developed a simple mixed integer programming formulation of the train holding problem. By designing a two-step solution procedure, we addressed the tradeoff between minimizing a linearly approximated cost function in real-time and large solution times for the non-linear program formulation. The running time of this procedure was comparable to the solution times obtained for a linearly approximated objective function.

Furthermore, results from the model implementation suggested that control strategies which minimize the non-linearized cost function are sufficiently simple to be implemented by transit practitioners.

We also presented a general analysis of the short-turn control strategy and differentiated two types of short-turning: short-turning ahead of the blockage and short-turning behind the blockage.

It was shown that the short-turn ahead strategy is generally the simplest to assess and that the holding model developed in this paper can be used to determine the complementary holds that optimize the benefits of any given short-turn ahead decision. The short-turn behind strategy was shown to be more difficult to assess for it involves many tradeoffs that need to be made simultaneously and does not lend itself to a simple use of the holding model.

To remedy this shortfall, a model based on modified headways similar to the holding model could be developed, but the difficult problem of train reordering must be addressed for this purpose. Since such a model is likely to use additional integer variables, methods based on simple logical considerations similar to the ones developed in the holding model's two-step solution procedure could be effective in pruning the solution tree and reducing the solution times.

References

Barnett, A. (1978). Control strategies for transport systems with nonlinear waiting costs. *Transportation Science*, **12**(2), 119–136.

Lin, T.-M. and Wilson, N. (1992). Dwell-time relationships for light rail systems. *Transportation Research Record*, **1361**, 287–295.

O'Dell, S. and Wilson, N. (1999). Optimal real-time control strategies for rail transit operations during disruption. In N. Wilson, editor, *Lecture Note in Economics and Mathematical Systems*, volume 471, pages 299–323, Berlin. Springer.

Shen, S. and Wilson, N. (2001). An optimal integrated real-time disruption control model for rail transit systems. In S. Voß and J. Daduna, editors, *Computer-Aided Scheduling of Public Transport*, volume 505 of *Lecture Notes in Economics and Mathematical Systems*, pages 335–364, Berlin. Springer.

Welding, P. (1957). The instability of close interval service. *Operations Research Quarterly*, **8**(3), 133–148.

Wilson, N., Macchi, R., Fellows, R., and Deckoff, A. (1992). Improving service on the MBTA green line through better operations control. *Transportation Research Record*, **1361**, 296–304.

The Holding Problem at Multiple Holding Stations

Aichong Sun[1] and Mark Hickman[2]

[1] Pima Association of Governments, 177 N. Church St., Tucson, AZ 85712, USA
asun@pagnet.org
[2] University of Arizona, P.O. Box 210072, Tucson, AZ 85721-0072, USA
mhickman@engr.arizona.edu

Summary. Inherent stochasticity within the transit operating environment suggests there may be benefits of holding vehicles at more than one holding station on a route. In this paper, the holding problem at multiple holding stations considers holding vehicles at a given subset of stations on the route. By approximating the vehicle dwell time as the passenger boarding time, the holding problem at multiple holding stations can be modeled as a convex quadratic programming problem, with the objective function as a convex quadratic function subject to many linear constraints. This particular problem can be solved by a heuristic that decomposes the overall problem into sub-problems which can be solved to optimality. Also, a hypothetical numerical example is presented to illustrate the effectiveness of the problem formulation and heuristic.

1 Introduction

Traffic congestion has become increasingly common in central urban areas, and transit ridership has been continuing to grow. As a result, public transit service has become more subject to the on-street traffic environment, and transit agencies may find it more difficult to maintain the vehicle schedule. In order to reduce the impact of schedule disruptions and disturbances, transit agencies often employ control strategies to reduce overall system cost, from the perspectives of both operators and passengers. Among these strategies, holding control is the most commonly used strategy by transit agencies in practice. Holding involves keeping a vehicle at a station for a period of time, in order to improve the service performance.

Barnett (1974) developed a model for holding a vehicle at a chosen control point. He proposed a solution algorithm for constructing an approximately optimal dispatching strategy from the control point in terms of minimizing both at-stop and in-vehicle passenger delay. This strategy is a threshold: if the vehicle headway is less than the threshold, the vehicle is held until the threshold. If the vehicle headway is greater than the threshold, the vehicle is dispatched immediately. Barnett's algorithm was tested on actual operation data from a Boston subway line to propose service improvements. Abkowitz and Tozzi (1986) conducted a study to evaluate the sensitivity of headway-based holding control to varying boarding and alighting profiles, headways, and other characteristics of route operations. They found that profiles with passengers boarding at the middle and alighting at the end of a route produce the most significant passenger waiting time savings with holding control. Also, increases in the initial headway variation and the amount of parking permitted along a route leads to worse service reliability; thus, holding strategies can be more effective in these situations. At about the same time, Abkowitz *et al.* (1986) investigated the effects of a threshold-based holding control strategy on reducing the headway variation at stops downstream of the control point. Their simulation results indicate that the headway variation does not increase linearly along a route. Also, the study results showed that it is preferable to locate the control point just prior to a group of stops where many passengers are boarding. Also, the threshold headway is sensitive to the number of passengers onboard the bus at the control point. In addition, this study concluded that the optimal holding control could result in a 3-10 % reduction in total passenger waiting cost. Later, Abkowitz and Lepofsky (1990) conducted a before-after study to evaluate the effectiveness of the threshold-based holding strategy on several real-life bus routes chosen from the MBTA in Boston. The results from this study were not conclusive; however, it appeared that certain route segments might have benefited from the holding actions. O'Dell and Wilson (1999) developed a deterministic model of a rail system and mixed integer programming formulations for the holding and short-turning problems. Three holding strategies, holding each train at any station, holding each train at the first station it reaches after the disruption occurs, and holding each train at an optimally chosen station, were considered and formulated. Study results based on the MBTA Red Line showed that passenger waiting time can be significantly reduced by applying the controls.

With the advent of AVL (Automatic Vehicle Location) and APC (Automated Passenger Counting) technologies, real-time vehicle location information is incorporated by many researchers into their studies. Furth (1995) developed a strategy to deal with a vehicle operating behind schedule, given the existence of an intelligent system providing information about vehicle location, vehicle load, and number of passengers waiting at stops. In his study, the problem is formulated as a constrained, non-linear optimization problem to decide how many vehicles following the initially delayed vehicles should be held; the location at which each vehicle should be held; and, the amount by which each vehicle should be held. Study results showed that the optimal solution is a gradual increase in the overall headway from the first vehicle, whose headway is short, until the last vehicle, with headway returning back to the base headway. Ding and Chien (2001) formulated a real-time operational control model in which the vehicle departure time at each stop is optimized so that the headway variance, weighted by passengers at each stop, can be reduced. The proposed real-time control model was tested by simulation based on a high frequency light rail transit route in the city of Newark, New Jersey. The simulation results demonstrated that the average passenger waiting time can be significantly reduced by applying the proposed control model.

Hickman (2001) presented an analytical model for optimizing the holding time at a given control point in the context of a stochastic vehicle operations model. In this study, the single vehicle holding problem is a convex quadratic program in a single variable, and is easily solved using gradient or line search techniques. Eberlein *et al.* (2001) also formulated an analytic model using a rolling-horizon approach, using real-time AVL vehicle location information. The problem can be effectively solved by a proposed heuristic. The study results showed significant reductions of passenger waiting time at stops. Fu and Yang (2002) investigated both the threshold-based holding control model and an optimal holding control model by considering both a vehicle's preceding and following headways, with the assumption that the future bus arrival time at the control stop can be predicted with real-time location information. Based on a simulation, the study results indicated that: the control point should be placed at the bus stop with high demand and located close to the middle of the route; two control points are preferable to one; holding control is fairly robust with respect to the control parameter, control strength or headway threshold; and, real-time bus location information can help reduce passenger in-vehicle time and bus travel time when a number of control points are used.

Zhao *et al.* (2001) present a distributed control approach based on multi-agent negotiation (between bus agent and stop agent) for addressing the holding problem. The negotiation in this study is conducted based on the marginal cost and marginal benefit of a hold, negotiated between a vehicle and the set of stops on the route. Also, the comparison between the negotiation algorithm and other commonly used strategies was conducted through simulation, and study results indicated that the negotiation algorithm is robust to different transit operating environments.

From the literature review above, one may see that it is commonly concluded that holding can undoubtedly improve the performance of transit service by diminishing the vehicle headway variance and schedule deviation, and hence can reduce passenger cost, if the control location is judiciously selected. However, some of the previous studies also pointed out either explicitly or implicitly that the transit operating stochasticity still plays a role on the vehicle's trajectories downstream from the control point after holding is applied. Based on the equations developed in their study, Abkowitz *et al.* (1986) concluded that:

> The reduction in headway variation at points downstream of the control point is not uniform. The maximum benefits of the control strategy are accrued by passengers at stops immediately downstream of the control point. Stops that are far from the control point may not be impacted significantly. (pp. 78-79)

Furthermore, Turnquist and Blume (1980) showed that there might be multiple points qualifying as holding point candidates along the route. Though not clearly indicated in the study, choosing one qualified location as the control point does not imply that the others cannot still qualify as additional control points, even when some correlation certainly exists between the potential holding points. Abkowitz and Tozzi (1986), Abkowitz *et al.* (1986) and Fu and Yang (2002) all define desirable conditions for a control point to hold selected vehicles. A route may have favorable conditions on separate segments, which might justify multiple control points.

Moreover, it has been assumed by a majority of the previous studies that the transit vehicle trajectories downstream of the holding station can be predicted precisely with the currently available information, typically from AVL technology; or, the vehicle trajectories can be predicted by the best-fit probabilistic distribution calibrated with historical data, if they are subject to random variation. However, in reality, as the transit vehicle's running time and dwell time may be both subject to significant variability, it becomes fairly difficult, if not impossible, to precisely predict vehicle trajectories far downstream from the holding station.

Seneviratne and Loo (1986) have analyzed the vehicle travel time data from two transit routes in Halifax, Nova Scotia, Canada, and found that fundamental to a realistic analysis of a bus route is proper segmentation; that is, routes may be broken into route segments within which operations are fairly consistent. To examine this, a preliminary analysis of bus AVL data from Tucson, Arizona is shown in Fig. 1.

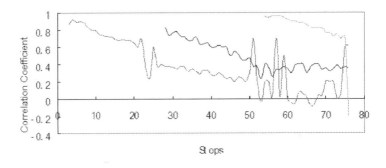

Fig. 1. Schedule Deviation Correlation at Stops

Fig. 1 is based on the AVL data collected by SunTran on Route 8 in Tucson, Arizona. The vertical axis represents the correlation coefficient between the schedule deviation at a specific upstream stop and all other downstream stops. In the figure, the bottom line, middle line and the top line represent the schedule deviation correlation between the 2nd, 28th and 54th stops and all downstream stops, respectively. For the purpose of clarity of presentation, only three lines are presented. Nonetheless, we can still see that the correlation coefficients drop abruptly at two stops, Stop 25 and Stop 53. Also, the correlation between the stops on route segments between these two stops and between Stop 25 and 53 and the terminals appears relatively strong. The reason behind this phenomenon in this particular example is due to the special nature of the two stops. Stop 25 is the downtown transit center, Stop 53 is a short-turn terminal, and a large amount of slack time has been built in the schedule at both stops for service recovery. However, one may see similar phenomena at other places, and the underlying reasons could be varying, e.g., different traffic conditions. This shows that the vehicle's schedule deviation can only be predicted accurately using the currently available information, typically the schedule deviation at the most recently passed stop, if this stop and the stops at which the schedule deviations need to be predicted are on the same route segment.

In summary, it appears possible that holding control can be implemented effectively at multiple stations, especially when the transit route is relatively long with many stops. This conclusion is based on the premise that separate route segments may need separate operations control actions.

Holding vehicles at multiple holding stations can essentially be seen as a three-dimensional decision problem: the vehicle holding time at a particular stop is one dimension, the vehicles to control are the second dimension, and the holding stations to use are the third dimension. Eberlein *et al.* (2001) presented a comparison of the benefits from holding vehicles at multiple holding stations versus only one holding station and concluded that holding a vehicle at more than one holding station did not show any significant advantages, using a numerical example based on a real-life transit route. However, the observation may not be conclusive due to the limitations of the selected passenger loading/boarding profile.

In this paper, we examine two dimensions of the problem: the holding times of multiple vehicles at a given set of holding stations. This problem is formulated, and a heuristic is proposed to solve for the optimal holding times.The remainder of the paper is organized into three sections. Section 2 formulates the general holding problem with either a single holding station or multiple holding stations. A heuristic based on an analytical model is also described in this section. Section 3 provides a hypothetical numerical example designed to demonstrate the effectiveness of the algorithm developed in Section 2. Finally, Section 4 concludes the study and presents the direction for future research.

2 Problem Formulation and Solution

As argued in Eberlein *et al.* (2001), the holding control problem can be formulated in the context of a deterministic model of transit operations. In a similar manner, the problem formulation in this study will also use a deterministic model.

2.1 Model Formulation

For the sake of simplifying the analysis that follows, several assumptions are made:

- The passenger boarding time dominates passenger alighting time at most stops or stations along the route. Therefore, the total passenger boarding time can be used as the vehicle dwell time.
- Vehicle overtaking is not a factor.
- The passenger arrival rate at any stop and vehicle average travel time between adjacent stops are given during the time period of interest.

- The number of alighting passengers at a stop is proportional to the number of passengers onboard.
- Vehicle capacity is not considered.

One may argue with the second assumption of no vehicle overtaking, but this assumption can be justified when:

- Transit service is provided at a high frequency, but the average headway is still relatively large, e.g., larger than five minutes.
- Traffic conditions do not change abruptly during the time period of interest, so that vehicle running times only differ randomly from one trip to another.

Therefore, no vehicle overtaking can be assumed in situations likely to satisfy the conditions above. In addition, holding control at multiple holding stations can help regularize vehicle trajectories, which greatly reduces the chance for vehicle overtaking to occur. This will be further discussed later in the paper.

Before we get to the problem formulation, major variables are defined below.

i, j, k	Indicators of the holding station, vehicle, and stop, respectively
$h_{j,k}$	Leading headway for the jth vehicle at Stop k
$d_{j,k}$	Departure time for the jth vehicle at Stop k
$a_{j,k}$	Arrival time for the jth vehicle at Stop k
$L_{j,k}$	Onboard passengers of the jth vehicle when it departs from Stop k
s_i	Index of the ith holding station, as a stop
H_{j,s_i}	Holding time for the jth vehicle at holding station s_i
$B_{j,k}$	Passengers boarding the jth vehicle at Stop k
$A_{j,k}$	Passengers alighting from the jth vehicle at Stop k
λ_k	Passenger arrival rate at Stop k
r_k	Vehicle running time between Stop k and Stop $k + 1$
q_k	Passenger alighting proportion at Stop k
$DWL_{j,k}$	Dwell time for the jth vehicle at Stop k
α, β	Parameters defining the passenger boarding process represented by $DWL_{j,k} = \alpha + \beta \cdot B_{j,k}$
b_i	Index of the earliest dispatched vehicle among those operating on the segment $(s_{i-1}, s_i]$
e_i	Index of the latest dispatched vehicle among those operating on the segment $(s_{i-1}, s_i]$
M	Total number of holding stations
N	Total number of stops on the route
P	Total number of vehicles on the route, indexed $\{1, 2, \ldots, P\}$. Vehicle P is the last vehicle, waiting to be dispatched at the terminal.
S	The set of holding stations on the route $\{s_1, s_2, \ldots, s_M\}$

Within an entirely deterministic context, it is meaningless to consider holding one vehicle at all holding stations within one decision-making cycle, because all effects resulting from the hold can be achieved by holding the vehicle at the first holding station to which it arrives. More specifically, with M holding stations available, the transit route can be divided into $M + 1$ segments, either bounded by two consecutive holding stations as $(s_i, s_{i+1}]$, or by a terminal and a holding station as $(1, s_1]$ or $(s_M, N]$. On each segment $(s_i, s_{i+1}]$, vehicles in the set $[b_i, e_i]$ are the vehicles to be considered for holding at station s_i. It is assumed that all vehicles within this set will only be held at this holding station s_i in one holding decision. Obviously, those vehicles operating on the segment $(s_M, N]$ are free of any control.

In short, the multiple holding station problem can be described as:

At any decision time, the holding times are determined only for vehicles at the immediate downstream holding station, where multiple holding stations are available.

With the assumptions and variable definition above, the holding problem can be formulated as follows.

$$Minimize \ Z = \frac{1}{2} \cdot \sum_{i=1}^{M} \sum_{j=b_i}^{e_i} \sum_{k=s_i}^{N} \lambda_k \cdot (d_{j,k} - d_{j-1,k})^2 \tag{1}$$

$$+ \frac{1}{2} \cdot \sum_{k=s_1}^{N} \lambda_k \cdot (d_{P,k} - d_{P-1,k})^2 + \sum_{i=1}^{M} \sum_{j=b_i}^{e_i} L_{j,s_i-1} (1 - q_{s_i}) \cdot H_{j,s_i}$$

In this objective function, the first two components represent the total passenger waiting time at stops, and the third term defines the delay experienced by the onboard passengers at the holding stations. Though not salient, it can be seen in the objective function that the departure times $d_{j,k}$ of vehicles $j \in [b_i, e_i]$ at each holding station $k = s_i$ are the decision variables.

Each vehicle's departure time at any stop other than the holding station to which it "belongs" (e.g., $j \in [b_i, e_i]$ belongs to holding station s_i) is entirely deterministic: the arrival time and dwell time at these stops can be determined directly, once the holding times are known. Also, the dwell time in turn is essentially defined by the time when the preceding vehicle departed as well as the passenger arrival rate at the stop.

If $k \notin S$ (k is not a holding stop) or if $k = s_i \in S$ but $j \notin [b_i, e_i]$ (j is not available for holding at s_i), then the departure time of j at k is given by:

$$d_{j,k} = (d_{j,k-1} + r_{k-1} + \alpha - \beta \cdot \lambda_k \cdot d_{j-1,k})/(1 - \beta \cdot \lambda_k) \tag{2}$$

Equation (2) can be directly derived from the relationship below:

$$d_{j,k} = d_{j,k-1} + r_{k-1} + \alpha + \beta \cdot \lambda_k \cdot (d_{j,k} - d_{j-1,k}) \tag{3}$$

Otherwise, i.e., for those vehicles at the holding stations, the vehicle holding time will together define the vehicle's departure time. However, any vehicle j cannot be held later than the time when vehicle $j + 1$ arrives, to avoid overtaking:

$$d_{j,k} = (d_{j,k-1} + r_{k-1} + \alpha - \beta \cdot \lambda_k \cdot d_{j-1,k})/(1 - \beta \cdot \lambda_k) + H_{j,k} \qquad (4)$$

$$d_{j,k} \leq d_{j+1,k-1} + r_{k-1} \qquad (5)$$

$$H_{j,k} \geq 0 \qquad (6)$$

The number of onboard passengers when a vehicle departs from a stop is determined by the number of passengers boarding and alighting at the stop and the number of onboard passengers when the vehicle arrived at the stop.

$$L_{j,k} = L_{j,k-1} + B_{j,k} - A_{j,k} \qquad (7)$$

The number of passengers boarding a vehicle is the product of the average passenger arrival rate and the vehicle's leading headway.

$$B_{j,k} = \lambda_k \cdot (d_{j,k} - d_{j-1,k}) \qquad (8)$$

The number of passengers alighting a vehicle is assumed to be proportional to the number of onboard passengers.

$$A_{j,k} = L_{j,k-1} \cdot q_k \qquad (9)$$

Equations (7), (8) and (9) can be combined into a single equation:

$$L_{j,k} = L_{j,k-1} \cdot (1 - q_k) + \lambda_k \cdot (d_{j,k} - d_{j-1,k}) \qquad (10)$$

In the model formulation above, the decision variables can be either the vehicle holding times at holding stations or equivalently the vehicle departure times at holding stations, due to the linear relationship between them. From now on, in this paper, the decision variables are the departure times of vehicles $[b_i, e_i]$ at each holding station s_i, and according to Equation (4) are modified into the following inequality.

$$d_{j,k} \geq (d_{j,k-1} + r_{k-1} + \alpha - \beta \cdot \lambda_k \cdot d_{j-1,k})/(1 - \beta \cdot \lambda_k) \qquad (11)$$

In the objective function, the holding time can be replaced by:

$$H_{j,s_i} = d_{j,s_i} - (d_{j,s_i-1} + r_{s_i-1} + \alpha - \beta \cdot \lambda_k \cdot d_{j-1,s_i})/(1 - \beta \cdot \lambda_{s_i}) \qquad (12)$$

Equations (2) – (12) together define the feasible region for each decision variable. Specifically, inequalities (11) and (5) together set the lower bound and upper bound, respectively, for the decision variables.

2.2 Proposed Heuristic

With the problem definition and formulation in the previous sub-section, one may see that the departure time of a vehicle within a control vehicle group $[b_i, e_i]$ at the stops on the downstream segment $[s_i, s_{i+1})$ is determined by a subset of the decision variables as follows.

$$d_{b_i,k} = f(d_{b_i,s_i}) \text{ if } s_i \leq k < s_{i+1} \tag{13}$$

$$d_{b_i+j,k} = f(d_{b_i,s_i}, d_{b_{i+1},s_i}, \ldots, d_{b_{i+j},s_i}) \text{ if } s_i \leq k < s_{i+1} \text{and } b_i + j \leq e_i \tag{14}$$

$f(\bullet)$ is a linear function of the decision variables. Furthermore, the departure times of vehicles $[b_i, e_i]$ at the stops further downstream of the subsequent holding station, say s_{i+m} , will be determined by more decision variables as follows.

$$d_{b_i+j,k} = f(d_{b_i,s_i}, d_{b_{i+1},s_i}, \ldots, d_{b_{i+j},s_i}, d_{k,s_{i+1}}) \tag{15}$$

$$\text{for } k \in [b_{i+1}, e_{i+1}], \ldots, [b_{s_m}, e_{s_m}]$$

With the variable description in (13) – (15), it becomes clear that the problem formulation has a general form of:

$$Minimize \ \ Z = F(\bullet) + f(\bullet) \tag{16}$$

$$\text{subject to: } g_j(\bullet) \leq C_j \quad \forall j$$

Herein, $g_j(\bullet)$ is also a linear function of decision variables; $F(\bullet)$ is a quadratic function of the decision variables; $f(\bullet)$ again is a linear function of the decision variables; C_j is constant; and, j varies from 1 up to double the number of vehicles upstream of the most downstream holding station, since each decision variable is subject to two constraints of the form of inequalities (11) and (5). Therefore, this problem formulation is essentially a convex problem with a convex objective function and a set of linear constraints. Such a problem can be solved to optimality by many classical techniques. However, the scale of the problem is not necessarily small when the route is long with many stops and many vehicles operating at the same time.

This paper presents a solution algorithm by decomposing the overall problem into several two-dimensional problems smaller in scale. Furthermore, the two-dimensional problem is further decomposed into one-dimensional problems, which eventually can be solved analytically.

Before getting into the details of the algorithm, a proposition regarding vehicle overtaking is presented.

Proposition 1 *Let h_2 and h_3 be the real headways of Vehicles 2 (the control vehicle's first following vehicle) and 3 (second following vehicle), respectively. If $h_2 \geq h_3 \cdot \beta \cdot \lambda_k/(1 - \beta \cdot \lambda_k)$ holds, the real objective value is always less than the model objective value on the route segment downstream of where vehicle overtaking occurs.*

The condition in the proposition is tighter than is needed. The proof of the proposition is presented in the Appendix.

Since the proposed model formulation does not explicitly include overtaking, this proposition states that a solution to the model formulation will have a larger (or higher) objective value than would occur if overtaking were included. In this way, our model formulation is more conservative, in that it will recommend holding actions that result in smaller improvements than if overtaking were included explicitly.

The following sub-sections start with the simplest problem, holding a single vehicle at a single holding station, then gradually add complexity to the problem to achieve the full problem solution for multiple vehicles at multiple stations.

Holding a Single Vehicle at a Single Holding Station (*PSS*)

The complexity of the holding problem lies in the fact that any adjustment to the departure time of one particular vehicle at a stop will in turn change this vehicle's trajectory downstream of the stop, and also affect many following vehicles' trajectories. Therefore, while considering holding one particular vehicle, it is also necessary to account for the following vehicles (impacted vehicles), as well as the leading vehicle, which functions as a boundary vehicle in the solution. If we expand the impacted vehicles up to the first non-dispatched vehicle P, all vehicles upstream of the holding station can be categorized into two groups:

- Holding Group: the vehicles within this group will be considered for holding.
- Non-Holding Group: the vehicles within this group will not be held, but define the conditions for the holding control decisions for the holding group.

For the problem of holding one vehicle at a single holding station, only one control vehicle is within the holding group, and the non-holding group consists of all other impacted vehicles, including the first non-dispatched vehicle and the boundary vehicle immediately ahead of the control vehicle. Accordingly, the PSS can be seen as a one-dimensional problem due to the unique decision variable.

Though presented for the overall problem, problem formulation (16) and (13) – (15) can still apply to the PSS problem. Obviously, all impacted vehicle trajectories downstream of the holding station can be derived with equations of the same form as (15). A univariate convex problem can be easily solved by many techniques. However, since the PSS problem solution is the core of the overall heuristic, an analytical solution is employed to solve the PSS problem in this particular study. The global optimal solution to PSS is either at the local optimal point of the objective function, if it exists, or at one of the extreme points.

Holding Multiple Vehicles at a Single Holding Station (*PMS*)

As more than one vehicle is included in the holding group for a single holding station, the holding problem becomes the PMS problem. For a particular holding station s_i, the set of vehicles $[b_i, e_i]$ constitutes the holding group, and all vehicles following the vehicle e_i up to the first non-dispatched vehicle P make up the non-holding group.

Equation (11) says that the decision variables are dependent on each other ($d_{j,k}$ is dependent on $d_{j-1,k}$). Therefore, for the general form of the problem (16), each of the linear constraints may include multiple decision variables. To make the concepts clearer and to simplify the problem, some special treatment is applied to the transit holding station.

Observing Equation (4), theoretically, holding control can be realized either by postponing the vehicle departure time for $H_{j,k}$ at the holding station, or by delaying the vehicle arrival time by an equivalent amount of time $H_{j,k} \cdot (1 - \beta \cdot \lambda_k)$.

If holding control is considered as a means to delay the vehicle's arrival time, the holding problem becomes an equivalent problem of how to optimize the vehicle arrival time at the holding station. As one may know, delaying one vehicle's arrival time at a stop would not affect the arrival times of other impacted vehicles. To clarify this idea, a simple treatment on the route and station is made by introducing a dummy stop to separate the vehicle arrival process and departure process at the real holding station. This dummy stop is inserted just upstream of the holding station to represent the vehicle arrival process, and will function as a surrogate for the original holding station, as shown in Fig. 2.

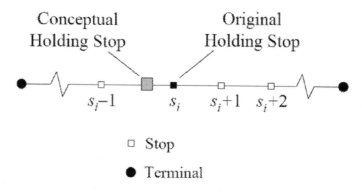

Fig. 2. Typical Transit Route with Multiple Holding Stations

With this "physical" treatment:

- The original holding station becomes a regular stop. Furthermore, it is assumed that all passenger boarding and alighting still occurs at the original control stop, with none at the dummy stop. The dummy link connecting the dummy stop and the original holding station has a length of zero.
- The dummy stop becomes the holding station, at which the vehicle arrival times are identical to the departure times if no control is implemented. The vehicle arrival times at the dummy stop then are independent of each other.
- The transit route operating process (the process of propagating arrival and departure times at downstream stops) remains the same as before any treatment is applied.
- The control vehicles' holding times are independent of each other, since no boarding and alighting occurs at the dummy stop and the interdependency of the holding times has to be realized through the passenger boarding and alighting process, as one may see from Equation (3).

However, it must be pointed out that the final observation only holds when the assumption that vehicle overtaking does not occur is strictly satisfied, because the dummy stop treatment can still result in vehicle overtaking at the original holding station. The dummy stop treatment itself does not change the essential nature of the problem, but adds a little more conceptual clarity. If the holding control at the dummy stop does not lead to vehicle overtaking at the original holding station, the holding times are certainly independent of each other at the original holding station even without the dummy stop treatment. However, as argued in Proposition 1, vehicle overtaking will occur only rarely in the given problem context.

With all treatments introduced above, the PMS problem still has a convex objective function with linear constraints. However, within the constraints, the decision variables are entirely independent of each other. With this additional characteristic, a solution algorithm for the PMS problem is developed. The solution algorithm basically decomposes the PMS problem into successive PSS problems, with each problem being to hold only one vehicle which can reduce the overall objective value the most. It finally converges at the point at which no additional holding control for any vehicle can reduce the objective value.

Step 1: Initialization.
    Set a threshold for algorithm convergence;
    Predict the current departure times at the holding station for all vehicles in the
        holding group, and set these current departure times as the Departure Time
        Lower Limit (DTLL). At the same time, DTLL will also function as the
        Departure Time Upper Limit (DTUL) for the preceding vehicles;
    Set the current departure times as the Solution 1;
    Compute the total passenger cost based on Solution 1, and set this passenger
        cost as the Previous Passenger Cost (PPC);
    Set $n = 2$.
Step 2: For iteration n:
    Optimize the departure time for each individual vehicle within the holding group
        $[b_i, e_i]$ by solving the PSS problem analytically for each vehicle sequentially,
        with all other vehicles' departure times the same as in solution $n - 1$.
Step 3: If all optimized vehicle departure times in Step 2 are earlier than, or the same as,
        in solution $n - 1$, go to Step 5;
    otherwise,
    Identify the departure time that leads to the minimum total passenger cost among
        all departure times;
    Update the corresponding vehicle departure time in solution $n - 1$ with this
        identified new vehicle departure time; and, set the minimum total passenger
        cost as the Current Passenger Cost (CPC); .
Step 4: Check the proximity of the CPC to PPC. If CPC is within the convergence threshold
        of PPC, go to Step 5; otherwise, PPC = CPC, $n = n + 1$, and go to Step 2;
Step 5: Stop.

Fig. 3. Algorithm H1

In more detail, solution algorithm H1 is described in Fig. 3. Following the steps of Algorithm H1, in each iteration, each vehicle's departure time is optimized conditional on other vehicles' departure times inherited from the last iteration, and H1 captures the most "efficient" vehicle's departure time to conclude the iteration. The interacting behavior between all control vehicles' departure times is hence realized by consecutive iterations.

Based on the Algorithm H1, Proposition 2 is introduced.

Proposition 2 *H1 solves the problem PMS to optimality.*

As has already been stated, the PMS problem is convex. It is also straightforward to show that the algorithm H1, by successive improvement of each departure time at each iteration, satisfies the Karush-Kuhn-Tucker (KKT) conditions in the final solution. A formal proof is given in Sun (2005).

Holding Multiple Vehicles at Multiple Holding Stations (*PMM*)

As a final extension of the previous two problems, the full problem is to hold multiple vehicles at multiple holding stations (PMM). As introduced earlier, holding multiple vehicles at multiple holding stations does not consider holding each vehicle at all downstream holding stations in one decision-making cycle. Instead, each vehicle is only considered to be held at the immediate downstream holding station. However, even with such a simplification, the problem becomes more complicated since the departure time d_{e_i,s_i} of the last control vehicle e_i of the downstream holding station s_i is always dependent on the departure time $d_{b_{i-1},s_{i-1}}$ of the first control vehicle b_{i-1} from its immediately upstream holding station s_{i-1}, and vice-versa. Recognizing this, heuristic H2 (see Fig. 4) is developed to search for a solution which can approximate the global optimum to the full problem.

This heuristic decomposes the overall problem into PMS problems first, then iterates to mimic the interaction among the control vehicles b_{i-1} and e_i at different holding stations. In more detail, the heuristic H2 is described below.

Always starting with the most downstream holding station in each iteration at Step 2, the heuristic solves the PMS problem for each holding station sequentially in descending order. As described in the heuristic, when the heuristic solves the PMS problem for a particular holding station s_i, all trajectories of the control vehicles belonging to all its upstream holding stations will function either as a boundary vehicle(s) or impacted vehicles. Certainly, the trajectories of the boundary vehicle(s) and impacted vehicles affect the solution of the PMS, and the revision of these trajectories is the essence of the iterative process in H2. The heuristic eventually converges at the point at which the objective cannot be improved significantly by changing any vehicle's departure time at the corresponding holding station.

Proposition 3 *If no vehicle $e_i; i = 1, ..., M - 1$, has a trajectory that is bound by the immediately following vehicle's arrival time, algorithm H2 solves the PMM problem to optimality.*

The proof of Proposition 3 follows a similar method as for Proposition 2, and is presented in Sun (2005).

Step 1: Initialization.

    Set a threshold for algorithm convergence;

    Check all en-route vehicles. Set $[b_i, e_i]$ as the holding group and all following
    vehicles up to the first non-dispatched vehicle in the non-holding group,
    for each holding station s_i;

    Predict all en-route vehicles' trajectories without holding, and set all vehicles'
    departure times at the corresponding holding stations together as Solution 1;

    Compute the total passenger cost based on Solution 1, and set it as the
    Previous Passenger Cost (PPC);

    Set $n = 2$;

Step 2: For iteration n.

    for $i = M$ to 1

        Solve the single holding station problem PMS by using H1 for holding station s_i,
        based on the solution $n - 1$.

        Update the corresponding terms in the solution $n - 1$ with the new optimized
        departure times for $[b_i, e_i]$ at holding station s_i.

    end

Step 3: Solution n = Solution $n - 1$;

    Compute the total passenger cost based on the solution n, and set it as the
    Current Passenger Cost (CPC);

    Compare CPC and PPC. If CPC is within the convergence threshold of PPC,
    go to Step 4; otherwise, PPC = CPC, $n = n + 1$, and go to Step 2.

Step 4: Stop.

Fig. 4. Algorithm H2

3 Numerical Example

In this section, using a hypothetical example, numerical results are given to demonstrate the problem formulation and solution. The test bus route is shown in Fig. 5.

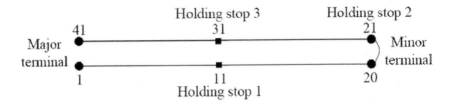

Fig. 5. Test Transit Route

The basic characteristics of this test route are:

- It has a major terminal and a minor terminal. Vehicle layover times occur only at the major terminal, and the minor terminal merely functions as an intermediate stop for the vehicle to turn around. Therefore, it is preferable to integrate the two directions since they are highly correlated from the operating perspective.

- There are a total of 40 stops (including terminals) on the transit route, 20 in each direction. Because the two directions are essentially treated as one continuous route in the following analysis, the major terminal will be double-counted as both the starting point and the end point. Therefore, a total of 41 stops will be shown in the analysis that follows.
- The one-way trip time is about one hour in each direction, and the average vehicle headway is ten minutes. Accordingly, there are twelve vehicles operating on the route at the same time.
- There are a total of three holding stations evenly spaced along the route, with one at Stop 11 (Station 1), another at Stop 21 (Station 2), and the last at Stop 31 (Station 3).

The passenger arrival profile is depicted in Fig. 6. This passenger arrival profile can result in a relatively even passenger loading profile along the route, provided that the headway is perfectly even everywhere.

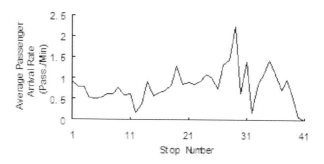

Fig. 6. Passenger Boarding Profile Along Route

Other parameters are given in Table 1.

Table 1. Operating Factors

Operating Parameters	Values
α, β (sec)	2, 2
Threshold Cost Value for PMS (Pass-Min)[1]	20
Threshold Cost Value for PMM with M Holding Stations (Pass-Min)[1]	$20 \cdot M$
Decision-Making Time Instant (Min)[2]	120

[1] The threshold cost values are set for the purpose of checking the convergence of algorithms H1 and H2.

[2] It is assumed that the first vehicle is dispatched at time 0; after 120 min the first vehicle is returning to the dispatch terminal.

The following analysis is only intended to demonstrate the problem formulation and solution. Therefore, only the results from one decision making at a specific time instant are given for illustration.

With this hypothetical route, at the time instant when the holding control decision is made ($t = 120$ minutes), the vehicle trajectories and the current locations are randomly generated: passenger boarding and alighting processes are deterministic, but the vehicle running time between adjacent stops is subject to variation with a coefficient of variation (COV) of 0.15. There are twelve vehicles operating on the route, and exactly three vehicles lie in the control vehicle group $[b_i, e_i]$ for each holding station s_i ($i = 1, 2, 3$), and the other three vehicles are operating on the segment downstream of holding station 3 (between stops 31 and 41).

By using algorithms H1 and H2, the estimated passenger cost reductions from holding vehicles at each one and at all of the holding stations are shown in Table 2.

Table 2. Passenger Cost Reduction Comparison

Holding Strategies	Passenger Waiting Cost Reduction (Pass-Min)
1	At All Holding Stations	1507
2	Only at Holding Station 1	965
3	Only at Holding Station 2	1120
4	Only at Holding Station 3	925

Again, it is emphasized here that the main purpose of this numerical example is to demonstrate the heuristics developed in this study. It is not meaningful to use the results in Table 2 to compare the performance of holding vehicles at each single holding station and at all holding stations for the following reasons:

- Across strategies, the passenger cost is counted based on different route segments and a different number of vehicles. For Holding Strategy 1, i.e., holding vehicles at all holding stations, the passenger cost is computed over three vehicles and the segment [11,41]; three vehicles and segment [21, 41]; and three vehicles and segment [31, 41]. In contrast, three vehicles and segment [11, 41] are involved for Strategy 2; six vehicles and segment [21, 41] are evaluated for Strategy 3; and, nine vehicles and segment [31, 41] are evaluated for Strategy 4. In short, the passenger cost reductions are not computed on a common basis.

- The results come from just one instance of a holding decision. However, in the context of a deterministic model, results from one decision-making cycle cannot give more than just a rough expectation, which may vary significantly from reality. How this deterministic model approximates the operational stochasticity is only realized by an adaptive decision-making process based on real-time information. Practically, instead of a single application of the PMM model, it would be applied frequently, with a decision made each time a vehicle arrives at a holding station.

- In this particular example, all vehicles are assumed not to be controlled previously on the route. This over-states the likely performance of the holding control at multiple holding stations. With more frequent application of holding along the route, it may not be as necessary to hold those vehicles which have been previously controlled at the upstream holding stations.

In this example, the expected passenger cost reduction from holding only three vehicles at Station 1 (11 stops from the terminal) can be expected to be 965 passenger minutes, which means that there is already significant vehicle headway irregularity when vehicles arrive at Station 1 from the major terminal, where vehicles are dispatched at perfectly even headways. Therefore, similar, or worse, vehicle headway irregularity may be observed at Holding Station 2 even after vehicles have been controlled at Station 1. Such headway irregularity may justify the placement of the second holding station, though there are fewer passengers downstream of Station 2 that can benefit from the control, and probably more onboard passengers will diminish the desirability of holding control at Station 2. Similar arguments can also apply to the third holding station.

The effectiveness of the model formulation and solution can also be illustrated by the vehicle trajectory change under the holding controls, as shown in Fig. 7. In the figure, the solid lines represent the vehicle trajectories after implementing the holding control. It can be easily seen in Fig. 7 that holding vehicles at multiple holding station does tend to regularize the vehicle headways more than a single holding station alone.

In more detail, a number of things can be seen. In the first graph, with no control, the vehicle headways become fairly uneven as vehicles proceed to the end of the route, and vehicle pairing tends to occur. In the second graph, as the single holding station is placed at Stop 11, only the last three vehicles are considered to be held there, and eventually, their trajectories along the remaining segment of the route (from Stop 11 to the end terminal) are regularized and their headways become more even than would be the case without holding control. Similarly, as the only holding station is placed further downstream (e.g., Stop 21 and Stop 31 in the third and the fourth graphs, respectively), the vehicle headway distribution can be improved by increasing the number of vehicles held, but only on a shorter segment of the route. From the fifth graph, multiple station holding control seems to be able to achieve the best tradeoff between the number of vehicles and the length of the route segment over which the vehicle headway distribution is improved.

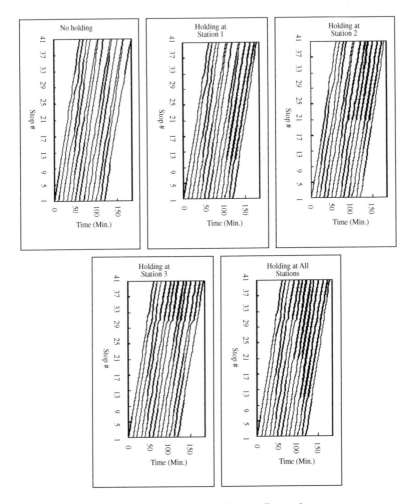

Fig. 7. Vehicle Trajectory Comparison

4 Conclusions

Multiple holding stations can offer more opportunities to regularize the vehicle headways, so that the overall passenger cost can be reduced further as compared to holding vehicles only at a single holding station. Certainly, the prerequisite of deploying multiple holding stations is that transit operation is subject to a certain level of variability.

The problem of holding multiple vehicles at multiple holding stations can be formulated as a convex problem with strictly convex objective function subject to linear constraints. Some classical techniques can solve this problem to optimality; however, this does not necessarily mean that the problem is small in scale. Therefore, heuristics are also developed in this study to solve this particular problem by decomposing the overall problem to sub-problems which can be tackled more easily. Respectively, the PSS problem can be solved analytically; the proposed H1 algorithm can solve the PMS problem to optimality; and, the H2 heuristic can also help to find the optimal solution to PMM problem if the assumption of no vehicle overtaking is strictly satisfied.

Though vehicle overtaking may be allowed within the problem formulation, it has been shown mathematically that the real objective is in most cases better than the objective derived merely from the model when vehicle overtaking does occur. This simply implies that the holding control decision made and the corresponding passenger cost reduction computed based on the model in this paper may be slightly conservative as vehicle overtaking occurs.

A hypothetical numerical example demonstrates the proposed heuristic, and shows further evidence to support the use of multiple holding stations even when transit operation variability is not very high (the coefficient of variation of travel time is 0.15 and the passenger boarding/alighting process is deterministic).

However, to demonstrate how holding control at multiple holding stations can outperform holding control at a single holding station, additional work is needed:

- One must judiciously select holding stations in terms of the number of holding stations and their locations; and,
- One may employ the model developed in this paper to make adaptive holding control decision based on a real-world example or a simulation study, by using the real-time information collected by AVL technology.

This work is of primary interest for our future study.

A Proof of Proposition 1

Though the problem formulation in this paper does not explicitly include vehicle overtaking, it essentially represents the vehicle overtaking as a negative headway. However, this still contributes positively to the objective since the headway item is always squared in the objective function. On the other hand, vehicle overtaking may not be allowed in practice. Without overtaking, the trajectory of the vehicle which tends to overtake the leading vehicle will intersect the lead vehicle's trajectory. Otherwise, if overtaking is allowed, these two vehicles may overtake each other alternately without ever deviating from each other much, and thus the two vehicles' trajectories can still be seen as intersecting. Therefore, a difference exists between the vehicle trajectories as formulated and the real vehicle trajectories when vehicle overtaking does occur, as shown in Fig. 8.

As shown in Fig. 8, as Vehicle 1 overtakes Vehicle 0, the trajectory of Vehicle 1 will follow the thin line after the overtaking point, according to the model. However, the solid line represents the real vehicle trajectories if overtaking is not explicitly modeled. Accordingly, H_1, H_2, H_3 are defined as the vehicle headways derived from the model formulation, and, in contrast, h_1, h_2, h_3 as the real vehicle

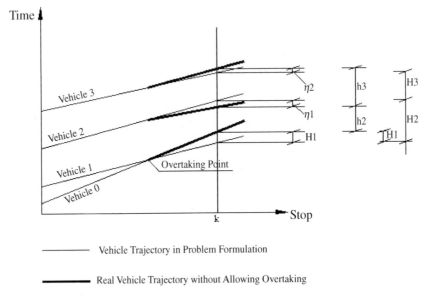

Time

Vehicle 3

Vehicle 2

Vehicle 1   Overtaking Point

Vehicle 0

η_2
η_1
H1

h3
h2

H3
H2
H1

k   Stop

—————— Vehicle Trajectory in Problem Formulation

—————— Real Vehicle Trajectory without Allowing Overtaking

Fig. 8. Comparison of Model Trajectory and Real Trajectory

headways ($h_1 = 0$ due to trajectory overlapping). For each stop k downstream where overtaking occurs, such headway patterns and the magnitude of η_1, η_2 (the difference between the model trajectory and real trajectory) can be easily seen and derived by mathematical induction based on Equation (2) as:

$$\eta_1 = \beta \cdot \lambda_k/(1 - \beta \cdot \lambda_k) \cdot H_1 \eta_2 = \beta \cdot \lambda_k/(1 - \beta \cdot \lambda_k) \cdot \eta_1 \qquad (17)$$

If only Vehicles 0, 1 and 2 are considered, it can be seen graphically that the real objective value is less than the model value. As Vehicle 3 is included, the model objective value can be expressed as:

$$H_1^2 + (h_2 + \eta_1 + H_1)^2 + (h_3 - \eta_1 - \eta_2)^2 \qquad (18)$$
$$= \omega + h_2^2 + h_3^2 + 2h_2 \cdot (\eta_1 + H_1) - 2h_3 \cdot (\eta_1 + \eta_2)$$

Herein, ω is a positive value. As we can see directly from Equation (18),

$$2h_2 \cdot (\eta_1 + H_1) - 2h_3 \cdot (\eta_1 + \eta_2) \qquad (19)$$
$$= 2h_2 \cdot (\eta_1 + H_1) - 2h_3 \cdot (\eta_1 + H_1) \cdot \beta \cdot \lambda_k/(1 - \beta \cdot \lambda_k)$$

Therefore, if

$$h_2 \geq h_3 \cdot \beta \cdot \lambda_k/(1 - \beta \cdot \lambda_k) \qquad (20)$$

it is always true that the model objective value is larger than the real objective value. The term $\beta \cdot \lambda_k/(1 - \beta \cdot \lambda_k)$ in the equation is essentially the departure time difference between Vehicle 2 and Vehicle 0 at Stop k. It is actually a very small number generally on the order of 0.1 or less. Therefore, unless the vehicle trajectory pattern is extreme, Inequality (20) always holds.

It would be always true that the model objective value is larger than the real objective value for the four vehicle case. Based on the same argument, it can be easily inferred that even when more vehicles are included, the proposition is still true.

References

Abkowitz, M. and Lepofsky, M. (1990). Implementing headway-based reliability control on transit routes. *Journal of Transportation Engineering*, **116**(1), 49–63.

Abkowitz, M. and Tozzi, J. (1986). Transit route characteristics and headway-based reliability control. *Transportation Research Record*, **1078**, 11–16.

Abkowitz, M., Eiger, A., and Engelstein, I. (1986). Optimal control of headway variation on transit routes. *Journal of Advanced Transportation*, **20**(1), 73–78.

Barnett, A. (1974). On controlling randomness in transit operations. *Transportation Science*, **8**(2), 101–116.

Ding, Y. and Chien, S. (2001). Improving transit service quality and headway regularity with real-time control. *Proceedings of the 80th Annual Meeting of the Transportation Research Board*.

Eberlein, X.-J., Wilson, N., and Bernstein, D. (2001). The holding problem with real-time information available. *Transportation Science*, **35**(1), 1–18.

Fu, L. and Yang, X. (2002). Design and implementation of bus-holding control strategies with real-time information. *Transportation Research Record*, **1791**, 6–12.

Furth, P. (1995). A headway control strategy for recovering from transit vehicle delays. *Transportation Congress: Civil Engineers–Key to the World Infrastructure*, **2**, 2032–2038.

Hickman, M. (2001). An analytical stochastic model for the transit vehicle holding problem. *Transportation Science*, **35**(3), 215–237.

O'Dell, S. and Wilson, N. (1999). Optimal real-time control strategies for rail transit operations during disruptions. In N. Wilson, editor, *Lecture Notes in Economics and Mathematical Systems No. 471: Computer-Aided Scheduling of Public Transport*, pages 299–323. Springer, Berlin.

Seneviratne, P. and Loo, C. (1986). Bus journey times in medium size urban areas. *Journal of Advanced Transportation*, **20**(3), 259–274.

Sun, A. (2005). AVL-based transit operations control. University of Arizona.

Turnquist, M. and Blume, S. (1980). Evaluating potential effectiveness of headway control strategies for transit systems. *Transportation Research Record*, **746**, 25–29.

Zhao, J., Dessouky, M., and Bukkapatnam, S. (2001). Distributed holding control of bus transit operations. *Proceedings of the 2001 IEEE Intelligent Transportation Systems Conference*, pages 978–983.

Part IV

Network Design, Fleet Sizing, and Strategic Planning

Models for Line Planning in Public Transport

Ralf Borndörfer, Martin Grötschel, and Marc E. Pfetsch

Konrad-Zuse-Zentrum für Informationstechnik Berlin, Takustr. 7, 14195 Berlin, Germany;
Email: {borndoerfer, groetschel, pfetsch}@zib.de

Summary. The *line planning problem* is one of the fundamental problems in strategic planning of public and rail transport. It consists in finding lines and corresponding frequencies in a public transport network such that a given travel demand can be satisfied. There are (at least) two objectives. The transport company wishes to minimize its operating cost; the passengers request short travel times. We propose two new multi-commodity flow models for line planning. Their main features, in comparison to existing models, are that the passenger paths can be freely routed and that the lines are generated dynamically.

1 Introduction

The *strategic planning* process in public and rail transport, i.e., the long and medium term design of the infrastructure and the service level of a transportation network, is usually divided into the following consecutive steps: *network design, line planning,* and *timetabling*. In each of these steps, operations research methods can support the planning decisions, see, e.g., the survey article of Bussieck *et al.* (1997a), which discusses the case of rail traffic. This article is about line planning in public transport. We start by briefly explaining the strategic planning process in this area to put our work into perspective.

All steps of strategic planning are generally based on so-called *origin-destination data* in the form of *OD-matrices*; each entry in an OD-matrix gives the number of passengers that want to travel from one point in the network to another point within a fixed *time horizon*. It is well known that such data have certain deficiencies. For instance, OD-matrices depend on the discretization used, they are highly aggregated, they give only a snapshot type of view, they are only valid when the transportation demand is fixed and does not depend on the service or price level, and it is often questionable how well the entries represent the "real" transportation demand. One can surely hope for better data, but gathering OD-matrices currently seems to be the best feasible choice for estimating transportation demand. Assembling such data is quite an art and rather costly. Public transportation companies do this routinely and employ OD-matrices as input for strategic planning.

Based on this demand data, the first step of the strategic planning process is the *network design problem*. It deals with the layout of the transportation system. Decisions are made about choosing streets/providing tracks of sufficient capacity to transport the number of passengers given by an OD-matrix such that construction costs are minimized. Typically, one considers extensions of existing, historically grown networks; designs from scratch, however, are also interesting, not only for the construction of completely new systems, but also for the evaluation of existing networks.

The *line planning problem* (LPP) that we discuss in this article is the second step in the strategic planning process for public transport. It consists of designing line routes and their frequencies in a given street or track network such that a given transportation volume, again given by an OD-matrix, can be satisfied. The lines include forward and backward directions, and they start and end at designated terminal points in the network. With each potential line we associate a certain transportation mode, such as tram, train, or different bus types, e.g., double-decker or kneeling bus. Each such mode has a capacity, and the capacity of a line is computed as the product of its mode capacity with an operating frequency; this frequency is supposed to indicate a basic timetable period. Restrictions on timetable periods, such as divisibility constraints and safety margins, may come up. Furthermore, the number of available vehicles for a mode may result in bounds on the frequencies. There are two competing objectives: on the one hand to minimize user discomfort and on the other hand to minimize the lines' operating costs. User discomfort is usually measured by the total passenger traveling time or the number of transfers during the ride, or both.

The third step is to refine the frequencies of a given line plan into a detailed *timetable*. The objective is either to minimize the number of necessary vehicles or to minimize the transfer times of the passengers. This timetable is the basis for the succeeding steps of *operational planning* such as vehicle scheduling, crew scheduling, rostering, and assignment, see, e.g., the survey article of Desrosiers *et al.* (1995).

In the recent literature on the LPP often a distribution of the passengers is estimated by a so-called *system split*. The system split fixes the traveling paths of the passengers *before* the lines are known, see Section 2. A second common assumption is that an optimal line plan can be chosen from a *line pool*, i.e., a precomputed set of lines. Third, maximization of *direct travelers*, i.e., travelers without transfers, is frequently considered as the objective. In such an approach, transfer waiting times do not play a role.

This article proposes two new multi-commodity flow models for the LPP. These models minimize a combination of total passenger traveling time and operating costs. The first model is compact in the sense that it uses arc variables for both lines and passenger paths; it can be used to compute lower bounds. The second model uses path variables for both lines and passenger paths; it is intended to deal with constraints on the line routes. The model also handles frequencies implicitly by means of continuous frequency variables. Both models allow for a dynamic generation of lines, and they allow passengers to change their routes according to the traveling times on the computed line system. In particular, they do not assume a system split, but compute a "best" passenger flow. These properties aim at line planning scenarios in public transport, where we see less justification for a system split and fewer restrictions in line design than one seems to have in railway line planning.

This paper is organized as follows. Section 2 gives an overview of the literature on the LPP. In Section 3 we describe and discuss our models. Section 4 discusses aspects of a column generation solution approach for the second model. We show that the pricing problem for the passenger variables is a shortest path problem. The line pricing problem turns out to be a longest path problem and it is, in fact, already $\mathcal{NP}$-hard to solve the LP relaxation of the second problem. However, if only lines of logarithmic length with respect to the number of nodes are considered, the pricing problem can be solved in polynomial time. We close with some final remarks in Section 5.

2 Related Work

This section provides a short overview of the literature for the line planning problem. More information can be found in the article of Ceder and Israeli (1992), which covers the literature up to the beginning of the 1990s; see also Odoni et al. (1994) and Bussieck et al. (1997a).

The first approaches to the line planning problem had the idea to assemble lines from shorter pieces in an iterative (and often interactive) process. An early example is the so-called skeleton method described by Silman et al. (1974), that chooses the endpoints of a route and several intermediate nodes which are then joined by shortest paths with respect to length or traveling time; for a variation see Dubois et al. (1979). In a similar way, Sonntag (1979) and Pape et al. (1995) constructed lines by adjoining small pieces of streets/tracks in order to maximize the number of direct travelers.

In the literature it is common to work in two-step approaches that precompute some set of lines in a first phase and choose a line plan from this set in a second phase. For example, Ceder and Wilson (1986) described an enumeration method to generate lines whose length is within a certain factor from the length of the shortest path, while Mandl (1980) proposed a local search strategy to optimize over such a set. Ceder and Israeli (1992) and Israeli and Ceder (1995) introduced a quadratic set covering model to choose among direct connections between destinations and transfer connections; they also proposed a heuristic to solve their model.

An important phase of development is related to the so-called *system split*, which distributes the passengers on paths in the transportation network *before* the lines are known. The system split is based on a classification of the transportation system into levels of different speed, as common in railway systems. Assuming that travelers are likely to change to fast levels as early and leave them as late as possible, the passengers are distributed onto several paths in the system, using Kirchhoff-like rules at the transit points. Note that this fixes, in particular, the passenger flow on each individual link in the network. The system split approach was promoted by Bouma and Oltrogge (1994), who used it to develop a branch-and-bound based software system for the planning and analysis of the line system of the Dutch railway network.

Recently, advanced integer programming techniques have been applied to the line planning problem. Bussieck *et al.* (1997b) (see also Bussieck (1997)) and Claessens *et al.* (1998) both proposed cut-and-branch approaches to select lines from a previously generated set of potential lines and report computations on real world data. They also both assume a homogeneous transport system, which can be assumed after a system-split is performed as a preprocessing step. Bussieck *et al.* (2004) extend this work by incorporating nonlinear components into the model. Goossens *et al.* (2004) and Goossens *et al.* (2002) show that practical problems can be solved within reasonable quality and time by a branch-and-cut approach, even for the simultaneous optimization of several transportation systems.

3 Two Models for the LPP

In this section we present two integer programming formulations for the line planning problem.

3.1 Notation and Terminology

We typeset vectors in bold face, scalars in normal face. If $\mathbf{v} \in \mathbb{R}^J$ is a real valued vector and I a subset of J, we denote by $\mathbf{v}(I)$ the sum over all components of $\mathbf{v}$ indexed by I, i.e., $\mathbf{v}(I) := \sum_{i \in I} v_i$.

In line planning, we are given an undirected multigraph $G = (V, E)$, which is supposed to model the topology of a transportation network; this graph is used to express line paths, which we assume to be undirected (or bidirectional). We consider also a symmetric directed version (V, A) of this graph, where each edge e in E is replaced by two antiparallel arcs $a(e)$ and $\bar{a}(e)$; the directed version is used to model passenger paths, which are not symmetric. We use the notation G to refer to both the directed or undirected graph depending on the context, i.e., for line paths we refer to the undirected version, while for passenger paths we use the directed version. If $a = (u, v)$ is an arc in the directed (multi)graph, we denote its antiparallel counterpart by $\bar{a} = (v, u)$ and by $e(a) = \{u, v\} \in E$ the undirected edge corresponding to a.

The nodes of G represent stops, stations, terminals (start and end points of lines), and origins or destinations of passenger flows (*OD-nodes* or "centroids" of certain traffic cells). The edges/arcs of G correspond to physical transportation links between two stations, to the formation or termination of lines at a terminal, or to the passenger in- and outflow between OD-nodes and stations. Associated with each edge e in E is a *mode* m_e of transportation, such as tram, train, double-decker bus, pedestrian traffic, etc.; we assume multiple edges between two nodes, one for each mode using the underlying link. We denote the set of all modes by $\mathcal{M}$ and by G_m the subgraph of G defined by the edges e with $m_e = m$. Furthermore, we have a traveling time τ_a for each arc $a \in A$, an (operating) cost c_e, and a capacity λ_e for each edge $e \in E$; all three, τ_a, c_e, and λ_e, are assumed to be nonnegative. The values λ_e bound the total frequency of lines using edge e, as will be explained below.

For each node pair $s, t \in V$ we assume a nonnegative *demand* d_{st} of passengers to be given that want to travel from s to t, i.e., (d_{st}) is the *OD-matrix*. We do *not* assume this matrix to be symmetric. We let $D := \{(s, t) \in V \times V : d_{st} > 0\}$ be the set of all *OD-pairs*, i.e., node pairs with nonzero demand. For such an OD-pair $(s, t) \in D$, an (s, t)-*passenger path* is a directed path in G starting at node s and ending at node t, which visits exactly two OD-nodes, namely, s and t. Since passenger paths will correspond to shortest paths with respect to some nonnegative weights, we assume them to be *simple*, i.e., without node repetitions. Let $\mathcal{P}_{st}$ be the set of all (s, t)-passenger paths, $\mathcal{P} := \bigcup \{p \in \mathcal{P}_{st} : (s, t) \in D\}$ the set of all passenger paths, and $\mathcal{P}_a := \bigcup \{p \in \mathcal{P} : a \in p\}$ the set of all passenger paths that use arc a. The *traveling time* of a passenger path p is defined as $\tau_p := \sum_{a \in p} \tau_a$.

For each mode m there is a set of *terminals* $\mathcal{T}_m \subset V$, where lines of mode m can start or end. Let $\mathcal{T} := \bigcup \{v \in \mathcal{T}_m : m \in \mathcal{M}\}$ be the set of all terminals. A *line* of mode m is an undirected path in G_m, starting and ending at a terminal from $\mathcal{T}_m$; we stipulate that the lines must be simple. Let $\mathcal{L}_m$ be the set of all lines of mode m, $\mathcal{L} := \bigcup \{\ell \in \mathcal{L}_m : m \in \mathcal{M}\}$ the set of all lines, and $\mathcal{L}_e := \bigcup \{\ell \in \mathcal{L} : e \in \ell\}$ the set of lines that use edge e. We assume that there are *fixed costs* C_ℓ and *capacities* κ_ℓ for one unit/vehicle/train of line ℓ, which depend only on the mode, i.e., $C_\ell = C_m$ and $\kappa_\ell = \kappa_m$ for $\ell \in \mathcal{L}_m$. We further associate a *frequency* f_ℓ with every line ℓ that is supposed to indicate the (approximate) number of times vehicles are employed to serve the demand over the underlying *time horizon* T. This not necessarily has to lead to a regular timetable period, but an estimate for such a period for line ℓ can be computed from this frequency as T/f_ℓ.

3.2 Service Network Design Model

In this section we present a model for the LPP in which lines are modeled as integer flows in the mode networks G_m; it is aimed at efficiently computing lower bounds. In order to achieve this goal, we have to circumvent several complications that are discussed at the end of this section. The model is related to a service network design model by Kim and Barnhart (1997).

We assume in this model a fixed finite set of possible frequencies $\mathcal{F} \subset \mathbb{R}_+$ for the lines of the transportation system. Furthermore, let Q be an upper bound on the number of lines that start and end in two given terminals. For mode m, let $R_m := \{(u, v, q, f) \in \mathcal{T}_m \times \mathcal{T}_m \times \{1, \ldots, Q\} \times \mathcal{F} : u < v\}$, and let $R := \bigcup \{R_m : m \in \mathcal{M}\}$. The set R represents all possible line-frequency combinations. For convenience, define $m_r := m$ and $r =: (u_r, v_r, q_r, f_r)$ for $r \in R_m$; r indexes the line numbered q_r of mode m with frequency f_r starting at u_r and ending in v_r. Moreover, we let $R'_m := \{(u, v, q) \in \mathcal{T}_m \times \mathcal{T}_m \times \{1, \ldots, Q\} : u < v\}$. We handle fixed costs by adding them to the costs on the arcs that emanate from the terminals $\mathcal{T}_m$.

There are two kinds of variables:

$y_a^{st} \in \mathbb{R}_+$: the flow of passengers from s to t $((s, t) \in D)$ using arc $a \in A$,

$z_a^r \in \{0, 1\}$: the flow of line numbered q_r (of mode $m_r = m_{e(a)}$) with frequency f_r, starting at u_r and ending at v_r, passing through arc $a \in A$.

The model is:

$$(\text{LPP}_1) \qquad \min \sum_{(s,t)\in D} \boldsymbol{\tau}^{\mathsf{T}} \mathbf{y}^{st} + \sum_{r\in R} \mathbf{c}^{\mathsf{T}} \mathbf{z}^r$$

$$\mathbf{y}^{st}(\delta^+(v)) - \mathbf{y}^{st}(\delta^-(v)) = \delta_v^{st} \qquad \forall\, v \in V \tag{1}$$

$$\sum_{(s,t)\in D} y_a^{st} - \sum_{r\in R} \kappa_{m_r}\, f_r(z_a^r + z_{\bar{a}}^r) \le 0 \qquad \forall\, a \in A \tag{2}$$

$$\mathbf{z}^r(\delta^+(v)|G_{m_r}) - \mathbf{z}^r(\delta^-(v)|G_{m_r}) = 0 \qquad \forall\, v \in V \setminus \{u_r, v_r\},\ r \in R \tag{3}$$

$$\mathbf{z}^r(\delta^-(u_r)) = 0 \qquad \forall\, r \in R \tag{4}$$

$$\mathbf{z}^r(A(W)|G_{m_r}) \le |W| - 1 \qquad \forall\, W \subseteq V \setminus \{u_r, v_r\},\ r \in R \tag{5}$$

$$\sum_{r\in R} f_r(z_a^r + z_{\bar{a}}^r) \le \lambda_{e(a)} \qquad \forall\, a \in A \tag{6}$$

$$\sum_{f\in\mathcal{F}} z_a^{(\mathbf{r}',f)} \le 1 \qquad \forall\, a \in A,\ \mathbf{r}' \in R_m' \tag{7}$$

$$z_a^r \in \{0,1\} \qquad \forall\, a \in A,\ r \in R \tag{8}$$

$$y_a^{st} \ge 0 \qquad \forall\, a \in A,\ (s,t) \in D \tag{9}$$

Here, $(A(W)|G_{m_r})$ are the arcs in G_{m_r} with both endpoints in $W \subseteq V$ and similarly for $(\delta^+(v)|G_{m_r})$.

The *passenger flow constraints* (1) and the nonnegativity constraints (9) model a multi-commodity flow problem for the passenger flow, where the commodities correspond to the OD-pairs $(s,t) \in D$. Here δ_v^{st} is zero except that $\delta_s^{st} = d_{st}$ and $\delta_t^{st} = -d_{st}$. This guarantees that the demand is satisfied. The lines are modeled as 0/1-flows in the z-variables for each $r \in R$: the *line flow conservation constraints* (3) ensure that every line that enters a non-terminal node also has to leave it. Constraints (4) ensure that the line-flow is directed from the start node u_r towards the end node v_r of the line indexed by r. The "subtour elimination" constraints (5) rule out isolated line circuits, i.e., circuits in the mode graphs G_{m_r} that are not connected to the terminal set $\{u_r, v_r\}$. The *frequency constraints* (6) bound the total frequency of lines using each edge. Constraints (7) ensure that at most one frequency for each line is used. The passenger and the line parts of the model are linked by the *capacity constraints* (2) in such a way that the total passenger flow on each arc is covered by lines of sufficient total capacity.

Formulation (LPP_1) models undirected line routes as directed paths in 0/1 variables, since this is the easiest way to model simple paths between terminals. Namely, it allows to eliminate isolated line circuits by constraints of the form (5). The model of Kim and Barnhart (1997), referred to above, does not incorporate terminals and can arbitrarily decompose any line flow into simple paths and circuits. It can therefore model lines using integer variables and does not need to resort to subtour elimination constraints. Note also that the discretization of the frequencies is used to linearize the capacity constraints (2).

Formulation (LPP_1) is of polynomial size except for the "subtour elimination" constraints. These constraints are well known from the traveling salesman problem and can be separated in polynomial time. By the equivalence of separation and optimization, see Grötschel *et al.* (1993), it follows that the LP relaxation of (LPP$_1$) can be solved in polynomial time to provide a lower bound for the line planning problem.

We also remark that the model is ready to accommodate a number of additional constraints. We mention as an example a restriction L on the total number of lines, which can be modeled as $\mathbf{z}(\delta^+(\mathcal{T})) \leq L$.

3.3 A Path Based Frequency Model

Our second model treats the lines by means of path and frequency variables.

There are three kinds of variables:

$y_p \in \mathbb{R}_+$: the flow of passengers traveling from s to t on path $p \in \mathcal{P}_{st}$,
$x_\ell \in \{0,1\}$: a decision variable for using line $\ell \in \mathcal{L}$,
$f_\ell \in \mathbb{R}_+$: frequency of line $\ell \in \mathcal{L}$.

This allows to model the cost of line ℓ of mode m directly as $x_\ell\, C_\ell + f_\ell\, c_\ell$. Here, $c_\ell := \sum_{e\in\ell} c_e$ is the total operating cost of line ℓ. Similarly, the capacity of line $\ell \in \mathcal{L}_m$ is $\kappa_\ell\, f_\ell = \kappa_m\, f_\ell$. The model is:

$$(\text{LPP}_2) \quad \min\ \boldsymbol{\tau}^\mathsf{T}\mathbf{y} + \mathbf{C}^\mathsf{T}\mathbf{x} + \mathbf{c}^\mathsf{T}\mathbf{f}$$

$$\mathbf{y}(\mathcal{P}_{st}) = d_{st} \quad \forall\, (s,t) \in D \tag{10}$$

$$\mathbf{y}(\mathcal{P}_a) - \sum_{\ell:e(a)\in\ell} \kappa_\ell f_\ell \leq 0 \quad \forall\, a \in A \tag{11}$$

$$\mathbf{f}(\mathcal{L}_e) \leq \lambda_e \quad \forall\, e \in E \tag{12}$$

$$\mathbf{f} \leq F\mathbf{x} \tag{13}$$

$$x_\ell \in \{0,1\} \quad \forall\, \ell \in \mathcal{L} \tag{14}$$

$$f_\ell \geq 0 \quad \forall\, \ell \in \mathcal{L} \tag{15}$$

$$y_p \geq 0 \quad \forall\, p \in \mathcal{P} \tag{16}$$

As in (LPP$_1$), the *flow constraints* (10) together with the nonnegativity constraints (16) guarantee that the demand is satisfied for each $(s,t) \in D$. The *capacity constraints* (11) link the passenger paths with the line paths to ensure sufficient transportation capacities on each arc. The *frequency constraints* (12) bound the total frequency of lines using each edge. Inequalities (13) link the frequency with the decision variables for the use of lines; they guarantee that the frequency of a line is 0 whenever it is not used. Here, F is an upper bound on the frequency of a line; for technical reasons, we also assume that $F \geq \lambda_e$ for all $e \in E$, see Section 4 for a detailed discussion.

The main advantage of (LPP$_2$) over (LPP$_1$) is that it is easy to incorporate additional constraints on the formation of individual lines such as length restrictions, as well as constraints on sets of lines, e.g., constraints on numbers of lines of certain types. As such constraints are important in practice, we are currently using (LPP$_2$) as the basis for the development of a branch-and-price algorithm. The disadvantage of the model is, however, that it is already $\mathcal{NP}$-hard to solve the LP relaxation, as we will show in Section 4.

3.4 Discussion of the Models

We discuss in this section advantages and disadvantages of the two models.

Objectives: Both models have objectives with two competing parts, namely, to minimize total passenger traveling time and to minimize operation costs. The models allow to adjust the relative importance of one part over the other by an appropriate scaling of the respective objective coefficients.

Passenger Routes: Previous approaches to the LPP often fixed the traveling paths of the passengers in advance by employing a system split. In contrast, our two models allow to freely route passengers in the line network in order to compute an optimal routing. To our knowledge, such routings have not been considered in the context of line planning before. Our models are targeted at local public transport systems, where, in our opinion, people determine their traveling paths according to the line system and not only according to the network topology.

Models (LPP_1) and (LPP_2) compute a set of passenger paths that minimize the total traveling times in the sense of a system optimum. However, in our case, with a linear objective function and linear capacities, it can be shown that the resulting system optimum is also a user equilibrium, namely, the so-called Beckmann user equilibrium, see Correa *et al.* (2004). We do not address the question why passengers should choose this equilibrium out of several possible equilibria that can arise in routing with capacities.

The routing in our models allows for passenger paths of arbitrary travel times, which may force some passengers to long detours. One approach to solve this problem is to restrict the lengths of passenger paths. For each OD-pair one computes the shortest path in G with respect to the traveling times in advance (every path is feasible independent of the line system) and modifies the model to only allow passenger paths whose traveling times are within a certain range from the traveling times of the shortest paths. This turns the pricing problem for the passenger variables into a constrained shortest path problem; see Section 4.1. Although this problem is $\mathcal{NP}$-hard, there are algorithms that are reasonably fast in practice. Note also that such an approach would measure travel times with respect to shortest paths in the underlying network (independent of any line system). Ideally, however, one would like to compare these to the shortest paths using only arcs covered by the computed line system.

Line Routes: The literature generally takes line routes as simple paths, with the exception of ring lines, and we do the same in this article. In fact, a restriction forcing some sort of simplicity is necessary to solve the line pricing problems, as otherwise the outcome will be a line that visits some edges back and forth many times consecutively; see Section 4.2. As a slight generalization of the concept of simplicity, one could investigate the case where one assumes that every line route is bounded in length and "almost" simple, i.e., when considering the sequence of nodes in a line route, no node is repeated within a given (fixed) number of nodes. It remains to be seen whether non-simple paths are useful in practice.

We consider lines as undirected, which implies that there are no one-way streets or tracks. However, it is easy to extend the model by including directed lines as they sometimes appear in ring lines.

Transfers: Transfers between lines are currently ignored in our models. The problem here are not transfers between different modes, which can be handled by setting up node disjoint mode networks G_m linked by appropriate transfer edges, which are weighted by the estimated transfer times. This does not work for transfers between lines of the same mode. The reason is that our models do not distinguish between lines of the same mode in the capacity constraints. In principle, this obstacle can be resolved by an appropriate expansion of the graph. However, this greatly increases the complexity of the model, and it introduces degeneracy; it is unclear whether such models have the potential of being solvable in practice.

Time horizon: An important consideration in any strategic planning problem is the time horizon that one wants to consider. In the LPP, it comes into play implicitly via the OD-matrix. Usually, such data are aggregated over one day, but it is similarly appropriate to aggregate, e.g., over the rush hour. In fact, the asymmetry of the demands in rush hours was one of the reasons to consider directed passenger paths.

Frequencies: In a real world line plan the frequencies have to produce a regular timetable and hence are not allowed to take arbitrary fractional values. Our first model takes this requirement into account. The second model, however, treats frequencies as continuous values. This is a simplification. We could have forced the second model to accept only a finite number of frequencies in the same way as in the first model, i.e., by enumerating lines with fixed frequencies. However, as the frequencies are mainly used to adjust the line capacities, we do (at present) not care so much about "nice" frequencies and view the fractional values as approximations or clues to "sensible" values. We note, however, that the approaches of Claessens et al. (1998), Goossens et al. (2004), and Goossens et al. (2002) are able to handle arbitrary finite sets of frequencies. This feature is clearly needed in future models that integrate line planning and timetable construction.

Additional Constraints: Several additional types of constraints can be added to the models, e.g., capacity constraints on the total number or on the frequencies of lines using an edge, on the number of lines of certain types, or other linear constraints.

4 Pricing Problems for (LPP_2)

In this section, we discuss the solution of the LP relaxation of (LPP_2). For this purpose, we have to analyze the pricing problems for the passenger and the line variables. Preliminary computational experience indicates that the LP relaxation gives a good approximation to an optimal solution of (LPP_2).

The LP relaxation of (LPP_2) can be simplified by eliminating the x-variables. In fact, since (LPP_2) minimizes over nonnegative costs, one can assume w.l.o.g. that the inequalities (13) are satisfied with equality, i.e., there is an optimal LP solution such that $Fx_\ell = f_\ell \Leftrightarrow x_\ell = f_\ell/F$ for all lines ℓ. Eliminating $\mathbf{x}$ from the system using these equations, we arrive at the following simpler LP (LP$_2$):

$$(\text{LP}_2) \qquad \min \; \boldsymbol{\tau}^\mathsf{T}\mathbf{y} + \boldsymbol{\gamma}^\mathsf{T}\mathbf{f}$$

$$\mathbf{y}(\mathcal{P}_{st}) = d_{st} \qquad \forall\, (s,t) \in D \tag{17}$$

$$\mathbf{y}(\mathcal{P}_a) - \sum_{\ell:e(a)\in\ell} \kappa_\ell f_\ell \leq 0 \qquad \forall\, a \in A \tag{18}$$

$$\mathbf{f}(\mathcal{L}_e) \leq \lambda_e \qquad \forall\, e \in E \tag{19}$$

$$f_\ell \geq 0 \qquad \forall\, \ell \in \mathcal{L} \tag{20}$$

$$y_p \geq 0 \qquad \forall\, p \in \mathcal{P} \tag{21}$$

Here, $\gamma_\ell = C_\ell/F + c_\ell$ denotes the cost of line ℓ resulting from the above substitution. After the elimination, (LP$_2$) contains inequalities $f_\ell \leq F$ for all lines ℓ. Since we have assumed that $F \geq \lambda_e$ for all $e \in E$, this exponential number of inequalities is dominated by inequalities (19) and can be omitted. Hence, (LP$_2$) contains only a polynomial number of inequalities (apart from the nonnegativity constraints (20) and (21)). We remark that the coupling between x_ℓ and f_ℓ by means of the equation $Fx_\ell = f_\ell$ is a typical weak point of IP models involving fixed costs.

Proposition 1. *The computation of the optimal value of (LP$_2$) with simple line paths is $\mathcal{NP}$-hard in the strong sense.*

Proof. We reduce the Hamiltonian path problem, which is strongly $\mathcal{NP}$-complete even for planar graphs, to (LP$_2$). Let (H, s, t) be an instance of the Hamiltonian path problem, i.e., $H = (V, E)$ is a graph and s and t are two distinct nodes of H.

For the reduction, we are going to derive an appropriate instance of (LP$_2$). The underlying network is formed by a graph $H' = (V', E')$, which arises from H by splitting each node v into three copies v_1, v_2, and v_3. For each node $v \in V$, we add edges $\{v_1, v_2\}$ and $\{v_2, v_3\}$ to E' and for each edge $\{u, v\}$ the edges $\{u_1, v_3\}$ and $\{u_3, v_1\}$, see Fig. 1. Our instance of (LP$_2$) contains just a single mode with only two terminals s_1 and t_3 such that every line must start at s_1 and end at t_3. The demands are $d_{v_1 v_2} = 1$ ($v \in V$) and 0 otherwise, and the capacity of every line is 1. For every $e \in E$, we set λ_e to some high value (e.g., to $|V|$). The cost of all edges is set to 0, except for the edges in $\delta(s_1)$, for which the costs are set to 1. The traveling times are set to 0 everywhere. It follows that the value of a solution to (LP$_2$) is the sum of the frequencies of all lines.

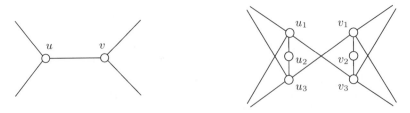

Fig. 1. Example for the Node Splitting in the Proof of Proposition 1

Assume that $p = (s, v^1, \ldots, v^k, t)$ (for $v^1, \ldots, v^k \in V$) is an (s, t)-Hamiltonian path in H. Then $p' = (s_1, s_2, s_3, v_1^1, v_2^1, v_3^1, \ldots, v_1^k, v_2^k, v_3^k, t_1, t_2, t_3)$ is an (s_1, t_3)-Hamiltonian path in H', which gives rise to an optimal solution of (LP$_2$). Namely, we can take p' as the route of a single line with frequency 1 in (LP$_2$) and route all demands $d_{v_1 v_2} = 1$ on this line directly from v_1 to v_2. As the frequency of p' is 1, the objective value of this solution is also 1. On the other hand, every solution to (LP$_2$) must have value at least one, since every line has to pass an edge of $\delta(s_1)$ and the sum of the frequencies of lines visiting an arbitrary edge of type $\{v_1, v_2\}$, for $v \in V$, is at least 1. This proves that (LP$_2$) has an optimal solution of value 1, if (H, s, t) contains a Hamiltonian path.

For the converse, assume that there exists a solution to (LP$_2$) of value 1, for which we ignore lines with frequency 0. We know that every edge $\{v_1, v_2\}$ $(v \in V)$ is covered by at least one line of the solution. If every line contains all the edges $\{v_1, v_2\}$ $(v \in V)$, each such line gives rise to a Hamiltonian path (since the line paths are simple) and we are done. Otherwise, there must be an edge $e = \{v_1, v_2\}$ $(v \in V)$ which is not covered by all of the lines. By the capacity constraints (18), the sum of the frequencies of the lines covering e is at least 1. However, the edges in $\delta(s_1)$ are covered by the lines covering edge e plus at least one more line of nonzero frequency. Hence, the total sum of all frequencies is larger than one, which is a contradiction to the assumption that the solution has value 1.

This shows that there exists an (s, t)-Hamiltonian path in H if and only if the value of (LP$_2$) with respect to H' is 1. $\qquad\square$

Note that Proposition 1 highlights a subtle, but important difference in the line planning parts of the LP relaxations of the two models (LPP$_1$) and (LPP$_2$). In the LP relaxation of (LPP$_2$), the line planning part optimizes over a convex hull of simple paths; Proposition 1 shows that this is $\mathcal{NP}$-hard. As the LP relaxation of (LPP$_1$) is solvable in polynomial time, its line planning part must be weaker and contain additional solutions which are not convex combinations of simple paths. For example, an isolated circuit C in some mode graph G_m gives rise to the vector $(|C| - 1)/|C| \cdot \chi(C)$, which fulfills all constraints of (LPP$_1$), in particular the subtour elimination constraints (5). But it is not a convex combination of simple paths.

By Proposition 1, we also know that at least one of the pricing problems associated with (LP$_2$) must be $\mathcal{NP}$-hard as well. In fact, it will turn out that the pricing problem for the line variables x_ℓ and f_ℓ is a longest path problem; the pricing problem for the passenger variables y_p, however, is a shortest path problem.

The pricing problems for the variables of (LP$_2$) are studied in terms of the dual of (LP$_2$). Denote the variables of the dual as follows: $\pi = (\pi_{st}) \in \mathbb{R}^D$ (flow constraints (17)), $\mu = (\mu_a) \in \mathbb{R}^A$ (capacity constraints (18)), and $\eta \in \mathbb{R}^E$ (frequency constraints (19)). The dual of (LP$_2$) is:

$$\text{(DLP)} \qquad \max \ \mathbf{d}^\mathsf{T}\boldsymbol{\pi} - \boldsymbol{\lambda}^\mathsf{T}\boldsymbol{\eta}$$

$$\pi_{st} - \boldsymbol{\mu}(p) \leq \tau_p \qquad \forall\, p \in \mathcal{P}_{st}, \ (s, t) \in D$$
$$\kappa_\ell\, \boldsymbol{\mu}(\ell) - \boldsymbol{\eta}(\ell) \leq \gamma_\ell \qquad \forall\, \ell \in \mathcal{L}$$
$$\boldsymbol{\mu}, \ \boldsymbol{\eta} \geq 0,$$

where

$$\boldsymbol{\mu}(\ell) = \sum_{e \in \ell} \left(\mu_{a(e)} + \mu_{\overline{a}(e)} \right).$$

4.1 Pricing of the Passenger Variables

The reduced cost $\overline{\tau}_p$ for variable y_p for $p \in \mathcal{P}_{st}$, $(s,t) \in D$, is

$$\overline{\tau}_p = \tau_p - \pi_{st} + \boldsymbol{\mu}(p) = \tau_p - \pi_{st} + \sum_{a \in p} \mu_a = -\pi_{st} + \sum_{a \in p} (\mu_a + \tau_a).$$

The pricing problem for the y-variables is to find a path p such that $\overline{\tau}_p < 0$ or to conclude that no such path exists. This can easily be done in polynomial time as follows. For all $(s,t) \in D$, we search for a shortest (s,t)-path with respect to the nonnegative weights $(\mu_a + \tau_a)$ on the arcs; we can, e.g., use Dijkstra's algorithm. If the length of this path is less than π_{st}, then y_p is a candidate variable to be added to the LP, otherwise we proved that no such path exists (for the pair (s,t)). Note that each passenger path can assumed to be simple: just remove cycles of length 0 – or trust Dijkstra's algorithm, which produces only simple paths.

4.2 Pricing of the Line Variables

The pricing problem for the line variables f_ℓ is more complicated. The reduced cost $\overline{\gamma}_\ell$ for a variable f_ℓ is

$$\overline{\gamma}_\ell = \gamma_\ell - \kappa_\ell \, \boldsymbol{\mu}(\ell) + \boldsymbol{\eta}(\ell) = \gamma_\ell - \sum_{e \in \ell} \left(\kappa_\ell \left(\mu_{a(e)} + \mu_{\overline{a}(e)} \right) - \eta_e \right).$$

The corresponding pricing problem consists in finding a *suitable* path ℓ of mode m such that

$$
\begin{aligned}
\overline{\gamma}_\ell < 0 &\Leftrightarrow \gamma_\ell - \sum_{e \in \ell} \left(\kappa_\ell \left(\mu_{a(e)} + \mu_{\overline{a}(e)} \right) - \eta_e \right) < 0 \\
&\Leftrightarrow C_\ell / F + c_\ell - \sum_{e \in \ell} \left(\kappa_\ell \left(\mu_{a(e)} + \mu_{\overline{a}(e)} \right) - \eta_e \right) < 0 \\
&\Leftrightarrow C_m / F + \sum_{e \in \ell} c_e - \sum_{e \in \ell} \left(\kappa_\ell \left(\mu_{a(e)} + \mu_{\overline{a}(e)} \right) - \eta_e \right) < 0 \\
&\Leftrightarrow C_m / F + \sum_{e \in \ell} \left(c_e - \kappa_m \left(\mu_{a(e)} + \mu_{\overline{a}(e)} \right) + \eta_e \right) < 0. \\
&\Leftrightarrow \sum_{e \in \ell} \left(\kappa_m \left(\mu_{a(e)} + \mu_{\overline{a}(e)} \right) - \eta_e - c_e \right) > C_m / F.
\end{aligned}
$$

This problem turns out to be a longest weighted simple path problem, since the weights $\left(\kappa_\ell \left(\mu_{a(e)} + \mu_{\overline{a}(e)} \right) \right) - \eta_e - c_e$ are not restricted in sign and the graph G is in general not acyclic. Hence, the pricing problem for the line variables is $\mathcal{NP}$-hard (even for planar graphs). Note that longest non-simple path problems will often be "unbounded", e.g., because of repeated subsequences of the form $(\ldots, u, v, u, \ldots)$, which will lead to paths of "infinite length". As discussed in Section 3.4, we therefore restrict our attention to simple paths. In the rest of this section, we explain how this problem can be solved in practice.

For the following we fix some mode $m \in \mathcal{M}$ and, for convenience, write $G = (V, E)$ for G_m and $\mathcal{T}$ for $\mathcal{T}_m$. We let $n = |V|$ and $m = |E|$. We are now given edge weights w_e ($e \in E$) as described above, which are assumed to be arbitrary (rational) numbers. The pricing problem amounts to finding a longest weighted path in G with respect to $\mathbf{w}$ from each node $s \in \mathcal{T}$ to each node $t \in \mathcal{T} \setminus \{s\}$.

For any fixed path-length $k \in \mathbb{N}$ we can solve the problem to find a longest simple path using at most k edges by enumeration in polynomial time. We want to give two arguments that lines in typical transportation networks are not too long. The first argument is based on an idea of a transportation network as a planar graph, probably of high connectivity. Suppose this network occupies a square, in which its n nodes are evenly distributed. A typical line starts in the outer regions of the network, passes through the center, and ends in another outer region; we would expect such a line to be of length $\mathcal{O}(\sqrt{n})$. Real networks, however, are not only (more or less) planar, but often resemble trees. In a *balanced* and preprocessed tree, such that each node degree is at least 3, the length of a path between any two nodes is only $\mathcal{O}(\log n)$.

We now provide a result which shows that the longest weighted *simple* path problem can be solved in polynomial time in the case when the maximal number of edges k occurring in a path satisfies $k \in \mathcal{O}(\log n)$. This result is a direct generalization of work by Alon *et al.* (1995). Their method works both for directed and undirected graphs.

The goal of their work is to find induced paths of fixed length $k-1$ in a graph. The basic idea is to randomly color the nodes of the graph with k colors and only allow paths that use distinct colors for each node; such paths are called *colorful* with respect to the coloring and are necessarily simple. Choosing a coloring $c : V \to \{1, \dots, k\}$ uniformly at random, every simple path using at most $k - 1$ edges has a chance of at least $k!/k^k > e^{-k}$ to be colorful with respect to c. If we repeat this process $\alpha \cdot e^k$ times with $\alpha > 0$, the probability that a given simple path p with at most $k - 1$ edges is never colorful is less than

$$\left(1 - e^{-k}\right)^{\alpha \cdot e^k} < e^{-\alpha}.$$

Hence, the probability that p is colorful at least once is at least $1 - e^{-\alpha}$. The search for such colorful paths is performed by dynamic programming, which leads to an algorithm running in $n \cdot 2^{\mathcal{O}(k)}$ time and provides the correct result with high probability. This algorithm is then derandomized.

We have the following result, which can easily be generalized to directed graphs.

Proposition 2. *Let $G = (V, E)$ be a graph, let k be a fixed number, and $c : V \to \{1, \dots, k\}$ be a coloring of the nodes of G. Let s be a node in G and (w_e) be edge weights. Then colorful longest paths with respect to $\mathbf{w}$ using at most $k - 1$ edges from s to every other node can be found in time $\mathcal{O}\left(m \cdot k \cdot 2^k\right)$, if such paths exist.*

Proof. We find the length of the longest such path by dynamic programming. Let $v \in V$, $i \in \{1, \dots, k\}$, and $C \subseteq \{1, \dots, k\}$ with $|C| \leq i$. Define $w(v, C, i)$ to be the weight of the longest colorful path with respect to $\mathbf{w}$ from s to v using at most $i - 1$ edges and using the colors in C. Hence, for each iteration i we store the set of colors of all longest colorful paths from s to v using at most $i - 1$ edges. Note that we do not store the set of paths, only their colors. Hence, at each node we store at most 2^i entries. The entries of the table are initialized with minus infinity and we set $w(s, \{c(s)\}, 1) = 0$.

At iteration $i \geq 1$, let (u, C, i) be an entry in the dynamic programming table. If for some edge $e = \{u, v\} \in E$ we have $c(v) \notin C$, let $C' = C \cup \{c(v)\}$ and set

$$w(v, C', i + 1) = \max \left\{ w(u, C, i) + w_e, \; w(v, C', i + 1), \; w(v, C', i) \right\}.$$

The term $w(v, C', i+1)$ accounts for the cases where we already found a longer path to v (using at most i edges), whereas $w(v, C', i)$ makes sure that paths using at most $i - 1$ edges to v are accounted for. After iteration $i = k$, we take the maximum of all entries corresponding to each node v, which is the wanted result. The number of updating steps is bounded by

$$\sum_{i=0}^{k} i \cdot 2^i \cdot m = m \cdot \left(2 + 2^{k+1}(k - 1) \right) = \mathcal{O}\left(m \cdot k \cdot 2^k \right).$$

The sum on the left side of this equation arises as follows. In iteration i, m edges are considered; each edge $\{u, v\}$ starts at node u, to which at most 2^i labels $w(u, C, i)$ are associated, one for each possible set C; for each such set, checking whether $c(v) \in C$ takes time $\mathcal{O}(i)$. The summation formula itself can be proved by induction (Petkovsek *et al.*, 1996, Exc. 5.7.1, p. 95). The algorithm can be easily modified to actually find a wanted path.   □

We can now follow the above described strategy to produce an algorithm which finds a longest weighted simple path in $\alpha \, e^k \, \mathcal{O}\left(mk2^k \right) = \mathcal{O}\left(m \cdot 2^{\mathcal{O}(k)} \right)$ time with high probability. Then a derandomization can be performed by a clever enumeration of colorings such that each simple path with at most $k - 1$ edges is colorful with respect to at least one such coloring. Alon et al. combine several techniques to show that $2^{\mathcal{O}(k)} \cdot \log n$ colorings suffice. This yields:

Theorem 1. *Let $G = (V, E)$ be a graph and let k be a fixed number. Let s be a node in G and (w_e) be edge weights. Then a longest simple path with respect to $\mathbf{w}$ using at most $k - 1$ edges from s to every other node can be found in time $\mathcal{O}\left(m \cdot 2^{\mathcal{O}(k)} \cdot \log n \right)$, if such a path exists.*

If $k \in \mathcal{O}(\log n)$, this yields a polynomial time algorithm. Hence, by the discussion above and the polynomial equivalence of separation and optimization, see Grötschel *et al.* (1993), applied to the dual LP, it follows that the LP relaxation (LP_2) can be solved in polynomial time in this case. On the other hand we have the following result.

Proposition 3. *It is $\mathcal{NP}$-hard to compute a longest path of length at most k, if $k \in \mathcal{O}\left(n^{1/N} \right)$ for fixed $N \in \mathbb{N} \setminus \{0\}$.*

Proof. Consider an instance (H, s, t) of the Hamiltonian path problem, where the graph H has n nodes. We add $(n^N - n)$ isolated nodes to H in order to obtain the graph H' with n^N nodes, which is polynomial in n. Let the weights on the edges be 1. If we would be able to find a longest simple path with at most $k = (n^N)^{1/N} = n$ edges starting from s, we could solve the Hamiltonian path problem for H.   □

5 Conclusions

In this paper, we presented two novel models for the line planning problem, which allow to compute optimal line routes and passenger paths, and investigated their LP relaxations. We started to implement the second model, solving the line route pricing problem by enumeration. Preliminary computational experience shows that this approach is feasible to solve the LP relaxation of this line planning model for a medium sized city. We are currently working on the solution of the integer program and on the evaluation of the practicability of our approach.

Acknowledgements: We thank Volker Kaibel for pointing out Proposition 3. This research is supported by the DFG Research Center MATHEON "Mathematics for key technologies" in Berlin.

References

Alon, N., Yuster, R., and Zwick, U. (1995). Color-coding. *Journal of the Association of Computing Machinery*, **42**(4), 844–856.

Bouma, A. and Oltrogge, C. (1994). Linienplanung und Simulation für öffentliche Verkehrswege in Praxis und Theorie. *Eisenbahntechnische Rundschau*, **43**(6), 369–378.

Bussieck, M. R. (1997). *Optimal Lines in Public Rail Transport*. Ph.D. thesis, TU Braunschweig.

Bussieck, M. R., Winter, T., and Zimmermann, U. T. (1997a). Discrete optimization in public rail transport. *Mathematical Programming*, **79B**(1–3), 415–444.

Bussieck, M. R., Kreuzer, P., and Zimmermann, U. T. (1997b). Optimal lines for railway systems. *European Journal of Operational Research*, **96**(1), 54–63.

Bussieck, M. R., Lindner, T., and Lübbecke, M. E. (2004). A fast algorithm for near optimal line plans. *Mathematical Methods in Operations Research*, **59**(2).

Ceder, A. and Israeli, Y. (1992). Scheduling considerations in designing transit routes at the network level. In M. Desrochers and J.-M. Rousseau, editors, *Computer-Aided Transit Scheduling*, volume 386 of *Lecture Notes in Economics and Mathematical Systems*, pages 113–136. Springer, Berlin.

Ceder, A. and Wilson, N. H. M. (1986). Bus network design. *Transportation Research*, **20B**(4), 331–344.

Claessens, M. T., van Dijk, N. M., and Zwaneveld, P. J. (1998). Cost optimal allocation of rail passanger lines. *European Journal of Operational Research*, **110**(3), 474–489.

Correa, J. R., Schulz, A. S., and Stier Moses, N. E. (2004). Selfish routing in capacitated networks. *Mathematics of Operations Research*, **29**, 961–976.

Daduna, J. R., Branco, I., and Paixão, J. M. P., editors (1995). *Computer-Aided Transit Scheduling*, volume 430 of *Lecture Notes in Economics and Mathematical Systems*. Springer, Berlin.

Desrosiers, J., Dumas, Y., Solomon, M. M., and Soumis, F. (1995). Time constrained routing and scheduling. In M.O. Ball, T.L. Magnanti, C.L. Monma, and G.L. Nemhauser, editors, *Network Routing*, volume 8 of *Handbooks in Operations Research and Management Science*, pages 35–139. North-Holland, Amsterdam.

Dubois, D., Bel, G., and Llibre, M. (1979). A set of methods in transportation network synthesis and analysis. *Journal of the Operational Research Society*, **30**, 797–808.

Goossens, J.-W. H. M., van Hoesel, S., and Kroon, L. G. (2002). On solving multitype line planning problems. METEOR Research Memorandum RM/02/009, University of Maastricht.

Goossens, J.-W. H. M., van Hoesel, S., and Kroon, L. G. (2004). A branch-and-cut approach for solving railway line-planning problems. *Transportation Science*, **38**, 379–393.

Grötschel, M., Lovász, L., and Schrijver, A. (1993). *Geometric Algorithms and Combinatorial Optimization*, volume 2 of *Algorithms and Combinatorics*. Springer, Heidelberg, 2nd edition.

Israeli, Y. and Ceder, A. (1995). Transit route design using scheduling and multiobjective programming techniques. In Daduna *et al.* (1995), pages 56–75.

Kim, D. and Barnhart, C. (1997). Transportation service network design: Models and algorithms. In N. H. M. Wilson, editor, *Computer-Aided Transit Scheduling*, volume 471 of *Lecture Notes in Economics and Mathematical Systems*, pages 259–283. Springer, Berlin.

Mandl, C. E. (1980). Evaluation and optimization of urban public transportation networks. *European Journal of Operational Research*, **5**, 396–404.

Odoni, A. R., Rousseau, J.-M., and Wilson, N. H. M. (1994). Models in urban and air transportation. In S. M. Pollock, M. H. Rothkopf, and A. Barnett, editors, *Operations Research and the Public Sector*, volume 6 of *Handbooks in Operations Research and Management Science*, pages 107–150. North Holland, Amsterdam.

Pape, U., Reinecke, Y.-S., and Reinecke, E. (1995). Line network planning. In Daduna *et al.* (1995), pages 1–7.

Petkovsek, M., Wilf, H. S., and Zeilberger, D. (1996). $A = B$. A. K. Peters, Wellesley, MA.

Silman, L. A., Barzily, Z., and Passy, U. (1974). Planning the route system for urban buses. *Computers & Operations Research*, **1**, 201–211.

Sonntag, H. (1979). Ein heuristisches Verfahren zum Entwurf nachfrageorientierter Linienführung im öffentlichen Personennahverkehr. *Zeitschrift für Operations Research*, **23**, B15–B31.

Improved Lower-Bound Fleet Size for Transit Schedules

Avishai Ceder

Civil and Environmental Engineering Faculty, Transportation Research Institute,
Technion-Israel Institute of Technology, Haifa 32000, Israel,
ceder@tx.technion.ac.il

Summary. This work describes a highly informative graphical technique for the problem of finding the lower bound of the number of vehicles required to service a given timetable of trips. The technique is based on a step function that has been applied over the last 20 years as an optimization tool for minimizing the number of vehicles in a fixed-trip schedule. The step function is called a Deficit Function (DF), as it represents the deficit number of vehicles required at a particular terminal in a multi-terminal transit system. The initial lower bound on the fleet size with deadheading (empty) trip insertions was found to be the maximum of the sum of all DFs. An improved lower bound was established later, based on extending each trip's arrival time to the time of the first feasible departure time of a trip to which it may be linked or to the end of the finite time horizon. The present work continues the effort to improve the lower bound by introducing a simple procedure to achieve this improvement that uses additional extension possibilities for a certain trip's arrival times.

1 Background on the Deficit Function

The minimum fleet size problem may be referred to with or without deadheading (DH) trips. When DH is allowed, we can reach the counterintuitive result of decreasing the required resources (fleet size) by introducing more work into the system (adding DH trips). This approach assumes that the capital cost of saving a vehicle far outweighs the cost of any increased operational cost (driver and vehicle travel cost) imposed by the introduction of DH trips.

1.1 Definitions and Notations

Let $I = \{i : i = l, \ldots, n\}$ denote a set of required trips. The trips are conducted between a set of terminals $K = \{k : k = l, \ldots, q\}$, each trip to be serviced by a single vehicle, and each vehicle able to service any trip. Each trip i can be represented as a 4-tuple (p^i, t_s^i, q^i, t_e^i), in which the ordered elements denote departure terminal, departure (start) time, arrival terminal, and arrival (end) time. It is assumed that each trip i lies within a schedule horizon $[T_1, T_2]$, i.e., $T_1 \leq t_s^i \leq t_e^i \leq T_2$. The set of all trips $S = \{(p^i, t_s^i, q^i, t_e^i) : p^i, q^i \in K, i \in I\}$ constitutes the timetable. Two trips i, j may be serviced sequentially (feasibly joined) by the same vehicle if and only if (a) $t_e^i \leq t_s^j$ and (b) $q^i = p^j$.

A deficit function is a step function defined across the schedule horizon that increases by one at the time of each trip departure and decreases by one at the time of each trip arrival. This step function is called a deficit function (DF) because it represents the deficit number of vehicles required at a particular terminal in a multi-terminal transit system. To construct a set of DFs, the only information needed is a timetable of required trips. The main advantage of the DF is its visual nature. Let $d(k, t)$ denote the DF for terminal k at time t for a given schedule. The value of $d(k, t)$ represents the total number of departures minus the total number of trip arrivals at terminal k, up to and including time t. The maximum value of $d(k, t)$ over the schedule horizon $[T_1, T_2]$, designated $D(k)$, depicts the deficit number of vehicles required at k.

1.2 DH Trip Insertion and Initial Lower Bound on the Fleet Size

This section follows Ceder and Stern (1981) and Stern and Ceder (1983). A DH trip is an empty trip between two termini that is usually inserted into the schedule in order to (i) ensure that the schedule is balanced at the start and end of the day, (ii) transfer a vehicle from one terminal where it is not needed to another where it is needed to service a required trip, and (iii) refuel or undergo maintenance.

Consider the example in Fig. 1. In its present configuration, according to the fleet size formula (Ceder and Stern (1981)), four vehicles are required at terminal a, 0 at terminal b, and 1 at terminal c for a fleet size of five. That is, $D(k)$, for all k, determines the minimum number of vehicles required at k. The dashed arrows in Fig. 1 represent the insertion of DH_1 trip from b to a and DH_2 from c to b. After the introduction of these DH trips into the schedule, the DFs at all three terminals are shown updated by the dotted lines. The net effect is a reduction in fleet size by one unit at terminal a. It is interesting to examine the particular circumstances under which this reduction was achieved. After adding an arrival point in the first hollow of terminal a before s_1^a, the maximal interval when using DH_1 is reduced by one unit, causing a unit decrease in the deficit at a. This arrival point becomes, therefore, e_1^a.

Since the DH_1 departure point is added in the middle hollow of terminal b, at e_1^b, it is necessary to introduce a second DH trip, which will arrive at the start of the second maximum interval of b. Fortunately, this DH_2 trip departs from the last hollow of c, where it could no longer affect the deficit at c. In general, it is possible to have a string of DH trips to reduce the fleet size by one unit: one "initiator trip" and the others "compensating trips."

The initial lower bound on the fleet size with DH trip insertions was found by Ceder and Stern (1981) to be the maximum of the sum of all DFs, $g(t)$, as shown in Fig. 1 by G. This initial lower bound is determined as 3 before inserting DH trips and becomes 4 after this insertion.

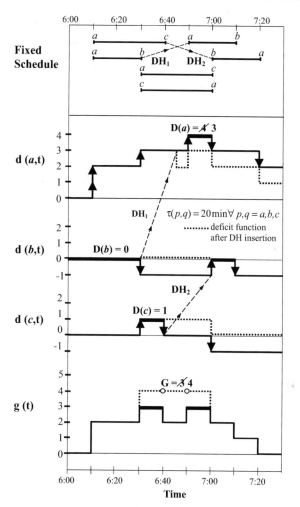

Fig. 1. Description of Six-trip, Two-terminal Example in Which the Fleet Size is Reduced by One Using a Chain of Two DH Trips (URDHC) and in Which $g(t)$ is Changed

2 Fleet Size Lower Bound

2.1 Overview and Example

An improved lower bound to that presented in Fig. 1 was established by Stern and Ceder (1983), based on extending each trip's arrival time to the time of the first feasible departure of a trip to which it may be linked or to the end of the finite time horizon. The direct calculation of the fleet size lower bound enables schedulers and transit decision-makers to ascertain more promptly how much the fleet size can be reduced by DH trip insertions and allowing shifts in departure times.

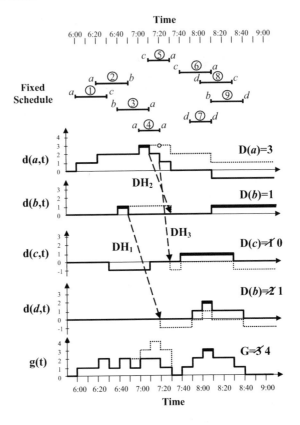

Fig. 2. Nine-trip Example With DH Trip Insertions for Reducing Fleet Size

Fig. 2 presents a nine-trip example with four terminals (a, b, c, and d). Table 1 shows the data required for the simple example used for demonstrating further improved lower-bound methods. Four DFs are constructed along with the overall DF. According to the next terminal (NT) procedure (see Ceder and Stern (1981)), terminal d (whose first hollow is the longest) is selected for a possible reduction in $D(d)$. The DH-insertion process selects two unit reduction DH chains (URDHC) in Fig. 2; i.e., $DH_1 + DH_2$, and the second DH_3. The result is that $D(c)$ and $D(d)$ are reduced from 1 to 0 and from 2 to 1, respectively; hence, $N = D(S) = 5$, and G is increased from 3 to 4 using three inserted DH trips.

2.2 Stronger Fleet Size Lower Bound

While Stern and Ceder (1983) extended each unlinked trip's departure time (i.e., one that cannot be linked to any trip's arrival time) to both T_1 and T_2, it is easy to show and prove that an extension only to T_2 is sufficient. The extension to the time of the

Table 1. Input Data for the Problem Illustrated in Fig. 2

Trip No.	Departure Terminal	Departure Time	Arrival Terminal	Arrival Time	DH Trips	
					Between Terminals	DH Time (same for both directions)
1	a	6:00	c	6:30	$a - b$	20 min
2	a	6:20	b	6:50	$a - c$	10 min
3	b	6:40	a	7:10	$a - d$	60 min
4	a	7:00	a	7:20	$b - c$	30 min
5	c	7:10	a	7:30	$b - d$	30 min
6	c	7:40	a	8:10	$c - d$	20 min
7	d	7:50	d	8:10		
8	d	8:00	c	8:30		
9	b	8:30	d	9:00		

first feasible departure time of a trip with which it may be linked, or to T_2, results in a schedule S' and an overall DF, $g'(t, S')$, with its maximum value $G'(S')$.

While S' is being created, it is possible that several trip-arrival points are extended forward to the same departure point that is their first feasible connection. However, in the final solution of the minimum fleet size problem, only one of these extensions will be linked to the single departure point. This observation provides an opportunity to look into further artificial extensions of certain trip-arrival points without violating the generalization of requiring all possible combinations for maintaining the fleet size at its lower bound.

Fig. 3 illustrates three cases of multiple extensions to the same departure point. Case (i) shows two extensions, Trips 1 and 2, both with the same arrival point b, which is their first feasible connection at point a of Trip 3. Because only one of the two trips will be connected to Trip 3, the question is, which one can be extended further? It is clear that Trip 1 has better DH chances to be connected to Trip 4 than to Trip 2 because of its longer DH time. Hence, Trip 1 can be further extended (2nd extension) to the start of Trip 4 if it is feasible. Case (ii), Fig. 3, shows that Trips 1 and 2 do not end at the same point and that Trip 4 has different points than in Case (i). The argument of Case (i) cannot hold here, since the DH time differs between each two different points. In this case, the second feasible connection for Trip 1 is T_2. By using the Case (i) argument, one can then create three possible chains [1], [2-3], [4], instead of two chains: [1-3], [2-4]. Case (iii) shows an opposite situation to that of Case (ii), with multiple extensions from different arrival points. If we link, in Case (iii), Trips 1 (longest DH time to the common departure point) and Trip 3 and extend Trip 2 to Trip 4, we have another multiple extension case like Case (i), this one concerning the start of Trip 4 (linked to Trips 2 and 3). Following the Case (ii) argument, Trip 3 will be linked to Trip 4, and Trip 2 will have its third extension. This results in three possible chains: [1-3-4], [2], [5], instead of two: [1-5] and [2-3-4]. Cases (ii) and (iii) show why it is impossible to apply any general rule to a multiple extension of different arrival epochs. Consequently further improvement of $G'(S')$ can be made only for Case (i) situations.

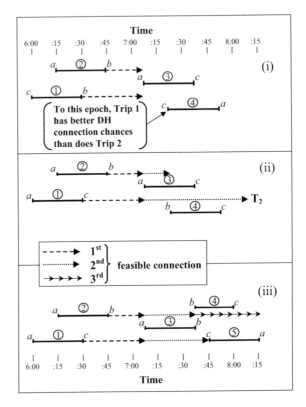

Fig. 3. Part (i) Shows Why One Should Select the Trip 2 Extension; Part (ii) Shows that the Argument in (i) Cannot be Used in Case of Multiple Connections from Different Terminals; Part (iii) Shows Another Case in Which Multiple Connections Cannot be Applied for Constructing the Lower Bound

Following is the procedure for finding a stronger fleet size lower bound.

1. Establish S'.
2. Select a case in which more than one extension is linked to the same departure time t^j_{sk} of trip j at terminal k. If no more such cases–STOP. Otherwise, select a group (two or more) of extensions with the same scheduled arrival terminal, u, and apply the following steps:

   2a. Find a trip that fulfills: $\min_{i \forall i \in E_u}(t^j_{sk} - t^i_{eu})$, where E_u = set of all trips arriving at u and extended to t^j_{sk}, and t^i_{eu} is the arrival epoch of trip i at terminal u;

   2b. Perform the second feasible extension for all trips $i \in E_u$, except the one selected in Step 2a. Go to Step 2.

Using this procedure, define the overall DF of the extended S' schedule by $g''(t, S'')$ with the maximum value $G''(S'')$. The following theorem and its proof establish that $G''(S'')$ is a stronger lower bound than $G'(S')$.

Theorem 1:   Let $N_o(S)$ be the minimum fleet size for S with DH insertions. Let $G'(S')$ and $G''(S'')$ be the maximum value of the overall DF for S' and S'', respectively. Then: (i) $G''(S'') \geq G'(S')$, and (ii) $G''(S'') \leq N_o(S)$.

Proof: (i) The new overall DF, $g''(t, S'')$, has more extensions than $g'(t, S')$; i.e., $g''(t, S'') \geq g'(t, S')$. Therefore, $G''(S'') \geq G'(S')$. (ii) According to the definition of S'', at any time t in which $g''(t, S'') = G''(S'')$, there exist $G''(S'') - g'(t, S')$ trip extensions over S'. The additional extensions in S'' represent multiple extensions (2nd, 3rd, ...), given that each extended trip is associated with another trip having the same arrival epoch and terminal, and has only one extension. In the optimal chain solution, a departure time t_s^* may or may not be linked to its nearest feasible arrival epoch (t_e^*) across all other points representing the same arrival terminal. Linkage to t_e^* complies with the procedure to construct S''. Otherwise, t_e^* in S'' is further extended either to another trip or to T_2 while t_s^* is linked to $t_e^{**} < t_e^*$. We should note that t_e^{**} is linked to t_s^{**} when using the procedure described. Because t_e^* to t_s^* is the shortest link, the additional extension of t_e^* cannot be linked to a trip that starts before t_s^{**} (otherwise, t_e^{**} too will be linked to it, and not to t_s^*). Therefore, the additional extension of t_e^* in the optimal chain solution, $N_o(S)$, results in a greater overlap between trips (when constructing $g''(t, S'')$). Hence, $G''(S'') \leq No(S)$. Q.E.D.

Fig. 4 presents the schedule of Fig. 2, with S' in its upper part, S'' in its middle part, and three overall DFs–$g(t, S)$, $g'(t, S')$, and $g''(t, S'')$–in the lower part. For S', it may be observed that Trips 3, 4, and 5 are extended to the same departure point as Trip 6 from the same arrival terminal a. According to the procedure for constructing S'', the extension of Trip 5 is selected, and Trips 3 and 4 are further extended to the departure time of Trip 9. These additional extensions create another multiple connection associated with Trips 3 and 4, in which Trip 4 is the selected extension and Trip 3 is further extended (3rd time). The initial lower bound is $G = 3$, the first improved lower bound is $G' = 4$, and the proposed improved lower bound is $G'' = 5$, which happens to be the optimal solution (see Fig. 2).

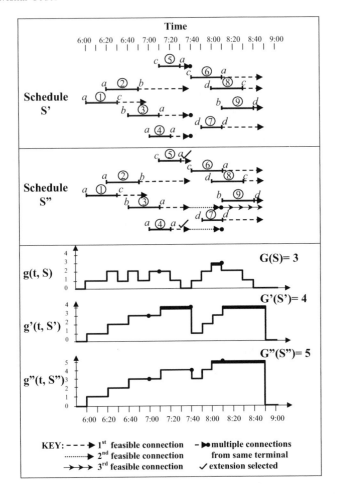

Fig. 4. Lower-bound Determination Using the Example Shown in Fig. 2, With the First and Second Improvement Procedures

References

Ceder, A. and Stern, H. I. (1981). Deficit function bus scheduling with deadheading trip insertion for fleet size reduction. *Transportation Science*, **15**, 338–363.

Stern, H. I. and Ceder, A. (1983). An improved lower bound to the minimum fleet size problem. *Transportation Science*, **17**, 471–477.

A Tabu Search Based Heuristic Method for the Transit Route Network Design Problem

Wei Fan and Randy B. Machemehl

Department of Civil Engineering, Ernest Cockrell, Jr. Hall, 6.9, University of Texas at Austin, Austin, TX 78712-1076, USA
{weifan, rbm}@mail.utexas.edu

Summary. Systematic tabu search based meta-heuristic algorithms are designed and implemented for the transit route network design problem. A multi-objective nonlinear mixed integer model is formulated. Solution methodologies based on three variations of tabu search methods are proposed and tested using a small experimental network as a pilot study. Sensitivity analysis is performed, a comprehensive characteristics analysis is conducted and numerical results indicate that the preferred tabu search method outperforms the genetic algorithm used as a benchmark.

1 Introduction

Public transit has been widely recognized as a potential way of reducing air pollution, lowering energy consumption, improving mobility and lessening traffic congestion. Designing an operationally and economically efficient bus transit network is very important for the urban area's social, economic and physical structure.

Generally speaking, the network design problem involves the minimization (or maximization) of some intended objective subject to a variety of constraints, which reflect system performance requirements and/or resource limitations. In the past decade, several research efforts have examined the *bus transit route network design problem* (BTRNDP). Previous approaches that were used to solve the BTRNDP can be classified into three categories: 1) Practical guidelines and ad hoc procedures; 2) Analytical optimization models for idealized situations; and 3) Meta-heuristic approaches for more practical problems. NCHRP Synthesis of Highway Practice 69 (1980) provides industry rule-of-thumb service planning guidelines. Furthermore, in the early research efforts, traditional operations research analytical optimization models were used. Rather than determining both the route structure and design parameters simultaneously, these analytical optimization models were primarily applied to determine one or several design parameters (e.g., stop spacing, route spacing, route length, bus size and/or frequency of service) on a predetermined transit route network structure. Generally speaking, these models are very effective in solving optimization-related problems for networks of small size or with one or two decision variables. However, when it comes to the transit route design problem for a network of realistic size in which many parameters need to be determined, this approach does not work very well. Due to the inherent complexity involved in the BTRNDP, the meta-heuristic approaches, which pursue reasonably good local optima but do not guarantee finding the global optimal solution, were therefore proposed. The meta-heuristic approaches primarily dealt with simultaneous design of the transit route network and determination of its associated bus frequencies. Examples of the general heuristic approaches can be seen in the work of Ceder and Wilson (1986), Baaj and Mahmassani (1992), and Shih *et al.* (1998). Genetic algorithm-based heuristic approaches that were used to solve the BTRNDP can be seen in Pattnaik *et al.* (1998), Chien *et al.* (2001) and Fan and Machemehl (2004).

However, the major shortcoming of most previous approaches is that they did not study the BTRNDP in the context of the "distribution node" (or bus stop) level and simply aggregate zonal travel demand into a single node. This precludes them as generally accepted applications for practical transportation networks because the frequency-based rule for the traditional transit trip assignment model based on this assumption is incorrect. Therefore, the BTRNDP should be considered in a more general real world situation. Furthermore, previous research efforts mainly centered on genetic algorithms and other potential heuristic algorithms such as tabu search methods are seldom used to solve the BTRNDP. To search for possibly good and/or better network solutions, these methods should be considered.

The objective of this paper is to systematically examine the underlying characteristics of the optimal BTRNDP in the context of the "distribution node" level. A multi-objective nonlinear mixed integer model is formulated for the BTRNDP. Characteristics and model structures of the Tabu Search (TS) algorithms are reviewed. A TS algorithm-based solution methodology is proposed. Three different variations of TS algorithms are employed and compared as the solution method for finding an optimum set of routes from the huge solution space. A genetic algorithm is also used as a benchmark to measure the quality of the TS methods. Numerical results including sensitivity analysis and characteristics identification are presented using an experimental network. The subsequent sections of this paper are organized as follows. Section 2 presents the model formulation of the BTRNDP from a systematic view. The objective function and related constraints are also described. Section 3 discusses general characteristics of the TS algorithms. Section 4 proposes the solution methodology for the BTRNDP, which contains three main components: an initial candidate route set generation procedure; a network analysis procedure and a TS procedure that guides the candidate solution generation process. Section 5 presents the applications of the proposed solution methodology to an experimental network and the numerical results are also discussed. Finally in Section 6, a summary concludes this paper.

2 Model Formulation

Essentially speaking, the transportation system is described in terms of "nodes," "links" and "routes." A *node* is used to represent a specific point for loading, unloading and/or transfer in a transportation network. Generally speaking, there are three kinds of nodes in a bus transit network system: (a) Nodes representing centroids of specific zones; (b) Nodes representing road intersections; and (c) Nodes with which zone centroid nodes are connected to the network through centroid connectors. Note that nodes could be real (identifiable on the ground) or fictitious. Furthermore, the term "distribution nodes" is introduced especially for the third kind of node. A *link* joins a pair of nodes and represents a particular mode of transportation between these nodes, which means that if two modes of transportation are involved with the same link, these are represented as two links, say walk mode and transit mode. This is natural since the travel time associated with every mode-specific link is different. A *route* is a sequence of nodes. Every consecutive pair of the node sequence must be connected by a link of the relevant mode. The bus line headway on any particular route is the inter-arrival time of buses running on that route. A *graph* (network) refers to an entity $G = \{N, A\}$ consisting of a finite set of N nodes and a finite set of A links (arcs) which connect pairs of nodes. A *transfer path* is a progressive path that uses more than one route. Note that a typical geographical zone system may be based upon census boundaries and all land areas are encompassed by streets or major physical barriers. The zone centroids are located somewhere near the centers of the zones and zone connectors are used to connect these centroids to the modeled network. Generally, the centroid node represents the "demand" center (origin and/or destination) of a specific traffic zone. Distribution nodes are the junctions of centroid connectors and road links and might physically represent bus stops. It should be pointed out that centroid connectors are usually fictitious and they are used as the origins and/or destinations for implementation of the shortest path and k shortest path algorithms. Furthermore, an important characteristic of these centroid connectors is the distances that transit users have to walk to get to the routes that provide service to their intended destinations. Note that the terms, "arc" and "link" are used interchangeably.

Consider a connected network composed of a directed graph $G = \{N, A\}$ with a finite number of nodes and arcs. The following notations are used.

Sets/Indices

$i, j \in N$ centroid nodes (i.e., zones)
$r_k \in R$ routes
$i_t \in N$ t-th distribution node of centroid node i
$tr \in R$ transfer paths that use more than one route from R

Data

R_{max} maximum allowed number of routes for the route network

D_{max} maximum length of any route in the transit network
D_{min} minimum length of any route in the transit network
d_{ij} bus transit travel demand between centroid nodes i and j
h_{max} maximum headway required for any route; (say, 60 minutes)
h_{min} minimum headway required for any route; (say, 5 minutes)
L_{max} maximum load factor for any route
P seating capacity of buses operating on the network
W maximum bus fleet size available for operations on the route network
C_v per-hour operating cost of a bus; ($/vehicle/hour)
C_m value of time; ($/minute)
O_v operating hours for the bus running on any route; (hours)
C_d value of one unsatisfied transit demand in dollars; ($/person)
C_i ($i = 1, 2, 3$) weights reflecting the relative importance of three components including the user costs, operator costs and unsatisfied total demand costs, respectively; note that $C_1 + C_2 + C_3 = 1$

Decision Variables

M the number of routes of the current proposed bus transit network solution
r_m the m-th route of the proposed solution, $m = 1, 2, \ldots, M$
D_{r_m} the overall length of route r_m
$d_{ij}^{r_m}$ the bus transit travel demand between centroid nodes i and j on route r_m
d_{ij}^{tr} the bus transit travel demand between centroid nodes i and j along transfer path tr
DR_{ij} the set of direct routes used to serve the demand from centroid nodes i and j
TR_{ij} the set of transfer paths used to serve the demand from centroid nodes i and j
$t_{ij}^{r_m}$ the total travel time between centroid node i and j on route r_m
t_{ij}^{tr} the total travel time between centroid node i and j along transfer path tr
h_{r_m} the bus headway operating on route r_m; (minutes/vehicle)
L_{r_m} loading factor in route r_m
T_{r_m} the round trip time of route r_m; $T_{r_m} = 2D_{r_m}/V_b$
N_{r_m} the number of operating buses required on route r_m; $N_{r_m} = T_{r_m}/h_{r_m}$
$Q_{r_m}^{max}$ the maximum flow occurring on the route r_m

Objective Function

The objective is to minimize the sum of operator cost, user cost and unsatisfied demand costs for the studied bus transit network. The objective function is as follows:

$$\min z = C_1 \cdot \left(\sum_{i \in N} \sum_{j \in N} \sum_{r_m \in DR_{ij}} d_{ij}^{r_m} t_{ij}^{r_m} + \sum_{i \in N} \sum_{j \in N} \sum_{tr \in TR_{ij}} d_{ij}^{tr} t_{ij}^{tr} \right) +$$

$$C_2 \cdot \frac{C_v}{C_m} \cdot O_v \cdot \left(\sum_{m=1}^{M} \frac{T_{r_m}}{h_{r_m}} \right) +$$

$$C_3 \cdot \frac{C_d}{C_m} \cdot \left(\sum_{i \in N} \sum_{j \in N} d_{ij} - \sum_{i \in N} \sum_{j \in N} \sum_{r_m \in DR_{ij}} d_{ij}^{r_m} - \sum_{i \in N} \sum_{j \in N} \sum_{tr \in TR_{ij}} d_{ij}^{tr} \right)$$

s.t.

$$h_{min} \le h_{r_m} \le h_{max} \qquad\qquad r_m \in R \text{ (headway feasibility constraint)}$$

$$L_{r_m} = \frac{Q_{r_m}^{max} \cdot h_{r_m}}{P} \le L_{max} \qquad r_m \in R \text{ (load factor constraint)}$$

$$\sum_{m=1}^{M} N_{r_m} = \sum_{m=1}^{M} \frac{T_{r_m}}{h_{r_m}} \le W \; r_m \in R \text{ (fleet size constraint)}$$

$$D_{min} \le D_{r_m} \le D_{max} \qquad\quad r_m \in R \text{ (trip length constraint)}$$

$$M \le R_{max} \qquad\qquad\qquad\quad \text{(maximum number of routes}$$
$$\text{constraint)}$$

$M, h_{r_m}, N_{r_m}, Q_{r_m}^{max}, d_{ij}^{r_m}, d_{ij}^{tr}$, are all integers.

The first term of the objective function is the total user cost (including the user cost on direct routes and that on transfer paths), the second part is the total operator cost, and the third component is the cost resulting from total travel demand excluding the transit demand satisfied by a specific network configuration. Note that C1, C2 and C3 are introduced to reflect the tradeoffs between the user costs, the operator costs and satisfied transit ridership, making the BTRNDP a multi-objective optimization problem. Generally, operator cost refers to the cost of operating the required buses. User costs usually consist of four components, including walking cost, waiting cost, transfer cost, and in-vehicle travel cost. The first constraint is the headway feasibility constraint, which reflects the necessary usage of policy headways in extreme situations. The second is the load factor constraint, which guarantees that the maximum flow on the critical link of any route r_m cannot exceed the bus capacity on that route. The third (fleet size) constraint represents the resource limits of the transit company and it guarantees that the optimal network pattern never uses more vehicles than currently available. The fourth constraint is the trip length constraint. This avoids routes that are too long because bus schedules on very long routes are too difficult to maintain. Meanwhile, to guarantee the efficiency of the network, the length of routes should not be too small. The fifth constraint is the maximum number of routes constraint, which reflects the fact that in solving the BTRNDP, transit planners often set a maximum number of routes, which is based on the fleet size. This has a great impact on the later driver scheduling work.

3 Tabu Search Algorithm

The TS algorithm has traditionally been used on combinatorial optimization problems and has been frequently applied to many integer programming, routing and scheduling, traveling salesman and related problems. The basic concept of TS is presented by Glover (1977), Glover (1986) who described it as a meta-heuristic superimposed on another heuristic. It explores the solution space by moving from a solution to the solution with the best objective function value in its neighborhood at each iteration, even in the case that this might cause the deterioration of the objective. (In this sense, "moves" are defined as the sequences that lead from one trial solution to another.) To avoid cycling, solutions that were recently examined are declared forbidden or "tabu" for a certain number of iterations and associated attributes with the tabu solutions are also stored. The tabu status of a solution might be overridden if it corresponds to a new best solution, which is called "aspiration." The tabu lists are historical in nature and form the Tabu Search memory. The role of the memory can change as the algorithm proceeds. Intensification strategies are based on modifying choice rules to encourage move combinations and solution features historically found good, and to initiate a return to attractive regions to search them more thoroughly. Diversification strategies are based on modifying choice rules to bring attributes into the solutions that are infrequently used, or to drive the search into new regions. Intensification and diversification are fundamental cornerstones of longer term memory in TS and reinforce each other. In many cases, various implementation models of the TS method can be achieved by changing the size, variability, and adaptability of the tabu memory to a particular problem domain. Basic versions of TS can be found in Glover (1989), Glover (1990), and variants ranging from simple to advanced can be found in Glover and Laguna (1997).

In all, TS is an intelligent search technique that hierarchically explores one or more local search procedures in order to search quickly for the global optimum. As one of the advanced heuristic methods, TS is generally regarded as a method that can provide a near-optimal or at least local optimal solution within a reasonable time for the BTRNDP. Details of our BTRNDP-specific TS algorithms are presented in Section 4.

4 Proposed Solution Methodology

The proposed solution framework consists of three main components: an *Initial Candidate Route Set Generation Procedure* (ICRSGP) that generates all feasible routes incorporating practical guidelines that are commonly used in the bus transit industry; a *Network Analysis Procedure* (NAP) that assigns the transit trips, determines the service frequencies on each route and computes many performance measures; and, a TS Procedure that combines these two parts, guides the candidate solution generation process and selects an optimum set of routes from the huge solution space. Fig. 1 gives the flow chart of the proposed solution framework. C++ is chosen as the implementation language in this research.

4.1 The Initial Candidate Route Set Generation Procedure (ICRSGP)

The ICRSGP configures all candidate routes for the current transportation network. It requires the user to define the minimum and maximum route lengths. The knowledge

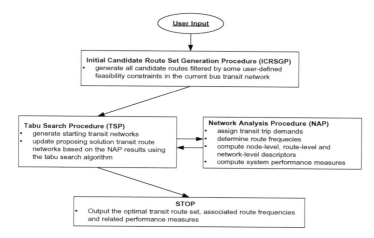

Fig. 1. Flow Chart of the Proposed Solution Methodology

of the transit planners has a significant impact on the initial route set skeletons, i.e., different user requirements result in different route solution space sets. ICRSGP relies mainly on algorithmic procedures including the shortest path and k-shortest path algorithms. Given the user-defined minimum and maximum length constraints, Dijkstra's shortest path algorithm (see Ahuja *et al.* (1993)) is used and Yen's k-shortest path algorithm (see Yen (1971)) is modified to generate all candidate feasible routes in the studied transportation network. Fig. 2 presents a skeleton for the ICRSGP.

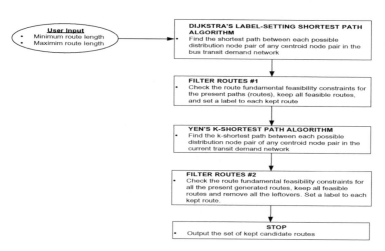

Fig. 2. Skeleton of the Initial Candidate Route Set Generation Procedure (ICRSGP)

4.2 The Network Analysis Procedure (NAP)

Fig. 3 shows the flow chart of the proposed network analysis procedure for the BTRNDP. Essentially, the NAP proposed in this paper is a bus transit network evaluation tool with the ability to assign transit trips between each centroid node pair onto each route in the proposed solution network and determine associated route frequencies. To accomplish these tasks for the BTRNDP, NAP employs an iterative procedure, which contains two major components, namely, a multiple transit trip assignment procedure and a frequency setting procedure, to seek to achieve internal consistency of the route frequencies.

Once a specific set of routes is proposed by the TS procedure in the overall candidate solution route set generated by the ICRSGP, the NAP is called to evaluate the alternative network structure and determine route frequencies. The whole NAP process can be described as follows. First, an initial set of route frequencies are specified because they are necessary before the beginning of the trip assignment process. Then, hybrid transit trip assignment models are utilized to assign the passenger trip demand matrix to a given set of routes associated with the proposed network configuration. The service frequency for each route is then computed and used as the input frequency for the next iteration in the transit trip assignment and frequency setting procedure. If these route frequencies are considered to be different from previous frequencies by a user-defined parameter, the process iterates until internal consistency of route frequencies is achieved. Once this convergence is achieved, route frequencies and several system performance measures (such as the fleet size and the unsatisfied transit demands) are thus obtained.

It should be noted that the trip assignment process considers each zone (centroid node) pair separately. Also, the transit trip assignment model presented in this paper adapts the lexicographic strategy (see Han and Wilson (1982)) and the previous transit trip assignment methods (see Shih *et al.* (1998)). However, several modifications have been made to accommodate more complex considerations for real world application. This model considers the number of transfers and/or the number of long walks to the bus station as the most important criterion. It first checks the existence of the 0-transfer-0-longwalk paths. If any path of this category is found, then the transit demand between this centroid node pair can be provided with direct route service and the demand is therefore distributed to these routes. If not, the existence of paths of the second category, i.e., 0-transfer-1-longwalk path and 1-transfer-0-longwalk paths are checked. If none of these paths is found, the proposed procedure will continue to search for paths of the third category, i.e., paths with 2-transfer-0-long-walk, 1-transfer-1-long-walk and/or 0-transfer-2-longwalks. Only if no paths that belong to these three categories exist, there would be no paths in the current transit route system that can provide service for this specific centroid node pair (i.e., these demands are unsatisfied). Note that at any level of the above three steps, if more than one path exists, a "travel time filter" is introduced for checking the travel time on the set of competing paths obtained at that level. If one or more alternative paths whose travel time is within a particular range pass the screening process, an analytical nonlinear model (i.e., the inverse proportional model) that reflects the relative utility on these

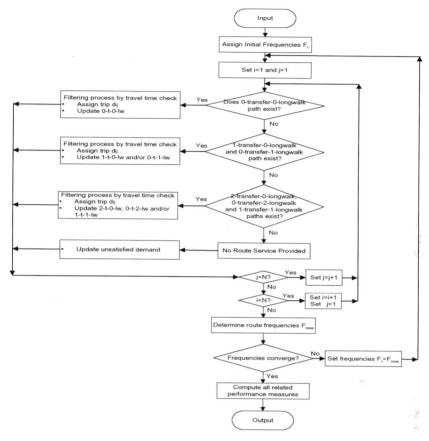

Fig. 3. Network Analysis Procedure (NAP) for the BTRNDP

competing paths is used to assign the transit trips between that centroid node pair to the network. In addition, policy headway and the demand headway are used together to determine the frequencies on each route in the frequency setting procedure. The whole process is repeated until all the travel demand pairs in the studied network are considered. Details of the transit trip assignment model can be seen in Fan and Machemehl (2004).

4.3 Tabu Search Procedure

Since the TS provides a robust search as well as a near optimal solution in a reasonable time, this approach is employed as one of the candidate solution techniques for BTRNDP. The following subsections present a systematic description for the TS algorithm-based implementation model for the BTRNDP.

Tabu Search Implementation Model: As with other heuristic algorithms, applying TS methods requires a significant amount of knowledge specific to the BTRNDP.

To make TS a potentially efficient algorithm for the BTRNDP, careful attention is required. Note that one of the significant contributions in this paper is using the TS algorithm to solve the BTRNDP. Since it is the first time for the TS methods to be applied for the BTRNDP, a detailed description of the BTRNDP-specific TS is presented.

Solution Representation: At any iteration t of the algorithm, let n represent the proposed solution route set size. A candidate bus transit route solution network can be represented by $X^t = (R_1^t, R_2^t, \ldots R_i^t, \ldots, R_n^t)$, where $R_i^t (i = 1, 2, \ldots, n)$ denotes the i-th bus route in the proposed solution set. Although the vector X^t is treated as ordered by the algorithm, it should be pointed out that X^t can also be treated as a set rather than a vector, and its ordering serves as a record keeping device for the algorithm rather than identifying a structural property of the solution itself. Let $f(X^t)$ represent the objective function as shown in the model formulation part for the proposed solution network defined by this n transit route network configuration $X^t = (R_1^t, R_2^t, \ldots, R_n^t)$.

Initial Solution: In this paper, all initial solutions for three different versions of the TS algorithms are randomly generated, with each solution being uniformly distributed in the solution space generated by the ICRSGP.

Neighborhood Structure: Undoubtedly, how to define the "neighborhood," i.e., the nearby solutions, might affect the quality of the transit route network solution. A different definition rule could result in a different solution of different quality. In this research, the neighborhood of a feasible solution route network set X^t is another feasible solution obtained by replacing one of the routes in the current proposed solution set, say the i-th route R_i^t to one of the routes that is next to R_i^t in the stored solution space. For route 1, the neighborhood can be defined as route 2 and route N, where N is the total number of routes in the stored solution space. For route N, the neighborhood can be defined as route 1 and route (N-1). The neighborhood of any route i $(1 < i < N - 1)$ that lies somewhere in the middle of the solution route space can be defined as the routes that are next to R_i^t. $Z(X_{ij}^t)$, the objective function value of a new solution X^{t+1} that is obtained from X^t by moving R_i^t to one of its neighbors R_j^t at generation t can be computed as follows: $Z(X_{ij}^t) = f(X^{t+1})$.

Moves and Tabu Status: As defined, a move consists of replacing a given route within X^t by one of its two neighboring routes that lie outside of X^t but within the stored solution space. It should be noted that both of these two neighboring routes are tried. At the beginning of this process, no move is tabu (i.e., forbidden). At any iteration with n number of routes in solution X^t, the algorithm executes the best non-tabu move out of $2 * n$ feasible moves to a feasible neighbor of the current solution. In addition, if a tabu move yields a worse solution which is, however, the best among all feasible neighbors of the current solution, it is also updated. Whenever a move is performed, the reverse move is declared tabu for m iterations, where m is either a user-defined parameter or a randomly generated one that follows a discrete uniform distribution in an interval $[m_{min}, m_{max}]$, where m_{min} and m_{max} are the user-defined minimum and maximum parameters of the algorithm. Comparisons of the model performance between these two strategies including the fixed and variable tabu tenure are performed in the numerical results part.

Diversification and Intensification: This part is developed to combine the diversification and intensification procedures to further explore the solution space for a possibly better solution. It starts from the best found solution route set and introduces a major perturbation by allowing q routes ($1 \leq q \leq n$) to move w positions up from their current solution location (say $q = 2$ and $w = 10$) in the stored solution space. Put another way, X^t is moved to another feasible solution by replacing q routes within X^t by q other routes that each of them go up w position from their current solution location in the stored solution space. This is called "diversification." Note that this is a "forced" movement no matter whether the solution improves or not, so that the solution space can be somehow traversed more evenly. To respect the original characteristics of the TS, this procedure is never applied more than once during a given operation (called "intensification"). Note that tabu moves are also applied to this situation. If this move is toward one direction (say increasing direction) of the current route, then moves toward to the opposite direction (i.e., decreasing direction) are prevented for a certain number of iterations (say using the same m). Model performance comparisons of the TS algorithms between using and not using this procedure are also achieved and the better approach will be identified in the numerical results part.

Implementation Model Summary: In all, the proposed TS algorithms for the BTRNDP in this paper include two main procedures described as follows.

Neighborhood Search Procedure: At iteration t, let $X^t = (R_1^t, R_2^t, \ldots, R_n^t)$ be a feasible solution of value $f(X^t)$. Let $N(X^t)$ be the set of feasible neighbors of X^t, as defined before. The best neighbor of X^t is a solution $X_{i^*j^*}^t \in N(X^t)$ obtained by replacing one given route $R_{i^*}^t$ within X^t to its best neighbor $R_{j^*}^t$ that is one of its two neighboring routes outside X^t but within the stored solution space. Similarly define the best feasible non-tabu neighbor of X^t as $X_{\overline{ij}}^t \in N(X^t)$. ($X_{i^*j^*}^t$ and $X_{\overline{ij}}^t$ may coincide). Let X^* be the incumbent (the best known feasible solution) and let $Z(X^*)$ be its value.

If $Z(X_{i^*j^*}^t) < Z(X^*)$, set $X^* = X^{t+1} = X_{i^*j^*}^t$ and $Z(X^*) = Z(X^{t+1}) = Z(X_{i^*j^*}^t)$. Declare the move of a route from $R_{j^*}^t$ to $R_{i^*}^t$ tabu for m iterations, where m can be a fixed user-defined parameter or is uniformly distributed with $m \in [m_{min}, m_{max}]$. If $Z(X_{i^*j^*}^t) > Z(X^*)$ and all moves defining the solutions of $N(X^t)$ are tabu, set $\delta = 1$ and return. Otherwise, set $X^{t+1} = X_{\overline{ij}}^t$ and $Z(X^{t+1}) = Z(X_{\overline{ij}}^t)$. Declare the move of a route from R_j to R_i tabu for m iterations, where m has the same definition as used before.

Diversification and Intensification Procedure: This procedure is the same as that in Neighbor Search but defines $N(X^t)$ differently. It allows q routes ($1 \leq q \leq n$) to move up to w more than the current solution location in the solution space (Note that in this paper, this procedure is called the "shakeup" procedure. Furthermore, for simplicity, q is set to n and w is set as a user-defined parameter). When a route is moved (i.e., replacing this route within X^t by another route that is w positions up/down from its current location in the stored solution space) in one direction (say the increasing direction), moving back in the opposite direction is declared tabu for m iterations, where m uses the same notation as before.

Tabu Search Algorithm for the BTRNDP:

Step 1 Randomly generate an initial feasible solution route network
  $X^t = (R_1^t, R_2^t, \ldots, R_n^t)$ with route size n in the proposed solution set.

Step 2 Set $\delta = 0, t = 1$ and $X^* = X^t$; While ($\delta = 0$ and $t \le$ MAX_Iterations)
  Apply Neighborhood Search to the solution X^t; $t = t + 1$.

Step 3 Apply the "Diversification and Intensification" procedure to X^*. Apply Neighborhood Search to the solution X^* until $\delta = 1$ or $t >$ MAX_Iterations.

Step 4 Output the current best solution found.

As mentioned before, since TS provides a robust search as well as a near optimal solution within a reasonable time, this algorithm is employed as the solution technique for the BTRNDP. Before implementing the TS algorithms, a set of potential routes, consisting of the whole solution space, has been generated by the ICRSGP. The objective of the TS algorithm presented here is to select an optimal set of routes from the candidate route set solution space with the sum of the total user, operator and unsatisfied demand cost being minimized.

A flow chart that provides the typical TS algorithm-based solution framework for the BTRNDP can be seen in Fig. 4. Note that the "neighborhood" for any route i is defined as the route left or right of route i stored in the solution space, as described before. At the beginning of the TS implementation, the initial solution is randomly generated. In the second (and later) generation, the TS procedure is used to guide the generation of the new transit route solution set and after it is proposed at each generation, the search process is started. The network analysis procedure is then called to assign the transit trips between each centroid node pair and determine the service frequencies on each route and evaluate the objective function for each proposed solution route set. For each iteration, if a solution route set is detected to improve over the current best one, the current best solution is updated. The new proposed solution sets are generated and are evaluated in the same way. If convergence is achieved or the number of generations is satisfied, the iteration for a specific route set size ends. Then, the proposed solution route set size is incremented and the processes are repeated until the maximum route set size is reached. The best solution among all transit route solution sets is adopted as the best solution to the BTRNDP for the current studied network.

Moreover, in this paper, three versions of TS algorithms are used: 1) TS without shakeup procedure (i.e., without the diversification procedure as defined before) and with fixed tabu tenures; 2) TS with shakeup procedure and fixed tabu tenure (i.e., the number of restrictions set for the tabu moves are fixed); and 3) TS with shakeup procedure and variable tabu tenure (i.e., the number of restrictions set for the tabu moves are randomly generated). The differences underlying each TS algorithm are self-explanatory by the names. All three variations of TS methods are implemented, sensitivity analysis for each version are presented, and algorithm comparisons are performed.

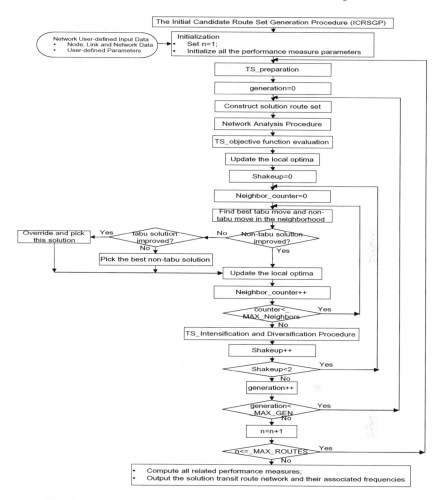

Fig. 4. A Tabu Search Model Based Solution Framework for the BTRNDP

5 Experimental Network and Numerical Results

5.1 Example Network Configuration

The TS algorithm-based solution methodology is implemented using a small example network as shown in Fig. 5. This example network contains seven travel demand zones and 15 road intersections. As noted before, the ICRSGP discussed in this paper first considers the BTRNDP under the "centroid" level. The network is processed as follows: 1) the zonal demands are distributed the same way as the highway network demand; and 2) if the same road link contains two or more demand distribution nodes from different zones, these distribution nodes are aggregated. After this preliminary process, 20 centroid distribution nodes, 35 nodes, and 82 arcs are obtained in this example network. The minimum and maximum route lengths are defined. In the example first phase, the ICRSGP generates 286 feasible routes whose distances satisfy two route length constraints as mentioned before.

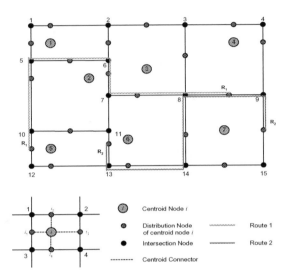

Fig. 5. A Small Network With Graphical Representations for Nodes, Links and Routes

5.2 Numerical Results and Sensitivity Analysis

It is noted that the performance of the proposed TS algorithms might greatly depend upon the chosen parameters such as the number of generations, the number of search neighbors, the number of tabu tenures and the shakeup number. Furthermore, note that since these parameters are basically continuous, one has to get the "nominally" optimal parameter through sequential testing. In addition, since the objective function is a multi-objective decision making problem, a commonly used weight set (0.4, 0.4 and 0.2) is assigned to each of the three objective function components (user cost, operator cost and unsatisfied demand cost), respectively, for demonstrating the sensitivity analysis here. Fig. 6 presents the sensitivity analysis of these parameters using the tabu algorithm without shakeup and with fixed tenures as an example. The effect of generations, tabu tenures and search neighbors are examined by varying these values within a specific range, and the results are given from Fig. 6.1 to 6.3, respectively. Details are described as follows.

Effect of Generations: Basically, "Generation" is a user-defined parameter which means how many iterations the transit planners want the developed solution algorithm run. It therefore can be varied from 1 to ∞. However, for efficiency, the effect of the number of generations is examined by varying this value from 5 to 100 and the result is given in Fig. 6.1. It can be seen from the figure that as the number of generations increases, the objective function value tends to decrease. It is also noted that the larger the chosen number of generations, the more the computation time. When the number of generations reaches 30, the optimal objective function is achieved, suggesting that 30 should be chosen as the optimal generations for the small network. Therefore, a generation of 30 was recommended.

Effect of Tabu Tenures: The effect of tabu tenures (i.e., the number of restrictions) is investigated by choosing this number ranging from 5 to 40 and the result is provided in Fig. 6.2. As can be seen, the least objective function value occurred with ten restrictions. Therefore, ten is chosen as the best number of tabu move tenures.

Effect of Search Neighbors: The effect of search neighbors is also studied by varying this value from 10 to 100. The result shown in Fig. 6.3 indicates that 20 might be the best value and as a result, it is recommended.

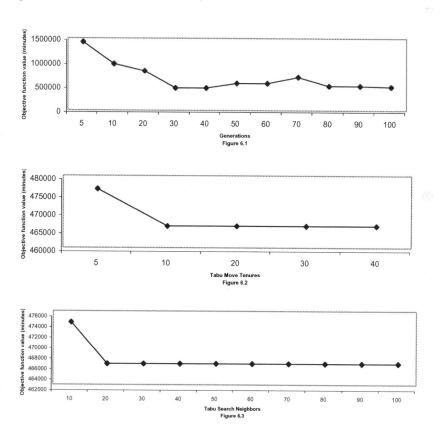

Fig. 6. Sensitivity Analysis for the Tabu Algorithm Without Shakeup and With Fixed Tenures

The above subsections presented the sensitivity analysis for tabu algorithm without shakeup and with fixed tenures using the example network. For sensitivity analysis regarding the other two developed TS methods including the tabu with shakeup/fixed tenures and that with shakeup/variable tenures, similar procedures can be followed. In addition, the genetic algorithm is used as a benchmark in this paper to examine the solution quality obtained from these three TS algorithms. The sensitivity analysis are also performed for the genetic algorithm using the same procedure (details about the genetic algorithm implementation model can be seen from Fan and Machemehl (2004)). Table 1 provides a summary of these sensitivity analysis for each algorithm for the BTRNDP. The best parameter set for each algorithm thus can be seen and chosen.

Table 1. Summary of Algorithm Sensitivity Analysis for the BTRNDP

	Genetic Algorithm	Population Size	30
		Generations	20
		Crossover Probability	0.8
		Mutation Probability	0.1
	Tabu w/o Shakeup and with Fixed Tenures	Generations	30
		Tenures	10
		Search Neighbors	20
Tabu Search	Tabu w/t Shakeup and Fixed Tenures	Generations	80
		Tenures	10
		Search Neighbors	10
		Shakeup Number	50
	Tabu w/t Shakeup and Variable Tenures	Generations	20
		Search Neighbors	40
		Shakeup Number	50

5.3 Multi-Objective Decision Making and Algorithm Comparisons

As mentioned, the model performance based on each proposed algorithm might greatly depend upon the chosen value of parameters inherent in that algorithm. In previous sections, a set of user-defined parameters associated with each algorithm is found by first assigning a commonly used weight set to each of the three objective function components and then running the developed programming codes based on that algorithm several times. The sensitivity analysis are then performed and the best parameter set is found by choosing those resulting in the least objective value from that algorithm. In this section, these chosen parameters for each algorithm are used and applied to the BTRNDP at different chosen weight levels. The objective is to see how the quality of these algorithms varies across different weight levels and one might therefore know which algorithm can be used to best solve the BTRNDP. The following sections compare the three employed TS algorithms to examine which variation is most suitable for the BTRNDP. Furthermore, the model performance is also compared to the genetic algorithm as a benchmark to examine the solution quality using TS algorithms from a multi-objective decision making perspective.

Fig. 7 presents numerical results for these comparisons using the example network. For each graph, the weight of total unsatisfied demand cost is set at a specific level between 0.1 and 0.8. The x-axis denotes the weight of total user cost and the y-axis is the objective function value measured in minutes. Note that each point shown for each algorithm in each graph is a decision making problem with a particular weight set for the three components contained in the objective function, where the weight of total operator cost can be obtained at each point by subtracting 1.0 from the weight sum of total unsatisfied demand cost and user cost. One can see from Fig. 7 that TS with shakeup and fixed tenures (i.e., fixed iterations) clearly seems to outperform other TS algorithms using the example network at any weight set level. Therefore, this tabu algorithm is chosen as the best TS algorithm for the BTRNDP.

It can also be seen from Fig. 7 that for each algorithm from any graph, as the weight of total user cost increases, the objective function value obtained by using that algorithm tends to increase. This is expected because the user cost is usually greater than the operator cost and the increase in total user cost due to a 0.1 unit increase in the weight of total user cost outweighs the decrease in total operator cost due to a 0.1 unit decrease in the weight of total operator cost. As a result, the total objective function value increases. One interesting phenomenon is that the genetic algorithm seems to be more variable than any TS algorithm (except the TS with shakeup and with variable tenures, which is also variable due to its inherent variable nature underlying the tabu tenures) in terms of the optimal objective function value (from Fig. 7.1 to 7.5.) This might suggest that, compared to TS algorithms, the Genetic Algorithm (GA) may largely depend on the chosen parameters at any particular level. If the chosen parameters inherent in the GA are fixed, the solution quality for the BTRNDP might be unstable. Therefore, to achieve the best solution network at each weight set level, one might need to run the program and get the optimal parameter set at that level although the computational burden would become larger. Furthermore, for each graph (i.e., for each weight level for the total unsatisfied demand cost), the TS with shakeup and fixed tenures seems to consistently outperform the GA in terms of the quality of solution (i.e., it always results in the least objective function value). This might allow the conclusion that compared to the GA, this TS method performs better for solving the BTRNDP. Furthermore, it can be seen that the local optimal solution obtained from this TS method can provide solution of very high quality because it is very near to the global optimum. The GA, however, seems to be the undesirable model. This might be possible because although the GA might achieve some better solutions by learning from the previous solutions through a genetic approach, it might take much more time inside the algorithm itself to look for this achievement, while it does not take much more effort looking for possibly better solutions from other "neighborhood" solutions in the candidate solution space (compared to the TS algorithms). Conversely, the TS with shakeup and fixed tenures not only can look for a good solution with a specific origin-destination node pair through "random search" in its early stage, but also can fully explore possibly better neighborhood solutions. Note that the tradeoffs between route coverage and the route directness might be well balanced between chosen shortest paths or k-th shortest paths between specific origin-destination node pairs. It is expected that this inherent characteristics of the TS algorithm might make it particularly suited for the BTRNDP and therefore outperform the GA.

5.4 Characteristics of the BTRNDP

The characteristics of the BTRNDP are very extensive due to its multi-decision making nature and the variety of parameters and procedures involved. These characteristics might depend upon the network size, the chosen parameters in the solution process, the chosen algorithm and the chosen weight level for each component of the objective function. In this sense, it is very hard to generalize all characteristics of the BTRNDP. However, it is expected that in most cases, the BTRNDP characteristics should be similar and the current comprehensive numerical results also show these similarities. Since the numerical results based upon weights of 0.4, 0.4 and 0.2 for the user cost, operator cost and unsatisfied demand cost, respectively, using the tabu algorithm without shakeup and with fixed tenures seem to be very representative, these are chosen here for presenting related BTRNDP characteristics.

The effect of the number of proposed routes in the transit network solution is investigated by varying it from 1 to 10 and the values of each performance measure of the optimal network at each route set size level including the user cost, the operator cost, the fleet size required, the unsatisfied demand cost, the percentage of the satisfied transit demand and the total objective function value are shown in Fig. 8.1 through 8.6, respectively. Generally speaking, as the number of routes provided in the network increases, more passengers will be served by transit and therefore, the satisfied transit demand increases. Furthermore, since the fixed transit demand is assumed, the percentage of satisfied transit demand also tends to increase as shown in Fig. 8.5. Also as a result, the unsatisfied demand cost decreases. However, the operator cost tends to increase because the fleet size required for the network generally increases. In addition, the user cost generally increases because more transit users travel and the total objective function value also increases. The reason might be that although service might be better in some sense (such as more passengers get direct route service) as more routes are provided, the headway might be longer on some routes. Therefore, the transit user cost as a whole might actually increase. In conclusion, the numerical results in Fig. 8 indicate that as a whole, as the route set size increases, the solution improved initially because more demand was satisfied and unsatisfied demand costs decrease. However, the least objective function value is achieved with two routes for this scenario and increases in the fleet size (i.e., operator cost) produces underutilization of routes and does not result in an improved objective function value. (Note that the optimal transit route network is shown in Fig. 5.)

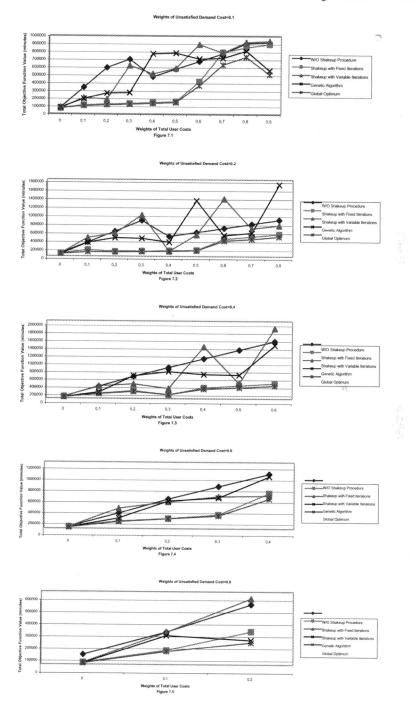

Fig. 7. TS and GA Comparisons for the BTRNDP

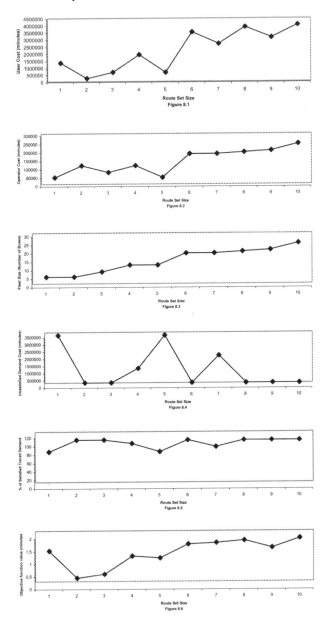

Fig. 8. Effect of Route Set Size on Objective Function and its Components for the BTRNDP

6 Conclusions

This paper uses TS algorithms to solve the optimal bus transit route network design problem at the distribution node level. A multi-objective nonlinear mixed integer model is formulated for the BTRNDP. The proposed solution framework consists of three main components: an Initial Candidate Route Set Generation Procedure that generates all feasible routes incorporating practical bus transit industry guidelines; and a Network Analysis Procedure that assigns transit trips, determines service frequencies and computes performance measures; and, a TS procedure that guides the candidate solution generation process. Three different variations of TS algorithms are employed and compared as the solution method for finding a hopefully optimal set of routes from the huge solution space. A C++ program is developed to implement the TS algorithms for the BTRNDP. A small example network is successfully tested as a pilot study. The model comparisons are performed and numerical results are presented. The TS with shakeup and fixed tenures is identified as the best TS method to solve the BTRNDP. A genetic algorithm is also used as a benchmark to measure the quality of the TS methods and numerical results clearly indicate that the preferred TS method outperforms the genetic algorithm using the example network. Furthermore, the local optimal solution obtained from this TS method can provide solutions of very high quality because it is very near to the global optimum. In addition, related characteristics and tradeoffs underlying the BTRNDP are also discussed.

BTRNDP is a really complex problem. One simple neighborhood rule can be the swapping of nodes. However, the link connectivity problem can make many routes resulting from swapping infeasible. Although one can always find routes to connect any two nodes to make it feasible, the efficiency can be a big problem. One option for future investigation is to examine a more flexible neighborhood definition that allows replacement by non-adjacent routes and the tabu status would then refer to forbidding the re-instatement of specific routes for a given period. Another possibility that may be worth mentioning is the investigation of a different type of short term memory that recent investigations have shown effectiveness (Glover and Laguna (1997)). Also, further application of this model to a very large network is under the way.

Acknowledgements: The authors want to express their deepest gratitude to two anonymous reviewers for their incisive and seasoned suggestions. The authors also appreciate the U.S. Department of Transportation, University Transportation Center through SWUTC to the Center for Transportation Research, The University of Texas at Austin for sponsoring this research by Projects 167525 and 167824.

References

Ahuja, R. K., Magnanti, T. L., and Orlin, J. B. (1993). *Network Flows: Theory, Algorithms and Applications*. Prentice Hall, Englewood Cliffs.

Baaj, M. H. and Mahmassani, H. S. (1992). Artificial intelligence-based system representation and search procedures for transit route network design. *Transportation Research Record 1358, Transportation Research Board*, pages 67–70.

Ceder, R. B. and Wilson, N. H. (1986). Bus network design. *Transportation Research*, **20B**(4), 331–344.

Chien, S., Yang, Z., and Hou, E. (2001). A genetic algorithm approach for transit route planning and design. *Journal of Transportation Engineering, ASCE*, **127**(3), 200–207.

Fan, W. and Machemehl, R. B. (2004). *A Genetic Algorithm Approach for the Transit Route Network Design Problem, CSCE 2004, 5th Transportation Specialty Conference*. Saskatoon.

Glover, F. (1977). Heuristics for integer programming using surrogate constraints. *Decision Sciences*, **8**(1), 156–166.

Glover, F. (1986). Future paths for integer programming and links to artificial intelligence. *Computers & Operations Research*, **5**, 533–549.

Glover, F. (1989). Tabu search, part I. *ORSA Journal on Computing*, **1**, 190–206.

Glover, F. (1990). Tabu search, part II. *ORSA Journal on Computing*, **2**, 4–32.

Glover, F. and Laguna, M. (1997). *Tabu Search*. Kluwer Academic Publishers.

Han, A. F. and Wilson, N. (1982). The allocation of buses in heavily utilized networks with overlapping routes. *Transportation Research*, **16B**, 221–232.

NCHRP Synthesis of Highway Practice 69 (1980). Bus route and schedule planning guidelines. Technical report, Transportation Research Board, National Research Council, Washington, D.C.

Pattnaik, S. B., Mohan, S., and Tom, V. M. (1998). Urban bus transit network design using genetic algorithm. *Journal of Transportation Engineering*, **124**(4), 368–375.

Shih, M., Mahmassani, H. S., and Baaj, M. (1998). Trip assignment model for timed-transfer transit systems. *Transportation Research Record 1571*, pages 24–30.

Yen, J. Y. (1971). Finding the k shortest loopless paths in a network. *Management Science*, **17**(11), 712–716.

Bus Tolling for Urban Transit System Management

Quentin K. Wan and Hong K. Lo

Civil Engineering, Hong Kong University of Science and Technology
cehklo@ust.hk

Summary. For transit services operated by competitive private companies, as in Hong Kong, the objectives of the companies are not to minimize the total traveler and/or infrastructure costs, but to optimize their profits. Other than engaging in a Bertrand Game, companies may also compete via their service frequencies. As evident in Hong Kong, the intense competition has led to a very visible phenomenon – companies putting more and more buses on major (profitable) corridors, leading to significant increases in congestion. This study aims to analyze externality pricing through bus tolling to manage the congestion caused by them. The result shows that bus tolling can be a promising tool.

1 Introduction

The bus system serves a crucial role in fulfilling the transportation needs of many transit-oriented cities. In Hong Kong, e.g., franchised buses and minibuses provided by private companies serve over 60% of the 11 million daily trips. The other 30% of these trips are carried by rail services, with the combined transit system serving over 90% of the daily trips. To ensure proper service provision, the Hong Kong government regulates bus operations, controlling their routes, fares, and minimum service frequencies. Within these regulations, private companies compete for revenue and market share in a rather profitable business. Recently Lo *et al.* (2003a) and Lo and Yip (2002) studied the possible outcomes of a competitive transit market based on the case of Hong Kong. The studies examined how private transit operators would act to maximize their own profits if their fares were fully deregulated. The results showed that all transit operators would simultaneously raise their fares; to exploit the situation, some would even double their fares. At the same time, higher transit fares encourage mode shifts to autos and taxis, which add to congestion and worsen network performance. The analysis showed that deregulated competition could lead to drastic changes in fares, network congestion, and social welfare.

In a market where both fares and routes of bus services are regulated, private companies would change their service frequencies to compete. The overall network congestion caused by buses is none of their concern, or, an externality. For demands on routes that are served by multiple transit operators, a simple strategy to increase one's market share and/or revenue is to operate more and more buses on profitable routes. Such a strategy would result in a net shift of rail users to the road network. In conjunction with the service frequency competition between bus operators, these factors lead to an oversupply of transit services and inefficient usage of the road space. The net effect is that significant congestion occurs on major corridors. Hence, it is important to incorporate this consideration on service competition into the transit system management strategy and closely monitor and regulate the bus operations.

The objective of this study is to examine the effect of bus tolling to price out the externality of excessive bus services. Essentially, a toll is charged for each additional bus in operation that is offered above the minimum frequency. The exact tolls are to be determined based on the locality of the route and its congestion level. The objective of this paper is to analyze how bus tolling would affect travelers, the competitive market, and overall system performance.

2 Modeling Bus Tolling and its Impacts

In a privately operated market, the ultimate objective of the transit operator is to maximize its profit. With fixed fares, the total revenue is simply the product of its fare and the number of passengers; whereas the total operating cost is determined by its marginal operating cost times the service frequency. As travelers choose their transport modes based on their perceived utilities or service qualities, in order to attract more passengers, an operator would improve their service quality as long as the improvement cost does not exceed the additional revenue generated. Consider a regulated bus market wherein only frequency is adjustable, the operator's problem can be formulated as:

$$\max_f \pi(f, d, \tau) = wd\rho - f\delta - [f - f_{min}]^+ \tau \tag{1}$$

$$\text{s.t. } w = \left\{ \sum_k \exp \theta[u_k(f) - u_i(f)] \right\}^{-1} \tag{2}$$

where π is the profit function; f is the bus frequency; f_{min} is the minimum bus frequency required by the terms of the franchised operation; d is total travel demand; ρ is the bus fare; δ is marginal operating cost; and τ is the bus toll. The bracket on the right hand side of (1) means that $[x]^+ = x$ if $x > 0$; zero otherwise. The terms on the right hand side of (1) are, respectively, the bus revenue, total operating cost, and total toll charge. (2) determines the market share on bus w using the standard logit model to capture travelers' choice behavior. The logit model is a popular member in the family of random utility models, the underlying principle of which is that passengers would choose the alternative with the maximum utility. The utility function for mode k is represented by $u_k(\cdot)$, with $k = 1$ for bus. The perceived utility parameter θ, whose reciprocal is sometimes known as the scale parameter, is a measure of the information content such that the homoscedastic variance of utility in the logit model is given by $Var(u_k) = \pi^2/6\theta^2$. The operator's problem in (1)–(2) is a maximization problem to determine the bus frequency f, subject to the equilibrium between market share w and utility function $u_k(\cdot)$ among the alternatives. In general, there is no closed form solution for the optimal bus frequency so determined. In terms of notation, we denote the solution to the operator's problem as:

$$f^*(d, \tau) = \underset{f}{\text{argmax}} \, \pi \tag{3}$$

As indicated in (3), the bus operator chooses to operate the service at different frequencies in response to the different demand levels and bus tolls. This decision by the bus operator not only affects its own service quality, revenue and cost, but also the patronage of the other transit modes, their service quality, and other users who share the roadway with the buses. That is, we study the effect of the bus toll τ on all travelers as well as the overall system performance.

3 Illustrative Case Study

We consider an illustrative case consisting of a major corridor connecting an origin and destination pair. Travelers choose between the bus service and the subway. This is fairly typical in a transit-oriented city such as Hong Kong. In the current study, we consider only a monopolistic bus service market provided by a single operator. This simple example is adequate to demonstrate how bus tolling can be used to manage the urban transit system. Without loss of generality, the bus tolling concept can be extended to oligopolistic and competitive bus service markets so as to consider explicitly the competition between different transit services. This we leave to a future study.

While the subway has exclusive rights of way and does not share congestion with others, buses operate on the road network and share congestion with other traffic such as trucks, company fleets, service fleets, and private vehicles. The amount of this background traffic is taken to be fixed at $x_0 = 1800$ pcu/hr, with an average occupancy of 1.5 prs/pcu. The practical capacity (defined as 75% of the maximum link capacity) of the roadway segment is $c = 1500$ pcu/hr. While the subway enjoys a constant travel time at 36 minutes for the OD pair, the bus travel time follows the BPR performance function:

$$t = t_0 \left[1 + 0.2 \left(\frac{x_0 + E_b f}{c} \right)^4 \right] \tag{4}$$

where t and $t_0 = 30$ minutes, respectively, are the actual travel time and free flow travel time between the OD pair on the road network. E_b is the equivalent passenger car unit (pcu) for buses. In order to consider the dissatisfaction from crowdedness on a transit mode, a discomfort function is used to modify the in-vehicle travel time (Nielsen (2000)). Generally, in transit studies conducted by the western world, as demand rarely exceeds vehicle capacity, the discomfort function usually does not impose any hard capacity constraint on the transit vehicle, similar to the case of the BPR function for roadway capacity (e.g., Lo et al. (2003b)). This may not be realistic in the current study, however, because overloading of the transit vehicle is not uncommon, which has implications on the frequency (and hence the line capacity) of bus services. Therefore, we adopt a function analogous to the Davidson volume delay function to adjust for the discomfort factor in a crowded transit vehicle. As a result, we define the congested time Γ as the travel time multiplied by a crowdedness factor ϕ, defined as:

$$\phi_i = \left[1 + \left(\frac{v_i}{C_i - v_i} \right)^2 \right]^{0.1} \tag{5}$$

where v_i denotes the average patronage per transit vehicle of mode i, with corresponding vehicle capacity C_i. We specify a homogeneous linear-in-parameter utility function that depends only on transit fare and the congested time as in (6):

$$u_k = \beta_i \rho_k + \beta_2 \Gamma_k \tag{6}$$

where $\beta_1 = -1$ and $\beta_2 = -\frac{2}{3}$ are the utility parameters and ρ_k is the transit fare on mode k in Hong Kong (HK) dollars[1]. These values imply a value of time (VOT) of HK\$40/hr, which is commonly adopted in local transportation studies. The transit fares are HK\$15 for bus, HK\$20 for subway. In addition, we adopt the perceived utility parameter $\theta = 0.1$ in the logit model as specified in (2). The marginal operating cost δ is assumed to be HK\$50/bus, and the bus fleet consists of identical vehicles with the capacity of 100 prs/vehicle. Referring to the objective function in (1), as the minimum frequency required, f_{min}, is a constant, one can drop this term without affecting the optimal result. In other words, it is the same as setting the minimum frequency to be zero. Though this problem is illustrated via a simple scenario, indeed some insights can be learned on the possible impacts of bus tolling.

3.1 The Impact of Bus Tolling

By varying the bus toll, we investigate how the following measures change: (i) *bus operation* – the profit, frequency, patronage, and load level; (ii) *transit congestion effect* – the congested time Γ on buses and the subway; and (iii) *system performance* – the crowdedness effect on both the total roadway travel time and congested transit time. For the representative case, we consider demand $d = 10,000$ prs/hr, with the pcu factor of buses fixed at $E_b = 3$. The capacity of the subway is $C_{subway} = 10,000$ prs/hr. We solve (3) for a range of bus tolls. That is, given a bus toll, the operator maximizes its profit by optimally setting its service frequency. The results are shown

---

[1] US\$1 is equivalent to HK\$7.8

in Figs. 1-3. In these figures, the effect of any change in parameter is presented in both absolute and relative terms: the left vertical axis shows the absolute scale and the right vertical axis the percentage change relative to the case without bus toll.

Bus Operation. The parameters are shown in Fig. 1. As expected, the optimal bus frequency drops with the bus toll. Figs. 1(b)-(d) show that the bus toll results in a lower service frequency; fewer travelers use the bus service but the load level per bus vehicle increases from around half-empty gradually to almost full. In this scenario, both the operator and the existing bus passengers suffer from the introduction of the bus toll. Therefore, from the perspective of the bus service alone, there is no winner.

Transit Congestion Effect. Fig. 2 shows the changes in transit congestion effect with the bus toll. We plot the congested times Γ on both the bus and subway services. They both show an upward trend. Less frequent bus services increase both the congested time on the buses as well as that on the subway, as travelers switch to the subway system. The increase is gradual at lower tolls but becomes more prominent at high tolls. The only winner is the subway operator, who gains in patronage and hence revenue without needing to improve its service.

Overall System Performance. Fig. 3(a) shows a gradual drop or improvement in the total roadway travel time as a result of the bus toll, as some buses are priced out of the system. Fig. 3(b) plots the total system congested time, which combines the congested time of all transit users (on both buses and the subway) as well as that of the background traffic including trucks, autos, etc. Initially the total system congested time descends to a global minimum at the bus toll of $\tau = HK\$85$ and then moves upward.

If one focuses on the profitability of the bus or transit users alone, bus tolling may not be attractive. In fact, its primary objective is to balance the supply and demand of bus services so that the entire system benefits, including all travelers. At low bus tolls, improvements in the travel time on the roadway more than compensate the slight deterioration in the congested time of the transit users, thereby driving down the total system congested time. At high bus tolls, however, the transit crowdedness associated with the frequency reduction outweighs the gain in the roadway travel time, leading to increases in the total system congested time. Thus, by applying the bus toll accordingly, one does have a way to strike the balance between different travelers, while at the same time allowing the bus company to set its own frequency policy to maximize its profit.

3.2 Optimal Bus Toll

As illustrated earlier, bus tolling can effectively mitigate the roadway traffic congestion, at the expense of transit service quality. Nonetheless, according to the result, we observe that with relatively low bus tolls, the deterioration in the transit system congestion is mild; whereas the overall system congested time can be substantially improved. By defining the objective to be the total system congested time, we can write the optimal toll τ^* as:

(a) Profit (10³ HK$)

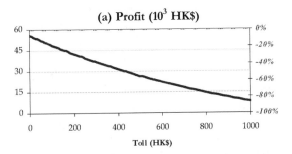

Toll (HK$)

(b) Frequency (hr⁻¹)

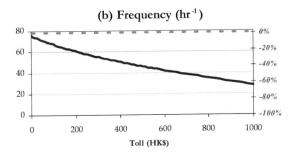

Toll (HK$)

(c) Bus Patronage (prs/hr)

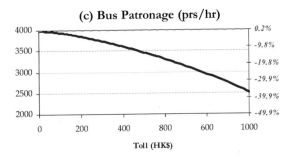

Toll (HK$)

(d) Load Level

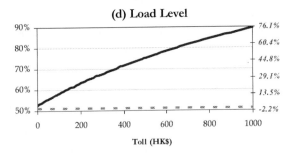

Toll (HK$)

Fig. 1. Optimal Bus Operation Parameters

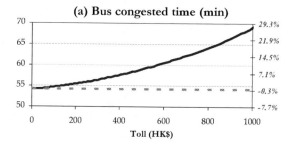

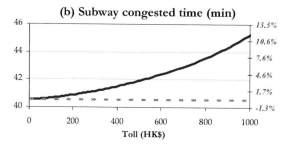

Fig. 2. Transit Congestion Effect

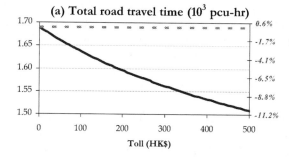

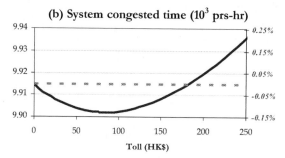

Fig. 3. System Performance

$$\tau^* = \underset{\tau}{\mathrm{argmax}} \sum_i x_i(f^*)\Gamma_i(f^*) \tag{7}$$

where x_i is passenger volume on mode i and f^* is obtained from (3). Together, (3) and (7) show the interrelated process in setting the optimal bus toll and optimal bus frequency. Given any bus toll τ, according to (3), the operator adjusts its service frequency f so as to maximize its profit. Fig. 1(b) shows how the optimal bus frequency f^* (optimal in the view of the operator, i.e., profit maximization) varies with the bus toll. Each instance of (τ, f^*) such determined will result in a certain total system congested time. By appropriately selecting the bus toll, while incorporating the reaction of the operator in adjusting its service frequency, one can achieve the objective of minimizing the total system congested time. In other words, one can consider this formulation as a leader-follower bi-level problem. The government acts as the leader, who sets the tolls so as to minimize the total system congested time (i.e., (7)), whereas the operator acts as the follower, who reacts to the toll and adjusts its service frequency so as to maximize its profit (i.e., (3)).

The optimal bus frequency (for profit maximization of the operator) on one hand depends on the bus toll; on the other hand, it affects the system performance which in turn affects the choice of the optimal bus toll (for total system congested time minimization). Though (3) cannot be expressed in closed form, it can be solved at different toll levels. Fig. 3(b) shows how the total system congested time varies with the toll. Indeed, for this case, the optimal bus toll is found to be around HK$85.

To study the sensitivity of the optimal bus toll to different traffic conditions, we numerically solve (7) and compare the results for different values of E_b and C_{subway}. Table 1 tabulates the optimal tolls and the corresponding frequencies for the fixed travel demand of 10,000 prs/hr. For the same subway capacity, one should charge a higher bus toll for bus operations with a higher pcu equivalent. A higher pcu equivalent occurs if a bus occupies more road space and/or operates in a less efficient manner than a passenger car. For example, a low speed bus with frequent stops will have a higher pcu equivalent. In other words, the policy of allowing buses to halt and wait for passengers at intermediate bus stops should be charged more. According to the results, doubling the pcu equivalent, say from 2 to 4, requires approximately a factor of 4 in the optimal toll charge. This indicates that the optimal bus toll is nonlinear to the pcu equivalent.

Table 1. Optimal Toll and Optimal Bus Frequency $(\tau^*, f^*)^2$ in Different Network Conditions [Demand at 10,000 prs/hr]

C_{subway}	E_b - Bus pcu-equivalent		
[prs/hr]	2	3	4
7,500	(0, 92.4)	(18, 79.1)	(37, 69.7)
10,000	(33, 83.1)	(84, 68.5)	(137, 58.8)
12,500	(73, 76.5)	(167, 60.9)	(279, 50.0)

---

[2] Toll in HK$, optimal bus frequency is hourly frequency.

Table 2. Optimal Toll at Different Demand [$C_{subway} = 12,500$ prs/hr and $E_b = 3$]

D [prs/hr]	7500	1000	12500	15000
τ^* [HK$]	272	167	91	40
f^* [hr^{-1}]	43.5	60.9	78.7	96.7
Bus load level	0.6267	0.6014	0.5931	0.5996

Table 2 shows the optimal bus tolls for different travel demands. The optimal bus toll declines with demand increases, which allows for more frequent services to cater to the higher demand. Interestingly, the load level remains roughly at 60% in all the cases. This indicates that an appropriate load level is essential in minimizing the total system congested time.

4 Concluding Remarks

We proposed bus tolling as a market-based strategy to address the supply of bus services to cope with demand in the presence of alternative transit services. In this strategy, the bus operator is free to choose its service frequency so as to maximize its profit. The government simply charges the bus toll based on the demand level and capacity of the alternative so that the system performance rests at the minimum total system congested time. The exact bus toll can be determined with the formulation developed herein.

We demonstrate in this study that bus tolling can be a flexible market-based strategy to strike a good balance between the objectives of transit users, for-profit operators, as well as the overall system performance, including other road users. This study is our first attempt to investigate the concept of bus tolling for managing the transportation system. Most of the results are based on the numerical study. In the future, we will examine if the results can be developed analytically. Many dimensions of this study can be extended, such as introducing the competition between multiple bus companies, extending the study to the case of a network, and considering bus route bundling in the competition.

Acknowledgement: This study is sponsored by the Competitive Earmarked Research Grants HKUST 6083/00E and HKUST6161/02E of the Hong Kong Research Grant Council.

References

Lo, H. K. and Yip, C. W. (2002). Fare deregulation of transit services: winners and losers in a competitive market. *Journal of Advanced Transportation*, **35**, 215–235.

Lo, H. K., Yip, C. W., and Wan, Q. K. (2003a). Modeling competitive multi-modal transit services. In W. H. K. Lam and M. G. H. Bell, editors, *Advanced Modeling for Transit Operations and Service Planning*, pages 231–256. Elsevier, Oxford.

Lo, H. K., Yip, C. W., and Wan, K. H. (2003b). Modeling transfers and non-linear fare structure in multi-modal network. *Transportation Research B*, **37**, 149–170.

Nielsen, O. A. (2000). A stochastic transit assignment model considering differences in passengers utility functions. *Transportation Research B*, **30**, 377–402.

Sensitivity Analyses over the Service Area for Mobility Allowance Shuttle Transit (MAST) Services

Luca Quadrifoglio and Maged M. Dessouky

Daniel J. Epstein Department of Industrial and Systems Engineering, University of Southern California, Los Angeles, California 90089-0193
maged@usc.edu

Summary. A Mobility Allowance Shuttle Transit (MAST) system is an innovative concept that merges the flexibility of Demand Responsive Transit (DRT) systems with the low cost operability of fixed-route bus systems. It allows vehicles to deviate from the fixed path so that customers within the service area may be picked up or dropped off at their desired locations. In this paper, we summarize the insertion heuristic presented by Quadrifoglio *et al.* (2007) for routing and scheduling MAST services, and we carry out a set of simulations to show a sensitivity analysis of the performance of the algorithm and the capacity of the system over different shapes of the service area. The results show that a *slim* service area performs better in general, but also that the positive effects of a proper setting of the control parameters of the heuristic is much more evident for wider service areas. In addition, a performance comparison shows that MAST systems can provide a better service to customers than fixed-route ones even for a slim service area.

1 Introduction

The Mobility Allowance Shuttle Transit system is an innovative concept in transportation that merges the flexibility of Demand Responsive Transit systems with the low cost operability of fixed-route bus systems, in order to satisfy the current needs of transit agencies, which are seeking ways to improve their service flexibility in a cost efficient manner. A MAST system is characterized by one or more vehicles driving along a base fixed-route covering a specific geographic zone, with one or more mandatory checkpoints conveniently located at major transfer points or high demand density zones. Given a proper amount of slack time, vehicles are allowed to deviate from the fixed path to serve (pick-up and/or drop-off) customers at their desired locations, as long as they are within a service area. Customers can make reservations before or during the service, thus the MAST system works under a dynamic environment.

Line 646 of the Metropolitan Transit Authority (MTA) of Los Angeles County offers a MAST *nightline* service. The vehicle drives nine times back and forth between two terminal checkpoints, passing by a third intermediate checkpoint in each trip. The vehicle is allowed to deviate from the fixed-route to serve customers as long as their service stops are within half a mile from either side of the main route. The demand of Line 646 is currently low enough to allow the bus operator to make all the decisions concerning accepting/rejecting requests and routing the vehicle. Quadrifoglio et al. (2007) developed a customized insertion heuristic algorithm to handle heavier demand in a potential daytime MAST operation and several requests for deviations. The vehicle's route and schedule are updated shortly after each request and customers are notified whether their request has been accepted and are provided with a time window for their pick-up and/or drop-off stops. The main characteristic of their algorithm is the development of efficient control parameters as a function of the future expected demand that, if properly set, significantly enhances the performance of the algorithm.

The purpose of this paper is to evaluate the sensitivity to the shape of the service area of the performance of MAST systems and of the effectiveness of the control parameters of the above mentioned algorithm. In particular we will show how a proper setting of those parameters is able to raise the saturation demand level in each configuration, allowing the system to serve more customers with a comparable service level. In addition, we perform a simulation comparison to test the competitiveness of hybrid systems like MAST versus conventional fixed-route types of services in a *slim* service area, apparently more suitable for the latter services.

Hybrid types of transportation systems have just lately been approached by researchers. Daganzo (1984) describes a checkpoint DRT system that combines the characteristics of both fixed route and door-to-door service. Malucelli et al. (1999) provide a general overview of flexible transportation systems. Crainic et al. (2001) incorporate the hybrid fixed and flexible concept in a more general network setting. Zhao and Dessouky (2004) study the optimal service capacity of a MAST system through a stochastic approach. Quadrifoglio et al. (2006) look at MAST systems from a design point of view, evaluating the relationship between the longitudinal velocity of the vehicle and the demand density, in order to allocate slack time and set other system parameters.

Some work approached hybrid systems in which different vehicles perform the fixed and variable portions. Aldaihani et al. (2004) develop a continuous approximation model for designing such a service. Scheduling heuristics based on a hybrid system include the decision support system of Liaw et al. (1996), the insertion heuristic of Hickman and Blume (2001) and the tabu heuristic of Aldaihani and Dessouky (2003). Another work studying a combination of fixed and flexible service can be found in Cortés and Jayakrishnan (2002).

Savelsbergh and Sol (1995), Desaulniers et al. (2000) and Cordeau and Laporte (2003) provide reviews on the Pickup and Delivery problem and Dial-a-Ride systems. Wilson et al. (1971) formulate the problem as a dynamic search procedure. Continuing work is presented by Wilson and Hendrickson (1980). Stein (1977), Stein (1978b), Stein (1978a) develops a probabilistic analysis of the problem and Daganzo (1978) presents a model to evaluate the performance of a Dial-a-Ride system. Theoretical studies of the problem case include the work by Psaraftis (1980),

Psaraftis (1983), Sexton and Bodin (1985a), Sexton and Bodin (1985b), Sexton and Choi (1986), Desrosiers *et al.* (1986) and Lu and Dessouky (2004).

Heuristics to solve multi-vehicle problems have been proposed by Psaraftis (1986), Jaw *et al.* (1986), Bodin and Sexton (1986), Desrosiers *et al.* (1988) and Madsen *et al.* (1995). Parallel insertion heuristics are proposed by Toth and Vigo (1997), Diana and Dessouky (2004) and Lu and Dessouky (2006). Diana (2006) assesses by simulation the effectiveness of the latter algorithm. Horn (2002a) develops an algorithm for the scheduling and routing of a fleet of vehicles that is embedded in a modeling framework for the assessment of the performance of a general public transport system with the latter being presented in Horn (2002b).

This paper is organized as follows. In Section 2 we describe the model for a MAST system. In Section 3 we briefly summarize the insertion heuristic algorithm described by Quadrifoglio *et al.* (2007), that we utilize to perform the simulation analysis described in Section 4, where a sensitivity over the shape of the service area is presented. Section 5 provides a MAST/fixed-route comparison and Section 6 the conclusions.

2 MAST System Model

The MAST system model is described by a service area shaped as a rectangular region L×W. C checkpoints are distributed along the x axis in the middle of the rectangle with a y coordinate W/2. Checkpoints 1 and C are at the extremities of the rectangle and the remaining C-2 checkpoints are within it (see Fig. 1). A single vehicle is assigned to this service area. A trip r begins at checkpoint 1 (or C) and ends at checkpoint C (or 1), after visiting in a predefined order all the intermediate checkpoints, which have fixed departure times. If R is the total number of trips, the total number of stops at the checkpoints is TC = (C-1)R+1. Hence, the initial vehicle's schedule is represented by an ordered sequence of stops from 1 to TC. We assume that the vehicle follows a rectilinear metric and has infinite capacity.

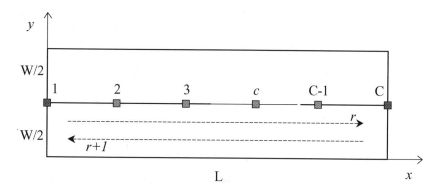

Fig. 1. MAST System Model

The demand is defined by a set of requests, which can be of three types: "hybrid" (having one service point at a non-checkpoint location in the service area and the other one at the checkpoints), "regular" (both service points at the checkpoints) or "random" (both service points located at non-checkpoint stops). We assume that the total demand rate θ is constant over time and that the non-checkpoint stops are uniformly distributed in the service area. At any moment a customer may call in (or show up at the checkpoints), specifying the locations of pick-up and/or drop-off points. "Regular" customers do not need a booking process to use the service.

In order to allow deviations from the main route to serve non-checkpoint requests, there needs to be a certain amount of *slack time* in the schedule. The initial slack time between any pair of consecutive checkpoints in the schedule is given by the difference between their scheduled departure times minus the time needed by the vehicle to travel from one to the other. The slack time is dynamically consumed by the insertion procedure when the demand arises. The amount of it to be allocated depends on the amount and type of demand and it may be adjusted properly to fit particular situations; see Quadrifoglio et al. (2006) and Zhao and Dessouky (2004) for more detailed analyses on the matter. In this paper we assume a slack time larger than the actual one in the MTA Line 646, where the demand is very low.

3 Algorithm Description

In this section we summarize the main features of the insertion heuristic algorithm described in Quadrifoglio et al. (2007) that will be utilized to perform the sensitivity analyses described in the following Section 4.

A *bucket* of a checkpoint c is the portion of the schedule beginning at one occurrence of c in the schedule and the following one. Since "hybrid" customers rely on a checkpoint c for either their pick-up or drop-off stop, the algorithm checks the schedule for possible insertion of their non-checkpoint stop "bucket by bucket" of c, until feasibility is found (for "random" requests, buckets are represented by trips). The following flowchart in Fig. 2 summarizes the insertion procedure.

All customers, once their request is placed in the schedule, are provided with time-windows for both their pick-up and drop-off stops. These time-windows depend on the current schedule at the time of the request and are naturally bounded by the hard time constraints of the checkpoints.

The cost function needed to select the best insertion among the feasible ones is given by

$$COST = w_1 \times \Delta t + w_2 \times \Delta RT + w_3 \times \Delta WT \qquad (1)$$

where Δt is the slack time consumed by the insertion. ΔRT is the sum over all passengers of the additional ride time, including the whole ride time of the requesting customer, caused by the insertion. In fact, a new inserted request would cause the passengers onboard to be delayed if the insertion takes place before and within the same pair of consecutive checkpoints of their drop-off. Also "regular" onboard passengers may be affected by this caused delay, because the *arrival* time at their

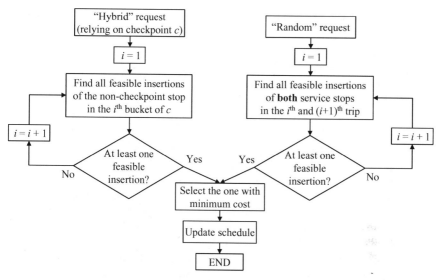

Fig. 2. Insertion Algorithm for MAST Systems, Quadrifoglio *et al.* (2007)

checkpoints is not fixed (the *departure* time at the checkpoints is) and depends on how much slack time is consumed in that portion of the schedule. ΔWT is the sum over all passengers of the additional waiting time caused by the insertion. In fact, customers that are already scheduled and are waiting for their pick-up at the time initially agreed might have to wait longer if the new insertion is placed before them and in between the same pair of consecutive checkpoints. w_1, w_2 and w_3 are the weights, which can be modified as needed to emphasize one factor over the others.

Insertion feasibility and control parameters   The "myopic" consumption of the slack time could prevent future requests to be properly satisfied, worsening the overall performance of the system. In order to prevent and solve this problem the heuristic makes use of two control parameters that are a function of the expected future demand and the relative position of the new request with respect to the already scheduled stops. The control parameter $\pi^{(0)} \leq 1$ is multiplied by the initial slack time and sets a cap on how much slack time each insertion may require. The BACK parameter (in miles) defines the maximum allowable backtracking distance available for each insertion. A proper setting of these two parameters (to be determined by simulation analysis) allows the system to control the consumption of slack time and improves the overall performance significantly, especially reducing the total mileage driven and allowing the system to serve more demand, raising the saturation level.

Thus, a candidate insertion is feasible if the customer precedence constraints are met, the slack time consumed is less than the current available and less than the maximum allowed (controlled by $\pi^{(0)}$), and the potential backtracking distance is less than the maximum allowed (controlled by BACK).

4 Sensitivity Over Service Area

In this section we perform a simulation analysis to observe the behavior of the system when modifying the shape of the service area, maintaining constant the total square mileage. In particular we want to observe the effect of the control parameters in each configuration over their saturation level.

The service area considered is described by Fig. 1. The time interval between the scheduled departure times of the two terminal checkpoints is assumed to be 50 minutes. We consider two different cases: C = 3, as for the MTA Line 646, therefore with only one intermediate checkpoint placed in the middle of the area (25 minutes between each pair of consecutive checkpoints) and C = 5, with three intermediate checkpoints (12.5 minutes between each pair of consecutive checkpoints). The initial slack time available between any pair of consecutive checkpoints will vary depending on the assumed proportion between W and L. With smaller L, the amount of slack time is larger because the checkpoints are closer.

The vehicle is riding back and forth between the two terminal checkpoints for a total simulated time of 50 hours, without interruption and therefore the total number of trips is R = 60. The simulation time has been chosen to ensure that the system reaches a steady state. The speed of the vehicle is assumed constant and equal to 25 miles/hour.

Demand is arising dynamically during the trip; we assume that the demand rate θ (customers/hour) is constant over time and that the customer types are distributed as shown in Table 1, as it is for MTA Line 646. In addition, we assume that check-point requests (P for pick-up and D for drop-off) are uniformly distributed among the checkpoints and that non-checkpoint requests (NP and ND) are uniformly distributed over the service area.

Table 1. Customer Type Distribution

Type	PD	PND	NPD	NPND
%	10%	40%	40%	10%

The weights in the COST function are $w_1 = w_2 = 0.25$, $w_3 = 0.5$, reasonably assuming that customers would rather stay onboard (w_2) than waiting (w_3) at the bus stop and assigning the same value to w_1 (slack time consumed) and w_2. These values can be modified accordingly depending on the objective function of the transit agency.

The main purpose of the analysis is to determine the demand saturation level of the system for each configuration, by running several simulation experiments: first, with no control (BACK = L and $\pi^{(0)} = 1$, which allow for any backtracking and any consumption of slack time, if available; therefore, giving the maximum freedom to the algorithm when checking for insertion feasibility); then, with the best setting of the control parameters that we could find, in order to maximize the saturation de-mand level. In addition, we compute the following performance parameters (directly related to the corresponding terms in the COST function) to compare the efficiency of the algorithm and the service level among the cases:

- M: total miles driven by the vehicle
- RT: average ride time per customer
- WT: average extra waiting time per customer

Configuration A: W = 1; L = 12 The first analysis is done over a *slim* service area with L = 12 and W = 1, both in miles. The distance between checkpoints is 6 miles and the slack time available between any consecutive pair is therefore about 10.5 minutes for C = 3 and 5 minutes for C = 5. The saturation levels of this system configuration with BACK = L and $\pi^{(0)} = 1$ (no control) and with the best setting of the parameters to maximize demand are shown in Table 2.

Table 2. Configuration A – Saturation Demand Levels: No Control / Best Control

C	3		5	
Control	None	Best	None	Best
BACK (miles)	L	0.2	L	0.2
$\pi^{(0)}$	1	0.3	1	0.6
θ (customers/hour)	18	21	15	18
WT (min)	0.99	1.43	0.34	0.46
RT (min)	25.33	25.42	27.04	25.97
M (miles)	1049.8	1018.2	1020.5	981.9

The system becomes unstable with θ greater than the values shown, that are approximately the saturation levels of these configurations.

For C = 3, the system is able to handle up to about 21 customers/hour, with a proper setting of the control parameters, namely BACK = 0.2 and $\pi^{(0)} = 0.3$. For C = 5 instead, the system capacity is about 18 customers/hour, with BACK = 0.2 and $\pi^{(0)} = 0.6$. The improvement on the capacity of the system is only 3 customers/hour for both cases (about 15-20% increase), but the improved efficiency of the algorithm is evident on the total mileage M as well, that has decreased by approximately 30-40 miles despite the increased demand. Note that the cases with C = 5 have lower capacities than the ones with C = 3, because of the additional constraints of the two extra checkpoints. From "None" to "Best" control cases, the ride time (RT) remains about the same, while the extra waiting time at stops (WT) slightly increases, due to the heavier demand that leads to an increased number of insertions and postponement of NP pick-ups. Also, the WT is lower for the cases with C = 5, because the number of possible insertions between consecutive checkpoints is smaller due to the checkpoints that are closer to each other and less slack time is allocated between each pair.

Configuration B: W = 2; L = 6 A similar analysis is performed over a service area with W = 2 and L = 6, always referring to the model in Fig. 1. The total square mileage is still 12 and all the other parameters of the system are kept the same. However, given the different shape of the area, checkpoints are closer to each other and therefore the initial slack time available between any pair of consecutive checkpoints is larger, namely equal to about 18 minutes for C = 3 and about 9 minutes for C = 5. Table 3 shows the figures for the saturation levels of this configuration.

Table 3. Configuration B – Saturation Demand Levels: No Control / Best Control

C	3		5	
Control	None	Best	None	Best
BACK (miles)	L	0.3	L	0.2
$\pi^{(0)}$	1	0.3	1	0.6
θ (customers/hour)	12	20	10	18
WT (min)	1.36	1.94	0.20	0.54
RT (min)	20.59	22.81	25.04	29.57
M (miles)	1054.5	933.5	909.8	917.8

In this case the improvement due to control parameter adjustment is more significant: the saturation level jumps from 12 to 20 customers/hour for C = 3 and from 10 to 18 for C = 5 (65-80% increase). The mileage (M) is reduced by about 120 miles for C = 3 and slightly increases for C = 5, even with the increased demand. The values of RT increase slightly more than in Configuration A.

Configuration C: W = 3; L = 4   We consider now a service area with W = 3 and L = 4. The total square mileage is again still 12 and all the other parameters of the system are kept the same, but checkpoints are even closer to each other and the initial slack time available between any pair of consecutive checkpoints is now about 20 minutes for C = 3 and about 10 minutes for C = 5. Results are in Table 4.

Table 4. Configuration C – Saturation Demand Levels: No Control / Best Control

C	3		5	
Control	None	Best	None	Best
BACK (miles)	L	0.5	L	0.2
$\pi^{(0)}$	1	0.5	1	1
θ (customers/hour)	12	18	10	15
WT (min)	1.73	1.68	0.38	0.51
RT (min)	17.37	22.17	21.62	24.86
M (miles)	1047.3	964.0	955.4	896.8

The increase in the saturation level due to control parameter adjustments is significant, from 12 to 18 customers/hour for C = 3 and from 10 to 15 for C = 5 (50% increase) and the mileage (M) also is reduced by about 80 and 60 miles, respectively. As for Configuration B, a more significant increase of the RT value is observed. Fig. 3 summarizes the findings shown in the previous tables.

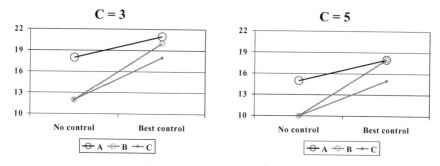

Fig. 3. Saturation Levels (Customers/Hour)

The analysis shows that a proper setting of the control parameters could significantly improve the performance of the system for every configuration. The results also show that the *slim* Configuration A performs better with or without the involvement of the control parameters, even though with different emphasis in the two cases.

With no control (BACK = L and $\pi^{(0)} = 1$) Configuration A outperforms Configurations B and C in terms of system capacity (18 vs. 12 customers/hour for C = 3 and 15 vs. 10 for C = 5), meaning that the insertion procedure is able to perform better in case of a slimmer service area and consequently a lesser amount of slack time. This is due to the fact that a "wild" consumption of the slack time is less likely to happen when there is a smaller amount of it available to begin with and the system is able to control itself better.

When properly setting the control parameters, every configuration benefits from it, but the improvements shown in Configurations B and C are much more evident than those in Configuration A, and while the *slim* case still performs better, the three "controlled" systems are comparable in terms of capacity and performance.

In addition, we note that the longitudinal velocity (along the x axis in Fig. 1) of the vehicle decreases with the widening of the service area (Configurations B and C), because of the increased amount of time needed by the vehicle to serve points along the larger width. Customers traveling to/from checkpoints could perceive this slowness unfavorably because on average they would experience ride times increasingly larger than the direct time needed to travel between their pick-up and drop-off. Therefore, only slimmer service areas, such as Configuration A, would be suitable for public transportation purposes where the longitudinal velocity of the vehicle is not much slower than a fixed route line traveling between checkpoints. However, configurations with wider service area could very well be appropriate for transportation of goods instead of people.

5 MAST/Fixed-Route Comparison

It could be noted that slimmer service areas, such as Configuration A, would be more suitable for a regular fixed-line service. For this purpose we perform a comparison between the MAST service (Configuration A, with C = 3) and a fixed-route bus service serving the same service area. Both systems serve the same demand of 21 customers/hour; with the distribution of Table 1. We assume the same vehicle speed v = 25 miles/hour and a service time of 18 seconds at each stop for both systems. The fixed-route line has C = 25 fixed stops evenly distributed along the x axis (one stop every 0.5 miles), therefore the headway is 72 minutes and the scheduled/actual travel time between two consecutive stops is 1.5 minutes. We assume that there is no variability in the travel time between two consecutive stops for the fixed line. The only variability for the MAST system is due to the random locations of the non-checkpoint demand. Fig. 4 illustrates the geometry and the features of the systems.

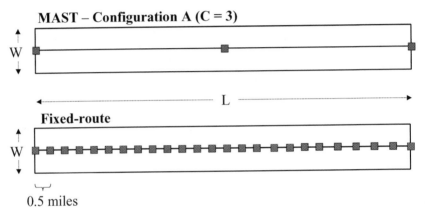

Fig. 4. MAST/Fixed-route Comparison

In order to perform the comparison we define WKT, being the average walking time per passenger (assumed walking speed = 3 miles/hour). While the MAST system serves its customers point to point and no walking occurs, a fixed-route system forces non-checkpoint requests to walk to/from the nearest fixed stop in order to use the service. Note that checkpoint requests could have a certain amount of walking time associated with it, but considering the same demand it would be equivalent for both systems and consequently we assume it to be zero.

We observe that for headways larger than 12-13 minutes the majority of the customers are aware of the schedule (Okrent (1974)) and this is true for all requests showing up at bus stops (for both systems). Therefore, we do not consider the waiting time until the pick-up as a valid parameter for this comparison. WT measures instead the extra waiting time that MAST customers have to wait at their stops, because of other insertions occurring after their requests.

Thus, the overall performance Z (in time units) is defined as follows:

$$Z = w_1 \times M/v + w_2 \times RT \times NC + w_3 \times WT \times NC + w_4 \times WKT \times NC \quad (2)$$

where NC is the total number of customers served by the system and the last term represents the contribution to Z of the amount of walking time. We assume that the weight for walking time (w_4) is conservatively equal to w_3 (even though customers would probably perceive walking time with more discomfort than waiting time at a bus stop, especially during nighttime for safety reasons). Hence the weights in Z are set as follows: $w_1 = w_2 = 0.25$ and $w_3 = w_4 = 0.5$.

We ran the simulations (using Common Random Numbers for the two systems) for 45 hours, so that for the fixed-route service R = 75 and for the MAST system R = 54 (since the headway is 100 minutes). The results are shown in Table 5.

Table 5. MAST/Fixed-route Comparison

θ (customers/hour)	21	
System	MAST	Fixed
	Conf. A (C = 3)	
WT (min)	1.56	0
RT (min)	25.53	16.6
WKT (min)	0	7.5
M (miles)	926.3	900
Z	6.804	7.831

The figures show that the MAST system compared to the fixed-route results has a small WT ($<$ 2 minutes) and a RT bigger by approximately 10 minutes, but M is lower and there is no walking for the customers as opposed to the fixed-route system where on average customers walk 7.5 minutes.

6 Conclusions

In this paper we summarize the insertion heuristic algorithm developed for the Mobility Allowance Shuttle Transit services presented by Quadrifoglio *et al.* (2007) and we utilize it to carry out a sensitivity analysis of its performance over the shape of the service area. The algorithm makes use of proper control parameters, aiming to cherish the consumption of the slack time. A proper setting of them allows the system to increase its capacity, maintaining an analogous service level for the customers. In particular, we show that this positive control effect is more evident in a wider service area with more slack time. The results also show that slimmer configurations perform better in terms of capacity and are more suitable for public transportation purposes. In addition, the findings show that MAST services are competitive with fixed-route ones and perform better under certain demand distributions, even for slim service areas.

Acknowledgement: The research reported in this paper was partially supported by the National Science Foundation under grant NSF/USDOT-0231665.

References

Aldaihani, M. M. and Dessouky, M. (2003). Hybrid scheduling methods for para-transit operations. *Computers & Industrial Engineering*, **45**, 75–96.

Aldaihani, M. M., Quadrifoglio, L., Dessouky, M., and Hall, R. W. (2004). Network design for a grid hybrid transit service. *Transportation Research*, **38A**, 511–530.

Bodin, L. and Sexton, T. (1986). The multi-vehicle subscriber dial-a-ride problem. *TIMS Studies in the Management Sciences*, **22**, 73–86.

Cordeau, J. F. and Laporte, G. (2003). The dial-a-ride problem (DARP): variants, modeling issues and algorithms. *4OR*, **1**(2), 89–101.

Cortés, C. E. and Jayakrishnan, R. (2002). Design and operational concepts of a high coverage point-to-point transit system. *Transportation Research Record 1783*, pages 178–187.

Crainic, T. G., Malucelli, F., and Nonato, M. (2001). Flexible many-to-few + few-to-many = an almost personalized transit system. In *TRISTAN IV, São Miguel Azores Islands*, pages 435–440.

Daganzo, C. F. (1978). An approximate analytic model of many-to-many demand responsive transportation systems. *Transportation Research*, **12**, 325–333.

Daganzo, C. F. (1984). Checkpoint dial-a-ride systems. *Transportation Research*, **18B**, 315–327.

Desaulniers, G., Erdmann, A., Solomon, M. M., and Soumis, F. (2000). The VRP with pickup and delivery. Technical report, Cahiers du GERARD G-2000-25, Ecole des Hautes Etudes Commerciales, Montréal.

Desrosiers, J., Dumas, Y., and Soumis, F. (1986). A dynamic programming solution of the large-scale single-vehicle dial-a-ride problem with time windows. *American Journal of Mathematical and Management Sciences*, **6**, 301–325.

Desrosiers, J., Dumas, Y., and Soumis, F. (1988). The multiple dial-a-ride problem. In *Computer Aided Transit Scheduling*, volume 308 of *Lecture Notes in Economics and Mathematical Systems*. Springer, Berlin.

Diana, M. (2006). The importance of information flows temporal attributes for the efficient scheduling of dynamic demand responsive transport services. *Journal of Advanced Transportation*, **40**(1), 23–46.

Diana, M. and Dessouky, M. (2004). A new regret insertion heuristic for solving large-scale dial-a-ride problems with time windows. *Transportation Research*, **38B**, 539–557.

Hickman, M. and Blume, K. (2001). A method for scheduling integrated transit service. In S. Voss and J. Daduna, editors, *Computer Aided Scheduling of Public Transport, Lecture Notes in Economics and Mathematical Systems 505*, pages 233–251. Springer, Berlin.

Horn, M. E. T. (2002a). Fleet scheduling and dispatching for demand-responsive passenger services. *Transportation Research*, **10C**, 35–63.

Horn, M. E. T. (2002b). Multi-modal and demand-responsive passenger transport systems: a modeling framework with embedded control systems. *Transportation Research*, **36A**, 167–188.

Jaw, J. J., Odoni, A. R., Psaraftis, H. N., and Wilson, N. H. M. (1986). A heuristic algorithm for the multi-vehicle advance request dial-a-ride problem with time windows. *Transportation Research*, **20B**(3), 243–257.

Liaw, C. F., White, C. C., and Bander, J. L. (1996). A decision support system for the bimodal dial-a-ride problem. *IEEE Transactions on Systems, Man, and Cybernetics*, **26**(5), 552–565.

Lu, Q. and Dessouky, M. (2004). An exact algorithm for the multiple vehicle pickup and delivery problem. *Transportation Science*, **38**, 503–514.

Lu, Q. and Dessouky, M. (2006). New insertion-based construction heuristic for solving the pickup and delivery problem with hard time windows. *European Journal of Operational Research*, **175**, 672–687.

Madsen, O. B. G., Raven, H. F., and Rygaard, J. M. (1995). A heuristic algorithm for a dial-a-ride problem with time windows, multiple capacities, and multiple objectives. *Annals of Operations Research*, **60**, 193–208.

Malucelli, F., Nonato, M., and Pallottino, S. (1999). Demand adaptive systems: some proposals on flexible transit. In T. Ciriania, E. Johnson, and R. Tadei, editors, *Operations Research in Industry*, pages 157–182. McMillan, London.

Okrent, M. M. (1974). Effect of transit service characteristics on passenger waiting time, MS thesis. Department of Civil Engineering, Northwestern University, Evanston.

Psaraftis, H. N. (1980). A dynamic programming solution to the single vehicle many-to-many immediate request dial-a-ride problem. *Transportation Science*, **14**, 130–154.

Psaraftis, H. N. (1983). An exact algorithm for the single vehicle many-to-many dial-a-ride problem with time windows. *Transportation Science*, **17**, 351–357.

Psaraftis, H. N. (1986). Scheduling large-scale advance-request dial-a-ride systems. *American Journal of Mathematical and Management Sciences*, **6**, 327–367.

Quadrifoglio, L., Hall, R. W., and Dessouky, M. M. (2006). Performance and design of mobility allowance shuttle transit services: Bounds on the maximum longitudinal velocity. *Transportation Science*, **40**, 351–363.

Quadrifoglio, L., Dessouky, M. M., and Palmer, K. (2007). An insertion heuristic for scheduling mobility allowance shuttle transit (MAST) services. *Journal of Scheduling*, **10**, 25–40.

Savelsbergh, M. W. P. and Sol, M. (1995). The general pickup and delivery problem. *Transportation Science*, **29**, 17–29.

Sexton, T. R. and Bodin, L. D. (1985a). Optimizing single vehicle many-to-many operations with desired delivery times: 1. Scheduling. *Transportation Science*, **19**, 378–410.

Sexton, T. R. and Bodin, L. D. (1985b). Optimizing single vehicle many-to-many operations with desired delivery times: 2. Routing. *Transportation Science*, **19**, 411–435.

Sexton, T. R. and Choi, Y. (1986). Pickup and delivery of partial loads with soft time windows. *American Journal of Mathematical and Management Sciences*, **6**, 369–398.

Stein, D. M. (1977). Scheduling dial-a-ride transportation systems: an asymptotic approach. Technical report, No. 670, Harvard University, Division of Applied Science.

Stein, D. M. (1978a). An asymptotic probabilistic analysis of a routing problem. *Mathematics of Operations Research*, **3**, 89–101.

Stein, D. M. (1978b). Scheduling dial-a-ride transportation problems. *Transportation Science*, **12**, 232–249.

Toth, P. and Vigo, D. (1997). Heuristic algorithm for the handicapped persons transportation problem. *Transportation Science*, **31**, 60–71.

Wilson, N. H. M. and Hendrickson, C. (1980). Performance models of flexibly routed transportation services. *Transportation Research*, **14B**, 67–78.

Wilson, N. H. M., Sussman, J. M., Wong, H. K., and Higgonet, B. T. (1971). Scheduling algorithms for a dial-a-ride system. Technical report, USL TR-70-13, M.I.T, Urban Systems Laboratory.

Zhao, J. and Dessouky, M. (2004). Optimal service capacity for a single bus mobility allowance shuttle transit (MAST) system. Submitted for publication.

Lecture Notes in Economics and Mathematical Systems

For information about Vols. 1–515
please contact your bookseller or Springer-Verlag

Vol. 562: T. Langenberg, Standardization and Expectations. IX, 132 pages. 2006.

Vol. 563: A. Seeger (Ed.), Recent Advances in Optimization. XI, 455 pages. 2006.

Vol. 564: P. Mathieu, B. Beaufils, O. Brandouy (Eds.), Artificial Economics. XIII, 237 pages. 2005.

Vol. 565: W. Lemke, Term Structure Modeling and Estimation in a State Space Framework. IX, 224 pages. 2006.

Vol. 566: M. Genser, A Structural Framework for the Pricing of Corporate Securities. XIX, 176 pages. 2006.

Vol. 567: A. Namatame, T. Kaizouji, Y. Aruga (Eds.), The Complex Networks of Economic Interactions. XI, 343 pages. 2006.

Vol. 568: M. Caliendo, Microeconometric Evaluation of Labour Market Policies. XVII, 258 pages. 2006.

Vol. 569: L. Neubecker, Strategic Competition in Oligopolies with Fluctuating Demand. IX, 233 pages. 2006.

Vol. 570: J. Woo, The Political Economy of Fiscal Policy. X, 169 pages. 2006.

Vol. 571: T. Herwig, Market-Conform Valuation of Options. VIII, 104 pages. 2006.

Vol. 572: M. F. Jäkel, Pensionomics. XII, 316 pages. 2006

Vol. 573: J. Emami Namini, International Trade and Multinational Activity. X, 159 pages, 2006.

Vol. 574: R. Kleber, Dynamic Inventory Management in Reverse Logistics. XII, 181 pages, 2006.

Vol. 575: R. Hellermann, Capacity Options for Revenue Management. XV, 199 pages, 2006.

Vol. 576: J. Zajac, Economics Dynamics, Information and Equilibnum. X, 284 pages, 2006.

Vol. 577: K. Rudolph, Bargaining Power Effects in Financial Contracting. XVIII, 330 pages, 2006.

Vol. 578: J. Kühn, Optimal Risk-Return Trade-Offs of Commercial Banks. IX, 149 pages, 2006.

Vol. 579: D. Sondermann, Introduction to Stochastic Calculus for Finance. X, 136 pages, 2006.

Vol. 580: S. Seifert, Posted Price Offers in Internet Auction Markets. IX, 186 pages, 2006.

Vol. 581: K. Marti; Y. Ermoliev; M. Makowsk; G. Pflug (Eds.), Coping with Uncertainty. XIII, 330 pages, 2006.

Vol. 582: J. Andritzky, Sovereign Default Risks Valuation: Implications of Debt Crises and Bond Restructurings. VIII, 251 pages, 2006.

Vol. 583: I.V. Konnov, D.T. Luc, A.M. Rubinov† (Eds.), Generalized Convexity and Related Topics. IX, 469 pages, 2006.

Vol. 584: C. Bruun, Adances in Artificial Economics: The Economy as a Complex Dynamic System. XVI, 296 pages, 2006.

Vol. 585: R. Pope, J. Leitner, U. Leopold-Wildburger, The Knowledge Ahead Approach to Risk. XVI, 218 pages, 2007 (planned).

Vol. 586: B.Lebreton, Strategic Closed-Loop Supply Chain Management. X, 150 pages, 2007 (planned).

Vol. 587: P. N. Baecker, Real Options and Intellectual Property: Capital Budgeting Under Imperfect Patent Protection. X, 276 pages , 2007.

Vol. 588: D. Grundel, R. Murphey, P. Panos , O. Prokopyev (Eds.), Cooperative Systems: Control and Optimization. IX, 401 pages , 2007.

Vol. 589: M. Schwind, Dynamic Pricing and Automated Resource Allocation for Information Services: Reinforcement Learning and Combinatorial Auctions. XII, 293 pages , 2007.

Vol. 590: S. H. Oda, Developments on Experimental Economics: New Approaches to Solving Real-World Problems. XVI, 262 pages, 2007.

Vol. 591: M. Lehmann-Waffenschmidt, Economic Evolution and Equilibrium: Bridging the Gap. VIII, 272 pages, 2007.

Vol. 592: A. C.-L. Chian, Complex Systems Approach to Economic Dynamics. X, 95 pages, 2007.

Vol. 593: J. Rubart, The Employment Effects of Technological Change: Heterogenous Labor, Wage Inequality and Unemployment. XII, 209 pages, 2007.

Vol. 594: R. Hübner, Strategic Supply Chain Management in Process Industries: An Application to Specialty Chemicals Production Network Design. XII, 243 pages, 2007.

Vol. 595: H. Gimpel, Preferences in Negotiations: The Attachment Effect. XIV, 268 pages, 2007.

Vol. 596: M. Müller-Bungart, Revenue Management with Flexible Products: Models and Methods for the Broadcasting Industry. XXI, 297 pages, 2007.

Vol. 597: C. Barz, Risk-Averse Capacity Control in Revenue Management. XIV, 163 pages, 2007.

Vol. 598: A. Ule, Partner Choice and Cooperation in Networks: Theory and Experimental Evidence. X, 202 pages, 2008.

Vol. 599: A. Consiglio, Artificial Markets Modeling: Methods and Applications. XV, 277 pages, 2007.

Vol. 600: M. Hickman, P. Mirchandani, S. Voß (Eds.): Computer-Aided Systems in Public Transport. XIV, 432 pages, 2008.

Vol. 601: D. Radulescu, CGE Models and Capital Income Tax Reforms: The Case of a Dual Income Tax for Germany. XVI, 168 pages, 2007.

Vol. 602: N. Ehrentreich, Agent-Based Modeling: The Santa Fe Institute Artificial Stock Market Model Revisited. XVI, 225 pages, 2007.

Vol. 603: D. Briskorn, Sports Leagues Scheduling: Models, Combinatorial Properties, and Optimization Algorithms. XII, 164 pages, 2008.

Vol. 604: D. Brown, F. Kubler, Computational Aspects of General Equilibrium Theory: Refutable Theories of Value. XII, 202 pages, 2008.